Reality in the Shadows

(or)

What the Heck's the Higgs?

S. James Gates, Jr.,
Frank Blitzer, and
Stephen Jacob Sekula

YBK Publishers
New York

Reality in the Shadows or What the Heck's the Higgs?

YBK Publishers, Inc.
39 Crosby Street
New York, NY 10013
www.ybkpublishers.com

ISBN:978-1-936411-39-9

Library of Congress Cataloging-in-Publication Data
Names: Gates, S. James (Sylvester James), author. | Blitzer, Frank, author. |
  Sekula, Stephen Jacob, author.
Title: Reality in the shadows, (or), what the heck's the Higgs? / S. James
  Gates, Jr., Frank Blitzer, and Stephen Sekula.
Description: New York, NY : YBK Publishers, Inc., [2017]
Identifiers: LCCN 2017054494| ISBN 9781936411399 (pbk. ; alk. paper) | ISBN
  1936411393 (pbk. ; alk. paper)
Subjects: LCSH: Physics--History.
Classification: LCC QC7 .G28 2017 | DDC 530.09--dc23
LC record available at https://lccn.loc.gov/2017054494

Manufactured in the United States of America for distribution in
North and South America or in the United Kingdom or Australia
when distributed elsewhere.

For more information, visit www.ybkpublishers.com

Contents

Acknowledgements

We thank our families for their indulgence and support for this book during the years of its preparation.

The Blitzer family, Frank Blitzer's lovely wife of 64 years, Arlene, whose many years of support has been Frank's inspiration, his children, Doug, Valerie, Debbie Jacobs, and grandson Ryan Blitzer, provided Frank with much support and input during the creation of *Reality in the Shadows*, especially while he was incapacitated for a brief period due to health issues. It was Frank's early efforts that created the concept and the framework upon which *Reality in the Shadows* was built.

Early on, Frank's daughter-in-law, Valerie Blitzer, Ryan's mother, suggested finding a teaching physicist to prepare a foreword for the book causing Frank to contact SJG after having viewed Gates' *Superstring Theory: the DNA of Reality*. This collaboration evolved well beyond creating just a foreword (which never came) as Gates joined with Blitzer as co-author.

The members of S. James Gates' family, Dianna, Delilah, and Sylvester, III provided a steadfast foundation and constant encouragement for his contributions to the project. They are sincerely acknowledged for the understanding and patience they selflessly gave. An especially big "shout-out" goes to wife and twin's mom, Dianna, who tolerated the absent-minded professor missing family events while working on the book.

Later, FB's grandson, Ryan, then attending Southern Methodist University, suggested one of his physics professors, Stephen Sekula, to be a contributor and reviewer of content. That suggestion happily evolved into and provided *Reality* with its third co-author.

Steve would like to thank his mother and father, Annetta and Steve, and his sister, Kate, for nurturing him as both a scientist and a writer. Added to by his wonderful wife, Steve has two families, and he would like to thank Ann, Rich, Jackie, Jerry, Jolene, and Deb for their support and patience. It's not easy having a physicist in the family!

A further driving motivation for Steve's contributions to the book was the thought that one day his wonderful nieces, Akira and Gwen, and his equally wonderful nephews, Andrew and Connor, might want to read this book because it stands as a door open onto the larger universe in which their uncle loves to play.

Most of all, Steve would like to thank his wife, Jodi, who not only provided intellectual input and feedback to him, but tolerated his working in the passenger seat of the car, writing chapters and editing text, when he should have been enjoying the road trip to Wisconsin during a winter vacation. However, Jodi, who is also a physicist, created some of the material for this book, so she cut him some slack. Her original contributions will be found in the sections on dark matter and dark energy.

Let it be noted that co-author, Stephen Sekula, and editor, Otto Barz, provided the final linchpins that made the completion of the project possible. The challenge of working within a group having such diverse backgrounds as formal theoretical physicist, rocket scientist, experimentalist, and demanding (yet extremely patient) editor, was an extraordinary yet generally entertaining experience for all. It is hoped that the result of this team effort will be one that readers will find informative, engaging, and not at all an effort to conclude.

In further acknowledgement, we would like to recognize the contributions of a few key individuals during the manuscript's early development who provided feedback from a personal standpoint using their diverse backgrounds as their crow's nest of observation:

Michael Barz, attorney; Donald Gendler, pharmacist; Dr. Tim Gutowski, mechanical engineer; Justin Kander, then-high school student; Glenn Opp, microbiologist; Dr. Roy Sandstrom, professor of ancient history; and Dr. John Wander, professor of American literature.

SS would also like to thank those who have been his scientific mentors during his career: Klaus Honscheid, Richard Kass, Yibin Pan, Michael Schmidt, Gabriella Sciolla, Sau Lan Wu, and Dick Yamamoto. He is also grateful to his faculty colleagues at SMU for their years of mentoring and support, and to all of his students for making him a better teacher.

SJG would like to recognize two close friends, Professor Arthur Popper (my "rabbi") and Mark Weinberg, Esq. for their stern—and accurate—critique of an early version of the manuscript. Finally, SJG is constantly in awe of the amazing support of his professional activities that is given by Ms. Mary Sutton.

Reality in the Shadows

(or)

What the Heck's the Higgs?

## Chapter 1

Cosmology in Motion

Reality in the Shadows chronicles the adventures and research of many who have sought to explore the question, "Just what is the universe?" They observed nature and wrote descriptions of their observations using mathematics, the only human language that is known to be capable of most completely and accurately summarizing their discoveries. While mathematics goes a great way toward quantifying the behavior of the universe, it is a means to *approximate* what is observed about the reality. These mathematical representations are like the shadows on the wall in Plato's *Allegory of the Cave,* in which prisoners face the wall of a cave, unable to turn their heads to see what it is that casts the shadows they see. Their descriptions of what causes these shadows are not always accurate. (Do a Google search on "allegory of the cave" for a more complete explanation.) Sometimes the mathematical statements present inadequate description and at other times they are completely incapable of describing the phenomena. However, each description, hit or miss, inches the observers forward toward a more accurate description of reality that is not yet fully emergent of its mathematical representations.

This process was initiated by a small number of ancient seekers who wanted to understand the universe. This continues today, spurring a relay race of discoveries that spans a period greater than 2,500 years. However, modern science got its real boost after the period of medieval Europe.

Old explanatory frameworks emerged from a questioning among philosophers and religious leaders who thought about life. They rationalized their existence through the worship of creator gods, founded astrological signs, and considered the orbits of the planets that they imagined guided their existence. This they called "natural philosophy."* As time went on, their successors furthered these questions by asking, "Why?"

These questioners became scientists by recording their observations of the heavens and the earth. Galileo Galilei, whom Albert Einstein, (the great physicist, about whom we will hear much more later) described as the father of all science said, "Philosophy is written in this grand book, the universe, which

---

* Today, natural philosophy has a different name. It is called "physics."

stands continually open to our gaze, but the book cannot be understood unless one first learns to comprehend the language and to read the alphabet in which it is composed. It is written in the language of mathematics..."

In seeking answers to the mysteries of its workings, men and women have endeavored over eons to unveil the universe as it slowly yields itself to study. In setting down the pages of their books, there were times when an expected conclusion resulted from an observation that was not anticipated—a point at which their vision of the universe failed to agree with their observations of it. This recording of expectation, observation, and consequence came to be codified as "the scientific method." It continues today, each time reaching a point when knowledge meets a stumbling block in the forwarding of its understanding of the universe. Science is dynamic, ever changing as the boundaries of knowledge are constantly expanded.

These early individuals were unique because they questioned what they saw as well as *why it happened*. They described these actions using mathematics and performed tests to clarify and support what they saw. Although the tools they devised captured ever more accurate information, the shadows continued to play on the walls of the cave, but the resulting output of this process produced theories (explanatory frameworks with predictive power that led to the discovery of new facts) about the grand working of the universe. All of science is a theory, but not in the sense of it being a collection of guesses. Science gives a most accurate description through the use of logical reasoning about structures that are observable.

Recent observations have determined the age of the universe to be 13.74 (±0.11) billion years; this, according to calculations from the Hubble telescope and WMAP (Wilkinson Microwave Anisotropy Probe) satellite observations.

Do we *really* know how old the universe is? Do we *really* know and understand what holds the universe together—where it is going, how it got here—and what is going to happen to it? Such questions often result in new questions. The WMAP satellite information has since been superseded by more powerful research called Planck, a project of the European Space Agency. Planck tells us that the universe is 13.807 (±.026) billion years old. Still, there is anticipated error in that number. That error factor is called "science," the going forward with continuing research to find even more accurate and precise information.

This book is intended to be a largely non-mathematical guidebook aimed at the more-engaged observer of nature to help in grasping the major ideas and concepts that bring us to the state of modern physics today. Mathematical equations that are used are intended to be illustrations to promote understanding of the manner in which arguments are framed. One need not have an understanding of the underlying mathematics. It is as with music, where one need not read the score of a great concert piece in order to appreciate the majesty of its composition.

The Geocentric Universe

In the time of Aristotle (384 B.C.–322 B.C.), people believed in the geocentric theory—that Earth is stationary and everything revolves around it. Fostered by Aristotle, it describes the universe as a series of concentric circles, each containing one of the five planets known at that time, all revolving around a stationary Earth.

Religion was the primary framework used by early searchers of the universe. Most people were illiterate and relied on learned religious leaders for guidance. Other, perhaps even more accurate, ideas were abandoned in favor of these teachings.

The geocentric concept was accepted because people believed that Earth was placed in a privileged state by the gods, believing that it was the only place in the universe where humans resided. This view remained popular for centuries because it coincided with the early understanding of the universe that was taken from naked-eye observation. It gained powerful support from the early Christian church due to the belief that humanity has a special place in creation. It seemed only logical that Earth was the center stage in the drama of creation. These teachings of the church combined with casual observation made this an enduring, albeit inaccurate, cosmology.

The geocentric theory was codified by Claudius Ptolemy (c. 90 A.D.–c. 168 A.D.), an Egyptian astronomer, mathematician, and geographer who studied the motions of the planets and described them in tables that were used to obtain past and future positions of the planets. He was able to see and plot Mercury, Venus, Saturn, Neptune, and Jupiter (all named after Roman gods). He wrote three treatises during his lifetime, the pertinent one on astronomy called the *Almagest*, "The Great Treatise." Contained in thirteen volumes, it is the only surviving comprehensive treatise on the universe of that time period.

Seeking to fit their motion into a geocentric theory, the planets were shown to traverse very complicated paths called epicycles. Due to a lack of technology, observations resulted in rudimentary data. Across time, many devices were invented to help to understand the planetary motions. Calendars were created to mark the days and years, and instruments were developed to study the stars and planets.

The Heliocentric View

So strong was the geocentric belief of astronomers, clergy, and philosophers in early history, that it was very difficult to propose or support any alternative to this belief. Scientists had a difficult time convincing the learned masters that other possibilities existed and were, perhaps, more accurate. Though the Aristotelian view predominated, there *were* dissenters.

Aristarchus of Samos (310 B.C.–230 B.C.) was the first person to propose a heliocentric universe—the concept that the planets revolve around the sun rather than the earth. Almost 300 years before Ptolemy, and about 1,800 years before Copernicus, he made careful measurements that gave him reason to believe that the earth and the planets revolved around the sun. Unlike his contemporaries, he realized the power that observation has in formulating theories about the stuff of reality. From his observation of eclipses he concluded that the radius of the moon was half that of Earth and that the distance from the earth to the moon was about 57 times that of the earth's diameter. That said, due to the inadequacies of his methods of observation, both of these answers were incorrect by a factor of about two, but the concept was accurate.

Eratosthenes of Cyrene (276 B.C.–194 B.C.), following the same kinds of logic as Aristarchus, deduced that the earth, like the moon and sun, was spherical in shape. Using geometry and calculating observations of the sun at the time of the summer solstice, he calculated the size of Earth to within only a few percent of today's measurements!

But the views of Aristarchus fell on deaf ears because most educated people continued to follow the teachings of those clergy and philosophers who believed strongly in the geocentric view. These remarkable achievements found little acceptance during the lifetimes of Aristarchus and Eratosthenes. Their critics were able to all but silence these dissenters.

Though their works were judged by their contemporaries to be incorrect, these scientists understood that using mathematics (in this case, geometry) in concert with observation could yield a more accurate view of the universe. This history holds a valuable lesson—as science advances through the work of individuals, the establishment will not necessarily embrace concepts that will produce progress . . . at least not at first. At this point, the Greeks and Romans were more mathematical philosophers than scientists.

Without a means of noting and cataloging planetary movement over long periods of time, it would be impossible to demonstrate that the heliocentric universe provided an accurate representation of the universe. Their ideas languished for centuries, largely unpursued by others.

Although there were others who came to the same conclusions as did Aristarchus and Eratosthenes, it was dangerous to assert beliefs different from the accepted religious edict. Giordano Bruno, a catholic priest, was martyred in 1584, in part for publishing his assertion that stars are other suns, circling around which were other planets like Earth, that might carry intelligent life. Those who denied arguments based on well-reasoned mathematical models created by observation of our cosmos, held sway for centuries—but the universe ultimately pays no heed to denialism.

A Major Scientific Revolution

Although early scholars had published heliocentric hypotheses centuries before Nicolas Copernicus (1473–1543), his publication of a scientific theory of heliocentrism, demonstrating that the motions of celestial objects can be explained without putting Earth in the center of the universe, stimulated further scientific investigation. It became a landmark in Western science known as the Copernican revolution. Enthusiasts focused on the quest to fathom the marvels of the universe. In doing this, a body of work emerged from the minds of a few contributors that set down the basic principles of the early universe. These contributions are accepted even today. These were the titans of science, who contributed significantly to our current understanding of the universe and made it possible for their followers to probe deeply into its workings.

Nicolas Copernicus

The geocentric theory continued to be believed until the 15th century, when Nicolas Copernicus appeared on the scene. He became the first astronomer to formulate a scientifically based heliocentric cosmology that removed Earth from the center of the universe. His book, *De Revolutionibus Orbium Coelestium* ("On the Revolutions of the Celestial Spheres"), is often regarded as the starting point for modern astronomy and the defining epiphany that began the scientific revolution. He published this work on his deathbed, fearing to be ridiculed (if not worse) during his lifetime. His research set down tables to show how the planets revolved around the sun in circular orbits, as illustrated in Figure 1.1. For the first time the world was given quantified evidence that Earth was not at the center of the universe, but just another planet that circled the sun.

This revelation sparked a major revolution around the world, greatly influencing interested observers and scientific experts such as Tycho Brahe, Johannes Kepler, Galileo Gallilei, and Isaac Newton, whose combined imaginations fueled the explosion of modern science.

Just as did other contributors to the Renaissance, Copernicus was multifaceted as an astronomer, physician, Catholic cleric, and more. Though one of his many studies, astronomy, was but an avocation, it made its mark on the world stage. It set down in great detail (for the first time) specific data

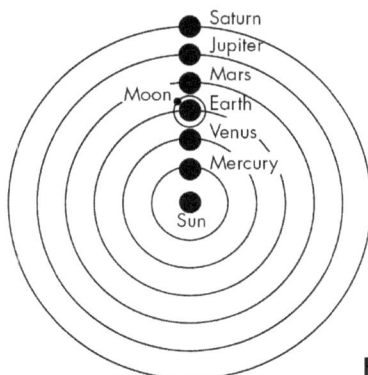

Figure 1.1 The Copernican Heliocentric Model

describing the motions of the planets in our solar system and their relationship to Earth's orbit.

Brahe and Kepler

Tycho Brahe (1546–1601) and Johannes Kepler (1571–1630), following on the heels of Copernicus, furthered the results of their predecessors. They advanced stellar observation and posted new findings of planetary motion even without the aid of a telescope, the invention of which was still to come. The results of the observations of Brahe and Copernicus were analyzed in detail by Kepler. This enabled Kepler to create a mathematical formalization of planetary motion that his predecessors had observed and set down. This was accomplished using mathematics to create a detailed analysis that concluded that the geocentric concept of the universe was incorrect; that Earth and the other planets revolved around the sun. Although these findings were formally published, their theories found no wide acceptance during their lifetimes.

Fascinated with astronomy at age thirteen, Tycho was coaxed by his uncle to enter college at the University of Copenhagen to study law and philosophy. At this impressionable age, an event took place that changed his life—a partial eclipse of the sun.

Brahe became obsessed with astronomy, putting aside the law and philosophy. He bought books and instruments and stayed up much of each night studying the stars. When he was seventeen he observed a special event—Jupiter and Saturn passing close to each other (August 17, 1563). He found on checking the data that the Alfonsine tables (which contained positioning information for the sun, moon, and planets, first published in 1483 and updated for about three hundred years after) were off by a month in predicting this event, and the Copernicus tables were off by several days. Tycho determined that this was unacceptable; that much better tables could be constructed through more accurate observation over an extended period of time. He decided that this was what he was going to do.

Tycho returned to Germany, falling in with some rich amateur astronomers in Augsburg. He persuaded them that what was needed was accurate observation. This required the use of large quadrants to obtain lines of sight on stars. This is shown in Figure 1.2, being a device having a nineteen-foot radius, probably made of logs, that defined one-quarter of a circle. It was graduated in sixtieths of a degree. There were 1350 divisions in each 22.5-degree sector of the quadrant. Thus, each division on the quadrant had a value of 0.01667 degrees of circumference—a *very* accurate protractor.

This device made it possible to make very accurate observations of the positions of the planets and other celestial bodies such as the moon by sighting along the lines marked out on the quadrant's circumference. From the data he developed, Brahe created his own model of the universe, which he published late in the 16th century.

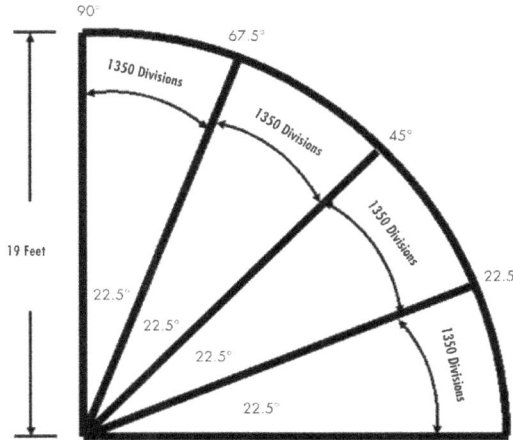

Figure 1.2   The Brahe Astronomical Quadrant

Johannes Kepler

Kepler, studying the detailed findings of Copernicus and Brahe, added his own data to their results. He then reduced it to simpler terms, analyzing that data to develop three laws:

First, the planets move in elliptical orbits (Figure 1.3) with the sun at one of its two foci.

Second, a line between the sun and the planet sweeps out equal areas in equal times (Figure 1.4). He noticed that as a planet approaches the sun, its speed increases. This increase causes the planet to traverse more distance in its orbit than it does at any other time.

Third, the ratio of the squares of the periods of revolution for two planets is equal to the ratio of the cubes of their semi-major axes. Expressed as an equation, Kepler's third law becomes, with P_1 and P_2 representing the two orbital periods and R_1 and R_2 the lengths of their semi-major axes, the mathematical relationship

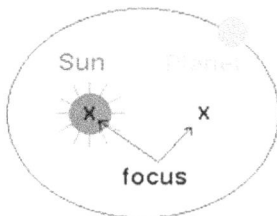

Figure 1.3   Kepler's 1st Law: Planetary Orbits are Ellipses with the sun at One Focus

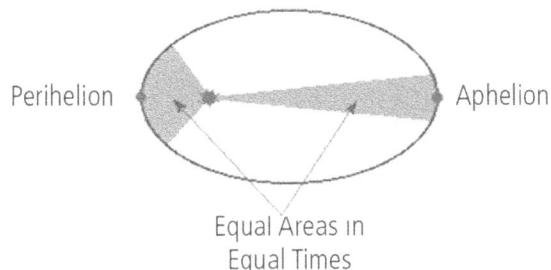

Figure 1.4   Kepler's 2nd Law: the Sun-Planet Line Sweeps out Equal Areas in Equal Times

shown in equation (1). If you know the length of the orbital periods of two planets, such as Earth and Mars, you can find their distance from the sun.

$$\frac{P_1^{\,2}}{P_2^{\,2}} = \frac{R_1^{\,3}}{R_2^{\,3}}$$

This geometrical conclusion, still valid today, was what Galileo proved with his telescope and what Isaac Newton captured in his seminal work on the mathematical description of the laws governing the universe, *Philosophiæ Naturalis Principia Mathematica.* Newton's work also created the basis for the development of the mathematics of calculus, so necessary to obtain an accurate description of the subject of classical mechanics.

The characteristics of a planet's orbit are shown in Figure 1.5. This figure also illustrates that two planets in different orbits have different orbital characteristics.

Orbital dimensions are different from each other. For example, Mars and Earth have different elliptical parameters—orbital periods, rotational speeds, semi-major and semi-minor radii, etc. We know that the earth's spin rate is 15 degrees per hour (360 degrees in 24 hours) and its orbital period is 365 days. Mars has different values for these two measurements. It spins once on its axis every 1.03 Earth days and makes one revolution around the sun every 687 Earth days. Figure 1.5 also shows that the sun is located at one of the two foci of the elliptical orbit of the planet in all cases. The other focus is located at the other end of the ellipse. These differences arise because the orbital parameters are different from each other. Their orbital periods differ because their semi-major and semi-minor axes have differing lengths. Thus, the orbital period of the larger ellipse is greater than that of the smaller one.

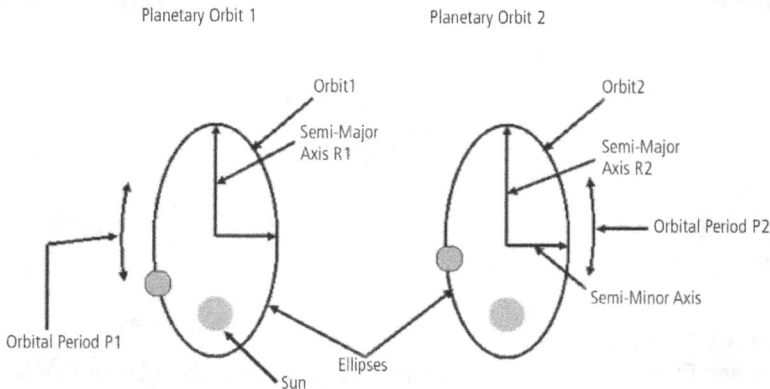

Figure 1.5 Comparing Two Planetary Orbits

These major contributors were the first to define a basic understanding of the universe resulting from their quantitative observations of the planets and stars. The work, developed over about a century, corroborated and reflected the first formalization of what is now understood about the universe. Of course, their knowledge was limited to the behavior of only the planets and stars within the universe as they knew it.

This work, although significant, reveals only a rudimentary understanding of the contents of the universe and its behavior. Without a telescope the universe was little understood in terms of its complete cast of characters—planets, stars, galaxies, and so on. The fact that Earth was part of the Milky Way galaxy was not on the horizon of their awareness. Other characteristics of the universe noted at that time were little understood. Without a magnifying capability it was impossible to observe, monitor, and record the behavior of other objects and events in space such as comets and eclipses.

Galileo and the Telescope

The telescope was invented in 1608 by Hans Lippershey, a Dutch lens maker, just before Kepler's publication of his three laws. One month after its invention, the telescope was taken to Italy and brought to Galileo, who was sought out to improve on its design. Galileo Galilei (1564–1642), one of the greatest scientific minds in history, made the needed refinements. He turned his improved telescope to the stars. After many observations of the planets with this new, higher-powered version (having a 30X capability—such that viewed objects appear to be 30 times closer or, said differently, 30 times larger than when not seen through the telescope), he embraced the Copernican heliocentric theory and published his findings. This came to the attention of the pope.

After reviewing Galileo's findings, the pope placed him under house arrest for the rest of his life because he had questioned church teachings. Galileo performed studies in many scientific areas, excelling in physics. He is considered to be the father of modern science. Among his achievements as a mathematician and astronomer are:

1. Improvements on the telescope for viewing the planets
2. Father of
   a. observational astronomy
   b. modern physics
   c. modern science
3. Development of the science of the motion of uniformly accelerated objects that is taught in nearly all high school and college physics courses.
4. Observation and analyses of sun spots, the moons of Jupiter, and the phases of Venus.
5. Improvements in the design of the magnetic compass.
6. The characterization of motion (kinematics).

The designation, "father of modern physics," was extended by none other than Albert Einstein who said "Pure logical thinking cannot yield us any knowledge of the empirical world; all knowledge of reality starts from experience and ends in it. Propositions arrived at by purely logical means are completely empty as regards reality. Because Galileo saw this, and particularly because he drummed it into the scientific world, he is the father of modern physics—indeed, of modern science altogether."

Galileo conferred onto physics yet another benefit, *precision timing*. In precisely timing the motion of pendula, Galileo hit upon the idea of using his pulse as a means to time the swings. Of course, this was not precise in the modern sense of the word as the human pulse has great variation, but by sitting calmly and quietly, he used his pulse to gain insight into motion at a level of precision not reached before that time. He developed very accurate water-based clocks for his experiments on motion.

Other Devices that Helped to Move Science from the Geocentric to the Heliocentric Understanding

Tools were developed to help understand what was being seen in the heavens. Though rudimentary, they provided a good deal of information about the universe. The abacus was such an early computational device that is still in use today. It is actually a form of digital computer as it works with beads that are counted to obtain a result. As the complexity of their questioning grew, the "computers" devised by these early scientists became more complex.

The Astrolabe

Devised around 150 B.C., the astrolabe evolved across several hundreds of years. Its operation uses the concept of projecting the celestial bodies onto a flat plane. It was one of the first instruments to give the user a means of locating positions on Earth.

By 800 A.D. the device was widely used in the Islamic world and was introduced to Europe from Spain in the early 12th century. It was the most popular astronomical instrument until about 1650, when it was replaced by more specialized navigation devices that were more accurate.

The device is a computer used to solve problems relating to time and the position of the sun and stars in the sky. Several types of astrolabes have been devised over the ages. The *planispheric* astrolabe (shown in Figure 1.6) designed by the French scientist and craftsman, Jean Fusoris in about 1400, depicts the celestial sphere projected onto a flat plane centered at the equator. It is made of brass and is about 6 inches (15 cm) in diameter.

Astrolabes depict the sky at a specific location and time. The sky is shown on the face of the instrument and marked with position locations. Its moveable

Figure 1.6   The Astrolabe

parts can be adjusted to specify a date and time. When set correctly, the sky at the user's location is presented on its face. The device can also be used to solve problems such as telling the present time or the time of a future event such as the summer solstice.

The Antikythera Mechanism

The Antikythera mechanism (Figure 1.7) is another unique mechanical analog computer devised and built more than 2000 years ago. Found around 1900, it astonished scientists. Greek sponge-fishers in the Kythera Sea sought refuge from bad weather, anchoring their boats off the island of Antikythera. While diving there they discovered an ancient shipwreck at a depth of about 125 feet. The ship had sunk, according to several documents, around 100 B.C. Two thou-

Figure 1.7   The Ancient Antikythera Mechanism

sand years of being in the salt sea corroded and destroyed its many artifacts. Among them were bronze and marble statues. Spurred on by these finds, continued searching produced a metal mechanical structure having multiple gears.

Various techniques were used to ascertain the nature and purpose of the severely corroded mechanism. The remains of the Antikythera mechanism are now part of the bronze collection of the National Archaeological Museum of Athens. The mechanism is so complex that it has taken many years to decipher its purpose and workings. It appears to have had the capability to track the celestial behavior of the sun, moon, and, possibly, other planets. In addition, it kept track of the dates of eclipses, Olympic Games events, and the annual and monthly calendars for a period of nineteen years. Its complexity amazed its discoverers as nothing like it has been found until the late middle ages when geared clocks began to appear.

This finding caused historians to reconsider the state of the technology of the Greeks of that period. Its design characteristics, according to Professor Michael Edmunds of Cardiff University, who led the most recent study in 2007, describes it in saying, " . . . This device is just extraordinary, the only thing of its kind. The design is beautiful, the astronomy is exactly right. The way the mechanics are designed just makes your jaw drop. Whoever has done this has done it extremely carefully . . . in terms of historic and scarcity value, I have to regard this mechanism as being more valuable than the Mona Lisa."

The beauty of the device is in the way that the mechanics were designed to present the celestial data in a very compact configuration. Several reproductions of the device were created beginning with the early designs of historian Derek de Solla Price, who spent three decades analyzing and reconstructing

Figure 1.8 Views of the Replicated Antikythera Mechanism

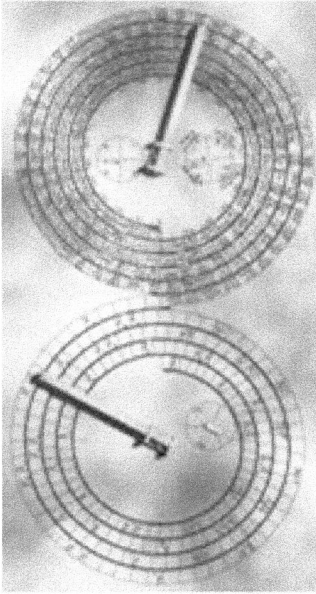

Figure 1.9 Rear Dial Face on the Antikythera Mechanism

the device as his knowledge of it increased. In a 1959 *Scientific American* article he described a replica of the device. The most recent re-creation by author Massimo Mogi Vicentini, in 2009, is shown in Figures 1.8 and 1.9.

Vicentini's replica shows the gearing arrangement and dials on the exterior of the instrument. The dials were manipulated to reveal the dates on which the Olympic trials would be held during each of the four-year intervals in that range.

The lunar calendar and the time for eclipses for a period of nineteen years can also be shown. Turning the crank allows the user to find the month, year, and day during any period of interest. The front view shows a crank and a single dial; the rear of the instrument has two dials. The mechanism was recreated in a clear plastic case to reveal the gearing of the mechanism that drives the dials. The mechanism may have had as many as seventy-two gears, yet it is remarkable in its miniaturization considering when it was designed. It is very compact and typical of 18th-century clocks that appeared a thousand years later.

The Antikythera mechanism was unknown to the person most often credited for ushering in the modern age of mechanical computers: Charles Babbage (1791–1871). Babbage was, among other things, a mathematician and an inventor. He conceived of two kinds of computer, an earlier one he called the "Difference Engine" and a later one he called the "Analytical Engine." The former was designed to solve a specific kind of problem—to find solutions to polynomial equations, those with unknowns raised to various powers (e.g., x^3, the cube of the unknown number, "x"). The analytical engine was in-

Figure 1.10   The Colossus Computer; Circa 1942

tended to be a more general kind of computational engine—much more like what we think of today as a "computer."

The mathematician Ada Lovelace (1815–1852), a contemporary of Babbage, was the first person to come to understand the true power of a machine like the analytical engine. She is credited as being the first computer "progammer"—one who envisioned a framework of instructions that machines could execute automatically. These visions of Babbage and Lovelace changed the way in which the Western world thought about mechanical computers and preceded the modern era of electronic computers capable of executing generic sets of instructions.

Electronic Computers

"Electronic" computers appeared in the early 1900s. These devices were electro-mechanical. In that period computers were primarily analog business machines, arithmetic devices used for making financial calculations. True electronic computers were devised concurrent with the development of vacuum tubes, making it possible to build more complex machines. By the late 1930s purely electronic devices became available. These devices could solve problems in a variety of disciplines, such as ballistics or measuring the heart rate of a patient, by modeling simple problems electro-mechanically or electrically.

Early computers utilized limited primitive electronics that took many electric circuits to make them work. They were subject to errors that were attributable to environmental factors such as temperature change and vibration. With the development of solid-state devices, computer devices came into existence that significantly reduced the amount of hardware. This came about through the use of the transistor, a quantum mechanical device that led to the total revolution of electronics during the late 1940s.

During the 1930s the electronic vacuum tube made the electronic computer possible and one of the first types of digital machines was invented by John

Atanasoff and Clifford Berry between 1939 and 1942 at Iowa State University. By the mid-1940s computer use was becoming more commonplace, and the first large-scale stored-program computer was developed.

The ENIAC, developed in the 1970s by Presper Eckert and John Mauchly, was the first digital device to be patented. While Atanasoff and Berry are considered by most historians to be the inventors of the first all-electronic computer, whole companies emerged then that were engaged in computer development.

The first all-digital computer, the Colossus (shown in Figure 1.10), was used during WWII to break German codes. Using 18,000 vacuum tubes, its processing power was sufficient to break German encryption. The computer room containing its electronic systems is on the right while the left photo shows the Input/Output (I/O) unit that enabled computer operators to communicate with it.

Modern digital computer architecture design and development began in 1944. The main figure in this work was a Hungarian, John von Neumann. Von Neumann was a mathematician who conceived of an architecture that is fundamental to present digital computers. That architecture has the means for communicating, memorizing, and performing mathematical operations. It is arranged so that its components can "talk" to each other while solving problems (see Figure 1.11).

Input and output devices enable the computer to interact with the outside world. They convert analog data (physical measurements, verbal data, numeric representations) that comes in from the outside world into digital form (a representation of that data as on-off pulses interpreted by a binary mathematical system) so that it can be digested and converted for output in summary form by the computer.

Originally, input and output circuits addressed the ALU (Arithmetic Logic Unit), that computed a process by responding to commands fetched from its

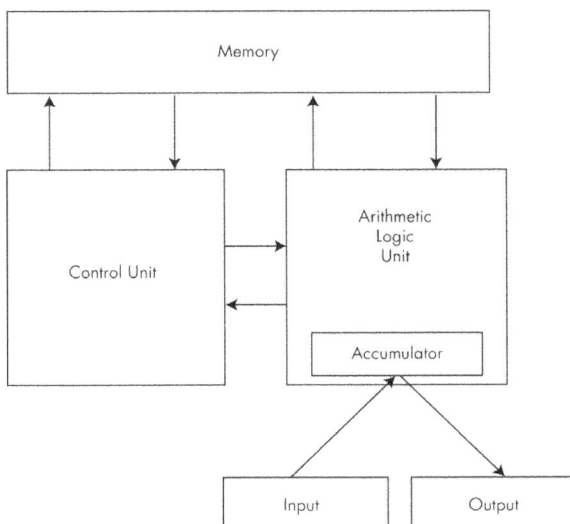

Figure 1.11 Von Neumann Computer Architecture

memory. The ALU performed the necessary computations and put the answers back into a different memory location where I/O (input/output) devices (monitors, data drives, keyboards, printers, etc.) responded to further directions from the memory device to deliver and display computed data in its final form.

The sophistication of computer architecture has grown markedly since then. Memory control and I/O units have grown in complexity as the demand for improved capability has required more speed, longer word lengths, and higher accuracy. Most importantly, computer language, software sophistication, and speed of response have moved computer design toward parallel processing, the use of more than one processor at a time to solve complex problems.

Of course, there is a galaxy of ingenious devices that have played enormously important roles in advancing the understanding of nature that we have not given due reference. Our reason for concentrating on computing devices will become clear later in our narrative.

Summary

This chapter has chronicled early views of the universe, showing the evolution of understanding from Aristotle's time to the sixteenth century. During this period humanity advanced its understanding of the universe, beginning with a geocentric model and progressing to a heliocentric model, revealing that the earth is a small planet that circles a sun along with five other planets. Because they lacked telescopes, it was not possible to search the heavens with more accuracy and greater resolution and thus they did not observe other objects and behaviors in the universe.

This lack of instrumentation and refined mathematics failed to enable these researchers to more accurately define the realities they observed. It is not surprising that what they described with their limited mathematics and tools produced limited results. These limited descriptions served to project an even greater number of shadows onto the wall.

Many more years would pass as researchers gathered more knowledge with more and better tools that came to hand. As we have shown, the march of technology—the development and improvement of the telescope; the construction of mechanical, electro-mechanical, and electrical computers—moves in concert with the march forward in the understanding of reality. As greater knowledge revealed greater refinement in understanding, previous views of their universe were clarified, but, still, they continued to project shadows that hid a reality that these probers knew was there. Thus, the interplay of technology, mathematics, and observation provides the basis for a continuing search for reality. In the midst of this, there appear insightful and dedicated thinkers who show the remarkable ability to use observation, reason, and mathematics as a means of escaping from a "prison of ignorance."

## Chapter 2

Emerging from the Shadows

The work of Copernicus, Brahe, Kepler, and Galileo began to place a conceptual basis on how the cosmology they observed actually works. Mathematical descriptions were introduced to provide the most detailed understanding of the phenomena humanly possible. They mathematically described properties of the orbits of the planets as they revolve around the sun. Building on this foundation, Newton would use mathematics to describe the motion. The shadows began to take on shape.

Without a telescope they were unable to identify the stars and planetary satellites. Only Galileo, using a telescope powerful enough to see the satellites of Jupiter, was first able to uncover other, now observable, parts of the universe. Using the telescope, he found that the sky was filled with faint stars not otherwise visible. Even so, he had no idea that more galaxies existed than our own and it was not until 1750 when Thomas Wright, an English astronomer, correctly guessed that we live on a rotating structure shaped like a pinwheel, the Milky Way, although this would not be observationally confirmed until the 1900s.

The telescope was a significant development that greatly enhanced our understanding of the structure of the universe. However, those discoveries that we today describe as simple facts required much input to reach a deeper understanding of what is going on in the universe. It is fortuitous that Isaac Newton felt compelled to enter the investigation.

Newton focused on the gravitational force. His developments in the fields of gravity and the laws of motion were widely celebrated and recognized accomplishments. Most often related about Newton, and known by every schoolchild, is that he believed that an apple falls from a tree because it experiences the gravitational force.

This description of gravity would be followed by a synthesis of past understandings by James Clerk Maxwell that led to electricity and magnetism being unified as a single electromagnetic force. It was later found that gravity and the electromagnetic forces are but two of four forces now known to influence the structure of the universe. Not only were the shadows taking shape, their potential sources were beginning to find focus.

The Forces of Nature

Sir Isaac Newton (1643–1727) was the first person to define the three fundamental laws of motion and formalize them in mathematical terms. He proved mathematically that all planetary bodies in motion under the influence of gravity traverse the sun in elliptical orbits. He is known for the development of Newtonian Mechanics, the law of Universal Gravitation, the development of the mathematics of calculus for use in precisely describing the laws of the motion of bodies, the development of classical optical theory, and he was suitably named the "Father of Classical Physics."

In a sense, Newton gave new definitions to time and space! Observed matter comes in granular or discrete form. There are chunks of matter scattered here and there throughout the universe, even down to the atom. Until the time of Newton one could argue that space and time might also exist in tiny, discrete chunks. In fact, to this day one of the approximations made to calculate the properties of space and time is to pretend that they are discrete because this makes it much easier to use computers to solve such calculations.

However, with Newton's calculus-based description of nature, the "chunky" concept became untenable. For calculus to work, one must be able to conceive of space and time in such a way that there is no smallest amount of either. Mathematically, this corresponds to a process called "taking a limit." The limit described is one where there is no lower bound on duration time or size of space. This is built into the definition of calculus. Thus, in this sense, Newton can be said to be the first person to understand space and time.

Newton was an English polymath as a physicist, mathematician, astronomer, natural philosopher, alchemist, and theologian. His *Philosophiæ Naturalis Principia Mathematica,* published in 1687, is said to be the greatest single work in the history of science. This work marked the birth of physics as a separate area of human thought, distinct from consideration of philosophy and nature; his exploration separating it from the older discipline called "natural philosophy."

Newton's discovery of natural laws laid the groundwork for classical mechanics, which mostly dominated the scientific view of the physical universe for the next three centuries and became the basis for modern engineering. A caveat has to do with the often overlooked work of Maxwell who definitively showed that "fields," in addition to Newton's clockwork view, were necessary. Newton showed that the motions of objects on Earth and the motions of the celestial bodies are governed by the same set of natural laws. Maxwell's work described mathematical fields of force that reach out through seemingly empty space.

Nobel Laureate physicist, Steven Weinberg (1933–) has said of Newton's work, "The unification of the celestial with the terrestrial—that the same laws that govern the planets in their motions govern the tides and the falling of fruit here on Earth—it was a fantastic unification of our picture of nature."

Newton's Early Years

In mathematics, Newton shares credit with Gottfried Leibniz for the development of differential calculus, one of the two basic forms of calculus; the other is known as integral calculus. He also demonstrated the generalized binomial theorem, developed "Newton's method" for approximating the zeroes of a function, and contributed to the study of infinite series. In a 2005 poll of the Royal Society about who had the greatest effect on the history of science, Newton was deemed more influential than Albert Einstein.

Newton attended The King's School, Grantham, where, from the age of about twelve until he was seventeen, he was the top student in the school. His signature can still be seen carved into a library window sill. He was later removed from school, and in October, 1659, he was to be found at Woolsthorpe-by-Colsterworth, where his mother, widowed now for a second time, attempted to make a farmer of him. It appears to have been Henry Stokes, master at the King's School, who persuaded his mother to send him back to school to complete his education. This he did at the age of eighteen, achieving an admirable final report.

In June, 1661, he was admitted to Trinity College, Cambridge. At that time, the teachings there were based on those of Aristotle, but Newton preferred to read the more advanced ideas of then-modern explorers such as Descartes, Galileo, Copernicus, and Kepler. In 1665, he discovered the generalized binomial theorem and began to develop a mathematical theory that would later become calculus. He obtained his degree in April of 1665.

Most modern historians believe that Newton and Leibniz had developed calculus independently, using their own unique notations. According to Newton's inner circle, Newton had worked out his method years before Leibniz, yet he published almost nothing about it until 1693, and did not give a full account until 1704. Meanwhile, Leibniz began publishing his methods in 1684. Leibniz's notation and differential method were universally adopted on the continent, and after 1820 or so, in the British Empire. Whereas Leibniz's notebooks show the advancement of the ideas from early stages until maturity, there is only the end product in Newton's notes. Newton claimed that he had been reluctant to publish his calculus because he feared being mocked for it.

Most of the modern notation for calculus is Leibniz's. Most surprising is that there is a third possible contender for the title of discoverer of calculus—Archimedes beat both Leibniz and Newton as the inventor of the integral form of calculus by more than seventeen hundred years!

Archimedes is regarded as one of the greatest mathematicians in history. One of his accomplishments was to provide the first accurate determination of pi. Pi is a special number in the world, the ratio of the circumference of a circle to its diameter; it's special because it has never been observed to repeat sequences of digits to any decimal place at any length to which it is computed. Although the significance of this number was known to ancient Egyptian, Babylonian, and Indian mathematicians before Archimedes, his first accurate determination of its value led some to call it "Archimedes' Constant."

Like a message in a bottle, the "Archimedes Palimpsest" began its reemergence in 1840. A palimpsest is a page from a book or scroll that has been scraped off and re-used. Many palimpsests are made from the well-processed hides of animals making them more durable than paper. In medieval Europe, when paper was rare, the practice of washing a palimpsest of its characters and reusing it for new text was common. A Biblical scholar, Constantine Tishendorf, visiting Istanbul in 1840, came across a palimpsest that contained Greek mathematical symbols. In 1906, Johan Heiberg had pages of that palimpsest photographed and published. Another scholar, Thomas Heath, translated the Greek and in 1971 an Oxford Professor, Nigel Wilson, realized that it was the lost Archimedes Palimpsest. The Archimedes Palimpsest now resides in Baltimore's Walker Art Museum. Eventually, using modern technological means, including a particle accelerator at the SLAC National Accelerator Laboratory (formerly known as the Stanford Linear Accelerator Center), and teams of scientists, it was found that the palimpsest contains seven mathematical works of Archimedes:

(a.) Equilibrium of Planes
(b.) Measurement of a Circle
(c.) On Sphere and Cylinder
(d.) On Floating Bodies
(e.) Methods of Mechanical Theorems
(f.) Spiral Lines
(g.) Stomachion

In "Methods," Archimedes describes techniques that are recognized today as the first use of integral calculus.

Archimedes was killed by a Roman soldier around 212 B.C. On his headstone is the inscription:

A solid sphere has 2/3 the volume of a circumscribed cylinder.

This inscription would cause one to conclude that this must be the resting place of a person who knew calculus before Newton or Leibniz.

The Gravitational Force

Newton was elected Lucasian Professor of Mathematics in 1669 at Cambridge University, the chair given to Paul A.M. Dirac in 1932, occupied until recently by Stephen Hawking, who retired in 2008, and currently held by Michael B. Green, one of the co-inventors of superstring theory.

At Cambridge, Newton described his understanding of the physics governing mechanics, the laws of motion, and the laws of optics, using calculus, which he invented for this purpose. Every high school student is familiar with Newton's law of gravitation— that two bodies under the influence of gravity are attracted to each other by a force that is proportional to the product of their masses and inversely proportional to the square of the distance between them.

He analyzed the physics of a body under the influence of gravity in orbit around the sun and determined the dynamical equation that describes an elliptical orbit due to the gravitational attraction between the planet and the sun. This mathematical demonstration supported Kepler's assertion, observing their motion to occur as elliptical orbits. This knowledge made it possible three centuries later for engineers to design and build the APOLLO spacecraft and send it to the moon in 1968. These simple laws have guided civilization for over three centuries and are still valid . . . at appropriate scales. . . for later, Einstein was to show that when objects move near the speed of light, Newton's "cathedral of thought" must be abandoned. Later, another group of physicists would show that when objects are too small, on the order of the size of atoms, Newton's edifice must once more be abandoned. Even today, in everyday situations, which includes most engineering, the work of Newton is a solid foundation.

As time moved forward, physicists worked independently in their individual areas of interest—heat, light, electromagnetism, fluid mechanics, etc. They characterized these phenomena and formalized their behaviors in companion scientific areas, developing and unifying the laws governing them as they sought to simplify the laws so that the mathematics would be elegant. It was difficult to perform experiments because experimental apparatus was mostly not yet developed. It was the same problem that hampered Copernicus, Brahe, and Kepler, but not Galileo, as he had access to the advanced technology of the then-new telescope. As we noted in the case of Brahe, when technology is not up to the needs of a scientist, they will often improve existing technology or invent their own to achieve their aims.

Newton's Law of gravitation was sacrosanct and remained so until Albert Einstein showed it to be incomplete and limited when he replaced it with his laws of relativity in 1905 and 1916. New devices were needed to examine the laws governing the universe. The technologies for such devices would emerge in the 19th and 20th centuries. It is technology that enables science to push back the shadows.

The Electromagnetic Force

By the 18th century, the world of physics was teeming with people searching for the laws that govern nature. They had what appeared to be a good foundation on gravity from Newton's work and they learned about electricity and magnetism through the work of scientists like Benjamin Franklin—who studied lightning, claiming it was electricity. Michael Faraday and James Clerk Maxwell were the two most influential people in the discovery and formalization of the electromagnetic force. Physicists of this era performed numerous experiments to develop an understanding of electricity and magnetism and defined the mathematical relationships and interactions between the two phenomena.

The fact that electricity and magnetism are entwined phenomena became apparent to Michael Faraday (1791–1867) who realized that electricity created magnetism and magnetism, in turn, created electricity. Using this knowledge he developed the principles of the electric induction motor. The electric induction motor demonstrates that electricity and magnetism work together to create electric currents and the forces of repulsion and attraction within wires and magnetic materials. He designed a motor that caused a rotating armature to spin. That motor could drive machinery and, when connected differently, would behave as a generator of electricity, as well. Although Faraday learned how these devices worked, he did not develop a formal mathematical explanation for them.

Maxwell's equations accurately defined how to harness energy to make motors, generators, radios, meters, industrial machinery, and many more things leading even to modern radar, X-ray machines, CAT-scanners, and MRIs. All of these behave according to his equations. To this date, no phenomenon associated with electricity or magnetism has ever deviated from the behavior predicted by Maxwell's equations.

Two American scientists (Richard Feynman [1918–1988] and Carl Sagan [1934–1996], respectively) have said of the work of Maxwell:

From a long view of the history of mankind—seen from, say, ten thousand years from now—there can be little doubt that the most significant event of the 19th century will be judged as Maxwell's discovery of the laws of electrodynamics. The American Civil War will pale into provincial insignificance in comparison with this important scientific event of the same decade.

Maxwell's equations have had a greater impact on human history than any ten presidents.

Although Newton and Einstein are better known today among the public than Maxwell, he is one of the three most highly regarded physicists in the history of the discipline. When Einstein was young he explicitly pointed to both as the models for which he strove to achieve his own accomplishment.

Maxwell had an impact on the thinking of physicists about the nature of reality. From the time that Newton completed his work, there had grown up a general philosophical viewpoint that the physical universe was a place very much like a clock. If one understood how the gears and levers worked, the operation of the clock could be understood as there are no invisible parts in a clock.

First Faraday, and later Maxwell, described aspects of the universe that are not seeable by establishing the concept of the "field." A field is thought of as being an intermediary quantity of physical significance that conveys the influence of one physical object through a region of space to another physical object with no apparent connection within the intervening region. Although a field may be invisible to the human eye, it is no less real than any physical object. It is the field of gravity that surrounds the earth that causes us to be able to distinguish up from down. The gravitational field points between these two directions. The Earth's magnetic field causes compass needles to point to the magnetic north pole. These invisible fields, while not obvious, influence human activity.

Instead of thinking of gravity as the sun magically reaching out to affect the orbit of a planet far away, a "field of gravity" is envisioned that surrounds the sun. It is this field that influences the path of the planet. The concept of the field of gravity solved one of the philosophical problems that had worried Newton. The ability of gravity to reach across space had been called by Newton, "action-at-a-distance." Though he was suspicious of this concept, it worked well enough to explain his rule for gravity and he accepted it.

Gravity and electromagnetism, two of the principal forces of nature, had been discovered and well-documented by the mid-1800s, but the world of the atom was still a great mystery to scientists who wanted to understand the link between tiny atoms and the massive universe. In the search for a unified theory, physicists believed that larger bodies in the universe must be built from smaller objects, atoms. They sought mathematical equations that would allow them to predict the universe's evolution from some set of initial conditions to any future condition—from the small atoms that form us, up to the planetary systems that surround us.

This was a strong continuation of Newton's vision of the universe as a clockwork system. It brought with it an inference that the key to a comprehensive understanding of the universe could be found by linking the small elements studied in the chemistry of that time with the methods that were successful in cosmology. This led to work by a group of scientists and especially chemists who searched for the minute particles that defined all matter.

The study of the atom, which began in the chemical industry, became the focal point of new research. The work of both physicists and chemists would contribute significantly to understanding the atom. In the 1700s, new work in chemistry led to the discovery of hydrogen by Henry Cavendish in 1756 and oxygen by Joseph Priestly in 1774.

In the 1860s, Dmitri Ivanovich Mendeleev (1834–1907) noted that the chemical elements had properties that correlated with their mass, leading him to create the periodic table of the elements in 1869, cataloging the first 63 atomic elements by their increasing mass, the number of electrons, and other chemical properties. The table has now been expanded to include more than 100 chemical elements.

The Structure of the Atom

In the 1800s little was known about the world of the small—the world of the atom. Chemists explored the behavior that describes the interactions among atoms. This was a kind of practical knowledge about the bulk behavior of atoms. It allowed the creation of new compounds and new materials having unexpected properties. These behaviors were related to how chemical elements combine by sharing electrons in their outer energy shells. The nuclei of atoms would eventually be known to contain protons and neutrons, accounting for the weight of substances, and electrons, the particles that share the atom's outer energy shells to create the compounds of chemistry.

Chemists were led to greater understanding of atomic behavior through the experiments they performed. By the 18th century, knowledge beyond gravity and electromagnetism gave way to the vigorous pursuit of atomic structure. Information about the nuclear forces of nature began to emerge from this research. By the end of the 19th century, both chemists and physicists believed that the atom was the fundamental particle of all matter—that it was indivisible. They would later learn that the atom was composed of smaller subatomic particles held together by very strong nuclear forces.

As chemists sought for a better understanding of the atom to help them design and build better products, physicists sought understanding of the atom from a purely scientific standpoint, believing that in understanding the lowest common particle of matter, the atom, it would aid them in unifying the laws of the universe. It would help them in understanding how the universe developed and of what it was composed.

Experiments performed in the late 1800s concerning the phenomenon of "natural radioactivity" revealed previously unknown instabilities in the atom that did not follow Newtonian law. The constituents of atoms—electrons, neutrons, and protons—would similarly be found to not behave like larger objects. While the electric charge of the electron is attracted to the positive charge of the proton, they are affected in their motion as well by having spin properties that need to be taken into account. These and similar unusual findings became the foundation of a new science that described particle interaction and behavior—particle physics.

This new field of research became an additional elusive character (a shadow) among the physical laws of nature. Work began in the early 1900s to make sense of these strange occurrences, and continued throughout the 20th century. Particle physics became a major thrust in the sciences of the 20th century, leading to nuclear science, atomic experimentation, and to the atomic bomb—research that truly changed history.

The Universe of the Small

When particle physics was in its infancy, hundreds of scientists were drawn to it. Many questions for research emerged from the experimental results derived as atomic and sub-atomic particles were subjected to various electric charges and magnetic forces.

When the equations of classical physics were used to describe an atom's behavior, the solutions did not yield the same results that actual experiments produced. There was difficulty in locating the position of a particle in space. It seemed as though sub-atomic particles, such as electrons, would rapidly move to many positions in the atom—sometimes the particle behaved like a billiard ball, and at other times like a wave or a force field. These non-conforming behaviors caused physicists to alter their mathematics to agree with the experimental measurements so that they could continue exploration. But, importantly, they did not know why these alterations were logically necessary.

These unexpected differences from expected classical predictions caused researchers to define new experiments that sought to resolve the anomalies. Significant development in what would become known as quantum mechanics began with Max Planck's research in 1900 that sought to understand black body radiation.

Classical physics predicted that, as one heated certain bodies, the intensity of re-radiated energy would become infinite. This was not compatible with observation. Planck instead considered the consequences of what would happen if such a body would not admit all possible energies to be absorbed. This resolved the problem. He was the first to find that energy would have to be "packaged" in discrete amounts—that is, to be "quantized."

In the world we see, when light falls on a body some of it is absorbed (thus heating that body) and some of it is reflected (otherwise it could not be seen). A black body is a hypothetical structure that absorbs all of the light that falls upon it. Therefore, exposing a black body to light will raise its temperature. This would be accurate regardless of the color of the light, extending to colors beyond the human ability to see. Planck found that each color, or frequency of light, corresponded to a particular temperature. However, in order to reach agreement between his mathematical description and experimental observation, Planck was led to the conclusion that, while the amount of energy carried by light depended on its frequency, more surprisingly, that energy had to come in discrete packets or quanta.

Planck concluded in a paper he published in 1900 that radiation energy is electromagnetic in nature and takes on discrete values (quanta) which are based on the product of its frequency, f, and a new quantity, h, which he defined as Planck's constant. Planck's constant imposes a discrete value on electromagnetic energy—a most profound discovery for which he was given the Nobel Prize in 1918.

Einstein's paper on the photoelectric effect in 1905 followed Max Planck's disclosure and underscored a new direction that would be needed to gain better understanding of quantum mechanics and particle theory. A new body of science would be needed to provide a description of the quantum behavior of particles. Particles undergo change in their energy levels due to the influence of the strong nuclear force, the electromagnetic force, and the weak nuclear force. This new theory would formalize the laws by which particles interact with each other and with the forces of nature; hence the term "quantum mechanics" to describe the characteristics of particles, just as classical mechanics governs the behaviors of the physical world.

Planck's and Einstein's research independently showed that an entirely new phenomenon was present in the world of the small—that these small particles behaved strangely and that confirming their unusual behavior required processes to occur in nature that had never before been considered. There had to be a "quantum mechanics process" at work in the shadows. A detailed understanding of these would require physicists to "invent" new mathematics! It would be necessary to further classify particles of atomic size according to such physical properties as their spin, charge, and mass.

The properties of charge and mass are widely known among non-scientists. However, "spin" is not so familiar. In 1922, two German physicists. Otto Stern and Walther Gerlach, did experiments that found that atoms, and even their components, have extra energy inside of them that resembles the kind of energy stored in, say, a spinning wheel. This is why this property is called spin. However, as was learned later, such particles do not actually spin. They simply possess internal energy that resembles the energy associated with a spinning object. The word "spin" is not to be taken literally for subatomic particles, just as we would not take literally the phrase, "he is on fire tonight!"

As you might guess, the quantum world doesn't fully resemble the classical macroscopic world, and such idioms as "spin" for quantum behaviors come, with differences to their familiar, human-scale counterparts. Unlike a spinning wheel, not all rotational speeds (and thus rotational energies) are possible.

In the world we see, wheels spin rapidly, slowly, or not at all, the transition being continuous across these states. In the world of spinning particles (at the atomic level and smaller) nature seems not to allow continuous transition in spin. Instead, particles must jump (go instantaneously) from one state to another. This is described as a "quantum leap."

Just as Planck had shown that the amount of energy carried by light comes in discrete packets or quanta, the rate of spin for elementary particles also comes only in discrete packets. These spin rates are multiples of the spin rate of the electron. They are described as half-integer or full-integer multiples. Bosons (force-carrying particles) are classified as particles having integer spin rates (0, 1, 2, 3, etc.). Particles with half-integer spin rates (0, 1/2, 3/2, 5/2, etc.) are fermions (matter particles). Particles can be described as spinning clockwise or counter-clockwise, relative to their direction of motion.. The physics community has given a special name to the rate of spin of an electron. It is called "h-bar."

Planck's discovery that black body radiation is constituted of electromagnetic waves that carry energy in packets was further strengthened by Einstein's paper in 1905 which defined the photon particle as a quantum unit of electro-magnetic force. The physicist, Louis de Broglie (1892–1987), following this line of investigation, hypothesized that if electromagnetic radiation could behave as both wave and particle (photon), then matter itself might also possess both wave and particle aspects. His hypothesis turned out to be correct, a fact that has stunning consequences, as we will explore later.

Einstein's 1905 paper showed that when light is shone on the atoms of any element, it increases that element's energy level causing an electron release to occur. He described this phenomenon as electrons interacting with light through what he called "photons," small quantized particles that carry the light's rays (an electromagnetic field). When an electron interacts with a photon, the electron can gain energy. In an atom, if the photon possesses enough energy, it can raise the overall energy level of the electron even while it remains part of the parent atom. With even more energy, a photon can "elevate" the electron from the atom, allowing it to flow freely through some other medium in the form of an electric current..

In these papers on special relativity and the photoelectric effect, Einstein showed with his equation, $E = mc^2$, that energy and mass are equivalent. Einstein further recognized that if one runs the equation backward—starting from matter and finding a way to convert it to energy—the conversion releases a tremendous amount of energy. Decades later, this simple observation would be understood as the seed of atomic weapons and atomic energy. Today, it is also the driving force that allows the discovery of new subatomic particles like the Higgs boson. It can also be used to save lives as through its application in the PET scan for non-invasively detecting the presence of cancer tumors.

Niels Bohr (1885–1962) was a Danish physicist who sought to define the structure of the atom. His studies were instrumental in the development of the first model of the atomic structure of hydrogen.

Physicists realized that subatomic particle characteristics and behavior needed to be unified in an atomic theory. Niels Bohr created an institute for this purpose. While at the Bohr Institute, Erwin Schrödinger developed the mathematics that governs the interaction of electrons in the atom. Schrödinger created his wave equation in 1925 and published the result in 1926. Werner Heisenberg (1901–1976), also working at the Bohr Institute, studied particle behavior and Schrödinger's equation. His studies showed that particles move in indefinite paths; that they might be found at a given point now, and are then observed at a different position an instant later.

This led him to define the "Uncertainty Principle," which he published in 1927. This paper showed that one can measure a particle's position precisely, but its momentum (the product of its mass "m" and its velocity "v") will be uncertain; or the reverse—that position is "smeared" when momentum is known. This effect is not experienced with large objects when observed in everyday situations, but small particles *do* exhibit this probabilistic behavior. Schrödinger developed a mathematical structure that could act as both particle and wave, but, according to his wave equation, one could find the position or momentum of a particle based on its wave function, Ψ(psi).

This body of work was instrumental in building the science that is quantum mechanics, the development of which proceeded through the 1930s and 1940s to culminate in the full mathematical theory of quantum electrodynamics (QED) that physicists use today. One foundational piece of this work was enhanced by Richard Feynman's work when he developed a mathematical diagraming technique that facilitates analysis of the quantum behavior of particles interacting in force fields. These accomplishments created a framework of mathematical rigor and formality for analyzing particle interaction with the electromagnetic force.

Schrödinger's equation accurately describes electrons only when they are moving at velocities that are small compared to the speed of light. Paul Dirac, on the other hand, sought to take advantage of Einstein's relativity and include these effects in the mathematics. The consequence of these differing responses formed two results: one based on Schrödinger's assumptions that did not include these effects; and Dirac's assumptions that did take these effects into account. Both solutions yielded similar results although Dirac's solution was more accurate at particle speeds approaching the speed of light. Furthermore, Dirac's mathematics made predictions due only to the requirement of consistency that did not follow from Schrödinger's equation. Schrödinger's equation is valid for particles possessing spin, but that property must be manually added into his equations; Dirac's formulation, on the other hand, predicted the existence and nature of spin as a consequence of uniting quantum mechanics with special relativity.

An additional prediction of Dirac's equation is quite widely known, even outside of science. This other prediction is that anti-matter should be seen in our universe. Anti-particles possess an opposite charge from ordinary matter (e.g. the anti-electron, or "positron," has a positive electrical charge) and should both be brought to the same location, this would cause both to be annihilated and replaced by an amount of energy precisely in accord with Einstein's equation.

Physicists demonstrated, as de Broglie had hypothesized, that sometimes particles behave as particles, and sometimes as waves. We can make this clearer by visualizing an electron gun that emits electrons toward two slits in a metal plate (see Figure 2.1). The electrons travel in the direction of the plate, some of them striking the metal plate while others pass through the slits.

Now cover up slit A. The electrons that go through slit B strike the back plate and are distributed along that plate. A sensor behind the slit counts the number of electrons that strike the plate. Their sum is shown as the distribution curve B. Should slit B be closed and slit A opened, and the test repeated, the result is found to create the same shape, just in a different location, as occurred during the first test.

Electron Distribution

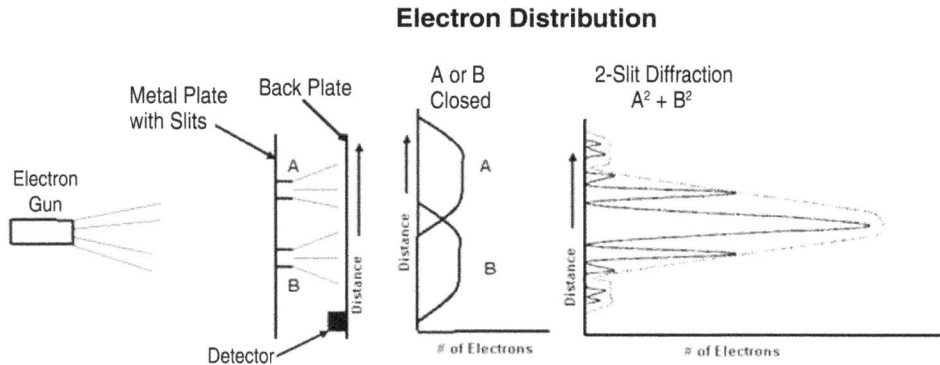

Figure 2.1 Wave–Particle Duality

These results show that each of the curves represents the number of electrons that exited through each slit and were distributed along the back plate acting as though the electrons were individual particles like bullets.

Now let's leave both slits open and repeat the test. Again, some of the particles strike the plate while others go through the slits. This time a single curve is obtained. These results are as though a wave had passed through the slits; an interference between the two electron waves that is much like water waves interfering similarly with each other when exiting two similar holes. One may conclude that the electrons are behaving as though they are waves, not particles.

In summary, research found that photons are quantum particles that carry light energy and behave both as particles and as waves. Light influences electron behavior and photons carry the electromagnetic force field in the form of quantum amounts of energy, confirming the quantum nature of particles and offering promise that these phenomena would reach into the understanding of how the universe works and help to reveal the laws of its operation. Even more stunning, what is true for light is true for matter. Fundamentally, all behavior is wave behavior, but it is noticed only at the scale of atoms. . . and smaller.

The Beginnings of String Theory

String theory, a new way to view the particle world, extrapolates our knowledge even farther beyond the billiard ball behaviors of early particle physics. Experimental results obtained in the late 1960s led to a conjecture that nuclear particles tend to behave as though they are spaghetti-like objects and exhibit spring-like properties. This line of thought ultimately failed and was replaced by the notions of quarks, gluons, and the mathematics of their interactions in the context of quantum chromodynamics (QCD). But the intial idea has not died, finding a more universal use since its birth.

In principle, a quantum theory of particles coupled with Einstein's theory of relativity, should have captured the fundamentals of the theory of the universe as it was understood in the early-to-mid 20th century. Einstein's Theory of General Relativity apparently accurately describes how the universe behaves at the scales of planets and larger (the macro-universe). In particular it does a remarkably accurate job of describing gravity. Physicists (at least some) believed for a long time that it should also work at very tiny scales equally well.

The "standard model" (created by large numbers of physicists working collaboratively) describes the world of atoms and smaller structures. The micro-universe is its domain of applicability. We will come to investigate it in great detail later.

But the two approaches, macro and quantum, could not be reconciled into one equation governing all characteristics of the universe because physicists could not find a way to include quantum gravity in the mix. This shadow persists into the 21st century, stalling efforts to move to a conclusion.

The quantum mechanics of particles is a probabilistic theory, which, when Einstein learned about it, said "God does not throw dice." What he meant is that the physics of the universe should be a deterministic process. The equations defining its laws should be deterministic in nature. He questioned the probabilistic view until his death, believing that an elegant, continuous solution is more likely than a probabilistic one. However, the quantum theory has been extremely successful in predicting particle behavior at the atomic level, even uncertainty seems fundamental to how the universe works.

The Einstein Revolution—The Universe in Turmoil

Einstein's revelations caused a great upheaval in the physics community in the 20th century. He became interested in the speed of light and how it affected the behavior of the universe. At first physicists questioned his equations, which predicted unusual behaviors in the universe. For example, his statement, "physical laws have the same value in every frame of reference," was a highly controversial assertion to those who adhered to Newton's laws because we observe that different frames of reference impose specific conditions on the objects that are placed within them. For example, one frame of reference might be at rest, while another might be moving relative to the first.

If you are in an airplane traveling at a constant speed of 500 miles per hour and you walk along the aisle, your speed, when viewed from the ground, will be 500 mph plus or minus your walking speed, depending on the direction in which you are walking. If you walk toward the rear of the plane your speed subtracts from the plane's speed, making your net speed 500 mph minus your walking speed. Your walking speed is no different from the speed at which you walk in your home. This is true because your reference frame is at rest in your home.

In another case, a plane coming toward you at a speed of 500 mph appears to pass you at a speed equal to the sum of your speed and the other plane's speed, or 1000 mph. Given this, Einstein asked the question: what happens when the speed of the frame of reference (the airplane) reaches a value close to the speed of light? He found that the two vehicles will appear to pass each other at a speed slower than the speed of light. This is strange because no matter how close to the speed of light the two are travelling, their speed of approach will always be smaller than the speed of light.

This startling result is because nothing can go faster than the speed of light—no matter its frame of reference.

Einstein showed that, at or near the speed of light, our present knowledge of space, time, and motion deviates considerably from those that were put forth by Isaac Newton. Einstein's initial insight into this problem and his resolution given in 1905 constitute his theory of special relativity. However, even that was restricted. Above we considered the case of a passenger walking to the front or the back of a plane in motion. Einstein's theory of special relativity is capable of giving precise and accurate description of this even if the passenger and the plane are moving at close to the speed of light. The theory of special relativity runs into trouble, however, if the plane and its passengers are immersed in a gravitational field. To handle this common situation, Einstein needed to generalize his notions of space, time, and motion. This led to the general theory of relativity.

A decade later, Einstein proposed that the mathematics of Georg Riemann (1826–1866), the father of non-Euclidean geometry, should be used for this more general situation. It had a marvelous capacity that could be used to describe the influence of gravitational behavior in our universe. Thus, Einstein declared the necessity to consider the curvature of space in defining gravity since this was intrinsic to the mathematics of Riemann.

The notions of space and time proposed by Isaac Newton had stood the test of time until Einstein published his paper on special relativity. The special theory of relativity does not include gravity, but considers that the speed of light is a universal constant at 3×10^8 m/sec in any frame of reference. By any frame of reference it is meant that this number is the same on Earth, in space, on Mars, in a spaceship and everywhere else, whether the frame of reference is in motion or static. . . with respect to a reference point. There are things that occur to objects in these frames of reference that vary according to the behavior of the particular frame of reference in which the object is found, but the speed of light remains a constant in any frame of reference.

In Einstein's general relativity paper the gravitational field is described as a space-time continuum that is warped by the presence of large bodies of matter and energy, making it appear as though the bodies are attracted to each other by a force (as Newton had described it) that is proportional to the product of the masses and diminishes with the square of the distance between them.

Einstein showed that time is a fourth dimension and the gravitational field is more correctly a space-time fabric that warps with the presence and motion of large bodies in the universe. As these large bodies move in the fabric, the warping moves with them. Because of this warping, the space-time fabric becomes curved, the density of the matter determining the amount of the curvature.

Light rays from distant stars are bent in the presence of a large body like the sun because light wants to travel on the shortest paths in space-time and those paths are bent as they move around massive objects. Einstein's prediction for the degree of this bending was proven in 1919 during an eclipse of the sun, by showing that the light from a star was bent as it passed near the large body of our sun.

In 1929 Edwin Hubble (1889–1953) discovered a shift in received light toward the red wavelengths (a Doppler shift), demonstrating that all galaxies were receding from Earth in all directions, dependent on their distance from Earth. This implies that the universe behaves as though it is on a bubble or it is on a balloon that is inflating.

One perceives a Doppler shift when the siren of an emergency vehicle is heard passing. At first the siren's sound is high-pitched, but lowers as it passes. This is a Doppler shift in the frequency of the sound. The same thing can happen to light waves. Light sources moving away from one are shifted to lower frequencies of light (more red), while those moving toward one are shifted to higher frequencies (more blue).

If the expansion continues forever, the Milky Way will eventually find itself alone and the temperature of space will become colder because the illuminated bodies that convey heat are moving away from each other (including Earth), causing the temperature of the universe to drop.

Enter the "Big Bang"

Physicists immediately began to alter their view of the universe after Einstein released his papers on relativity. Alexander Friedmann (1888–1925) and Georges Lemaitre (1894–1966), independently developed models of the universe, solving Einstein's equations, thus offering supporting evidence that the universe could be expanding or contracting. This work was mathematical and supported the idea of an expansion that Lemaitre suggested to have been a "primeval atom" from which the whole of the universe emerged.

Physicists today believe that a process, now generally referred to as "the big bang," rather than the primeval atom, is a more accurate description of this behavior. It was reasoned that if the process were a motion picture, and if the motion picture could be run backward, the universe would ultimately shrink to a point in time, $t = 0$, when it would be a very small entity at an extremely high temperature and density. This idea came to the fore because of the belief that the behavior of the universe is deterministic. It was further theorized that, at the point of the beginning of the universe there would have been a big bang—not actually a bang, but a starting condition at a singularity (a point before which there is no definable state of existence).

Today most of those then-new ideas are accepted, having been supported by observations over the years. Einstein's revolution is considered to be over and the world has accepted the fact that there are four dimensions, one being time. The concept of a space-time fabric is used throughout all of it. Satellite observations confirm that these kinds of fluctuations do occur in nature. When large bodies move, the space-time fabric warps in all dimensions. Recently, as we will show later, physicists have detected the rippling of the space-time fabric—gravitational waves—resultant of one of the most cataclysmic phenomena in the universe.

Atomic Experimentation Techniques

Considerable research was accomplished from the 1930s through the 1950s to develop particle theory and the quantum mechanics of particles. These efforts ascertained that atoms are not the most fundamental particles.

The cyclotron, developed in 1929 by Ernest Lawrence, was the primary instrument used to examine atomic behavior. The cyclotron accelerates particles to very high speeds within the cavity of an electron tube called a magnetron.

Cyclotrons

Cyclotrons are devices that allow an experimenter to impart great amounts of energy to particles. Electrons or protons (or any other type of particle) are inserted into the cavity of the cyclotron. Electric and magnetic fields are then applied to the cavity causing the particles to react to those fields. A particle having no electric charge will travel a different path through the device than one with an electric charge. Electric and magnetic field properties, such as alternating versus fixed fields, will cause particles to change direction, speed, and path.

In a cyclotron, electrons are created by heating the cathode of the magnetron using a filament (a wire that heats because of its high resistance to electric current flow, much as a light bulb's filament heats). Electrons created by this means are released from the cathode and are drawn to the anode (creating a current flow) by placing a high positive voltage between the cathode and the anode. This high voltage accelerates a stream of electrons (which are negatively charged particles) across the cavity toward the positively charged anode, following a path that can be made to curve by the influence of magnetic fields applied within the cavity.

The degree of curvature of the path varies with the strength of the applied magnetic field. As the electron stream follows this altered path it can be observed by instruments to determine how it reacts to the forces and charges that are applied to it.

Particle Accelerators

After World War II, particle acceleration science evolved rapidly. Particle accelerator design is quite different from that of the cyclotron. Particles are injected into a large circular tunnel. The particles are brought to a speed close to the speed of light. Two types of experiments are conducted in these systems. In the first type, two particle streams traveling in opposite directions are released into parallel circular accelerator tunnels where electric fields are used to speed up particles along the direction of their flight and magnetic fields are used to steer those particles to keep them traveling in a circle. Sped up to nearly the speed of light, the particles are caused to collide at a specific crossover point in the tunnels. Detectors are positioned there to observe the effects of the collision.

At the occurrence of these high energy collisions new particles are formed and appear in the collision debris, some of them decaying rapidly to other particles, while others do not. Their behavior indicates how particles might have behaved and interacted at a time just after the big bang. Figure 2.2 shows how such a collision might appear.

The energy level in the collider sets the conditions of the collision. These conditions determine the types of particles that will be produced. The collision produces an image like that shown in Figure 2.2. The debris is shown exploding in all directions in the detector. Hardware and software systems sift the debris, cataloging the kinds of particles produced and recording what takes place. Researchers monitor and later analyze this data, obtaining specific information about each particle involved in the experiment.

Figure 2.2 A proton-proton collision imaged by the ATLAS Experiment.

© CERN

The second approach is to accelerate a single stream of particles that is then aimed at a specific target (a metal object of a specific material). When the target is struck by the stream, you can both knock existing particles out of the target as well as create new ones. This data is then evaluated and analyzed.

A common unit of particle energy is the "electron-Volt" (eV). This is the amount of energy gained by a particle that experiences a one-volt potential difference. Electrons moving between the terminals of a AA battery will gain an energy of 1.5eV. Similarly, electrons going through a lamp that is plugged into a wall socket in the U.S. will experience a 120eV energy change. Today's colliders have energy levels as high as the Tera-electron-volt range (trillions of electron volts), expressed as a value, TeV.

The energy level at which collision is brought about determines the interactions that occur and are observed. Our universe currently exists in a state of relative low average energy density. At times close to the big bang, the average energy density was extreme and high. Thus, to observe behaviors our universe would have exhibited near to the time of the big bang, we have to create enormous energies that are concentrated in very small regions of space.

One of the secrets of our universe that the study of physics has revealed is that when higher energies are reached, the universe is more symmetrical. . . more symmetrical, meaning that it is easier and simpler to describe. For example, a basketball has a very symmetric shape. In being more symmetrical, it is easier to describe a basketball to another than it is to describe a random lump of mashed potatoes. The energy of a collider is successively increased upward until the forces holding the particles together are overcome and their force carriers can be observed as they separate into their more symmetrical conditions. From this, we can learn the character of the fundamental laws of nature.

It is possible to learn much from these collisions about the initial moments after the big bang and the force and particle interactions that occurred at that time. This work is ongoing at such facilities as the Large Hadron Collider (LHC) at CERN, near Geneva in Switzerland, and the Relativistic Heavy Ion Collider (RHIC) at The Brookhaven National Laboratory in Long Island, New York.

In the early 1980s, physicists in the United States proposed a proton collider to reach these higher energies that was hoped would reveal the then-mythical Higgs boson. This device was named the "superconducting super collider" or SSC. It was designed to collide at extremely high speed at energy levels as high as 20 TeV per single proton. It was hoped that such energy levels might create new interactions that would release heavier particles such as the Higgs particle. However, in 1993, the SSC project was cancelled. Although it is often said that the reason was cost, its estimated cost would have been equivalent to a few B-2 bombers. Even today, the lessons and technology of the LHC and SSC are used in the planning of bigger, more powerful future machines. From today's vantage point, another point is clear. As the SSC was designed to operate at about three times the initial energy of the LHC, the Higgs particle (discussed later) could have been discovered a decade earlier.

The energy level of one beam at the LHC is presently at 6.5 TeV. The ultimate energy per beam achievable at the LHC is 7 TeV. This machine has already made fundamental discoveries about the nature of the universe that we will discuss later.

Enter Hyperspace

Particle behavior corroborated by laboratory simulations and experimentation during the 20th century has given high confidence that certain interactions and behaviors occurred at the time of the big bang. These findings point to a period between the onset of the big bang until just minutes after its occurrence, a period during which much happened that prompted the expansion of primordial matter-energy into the formation of early particle matter and plasma. The period just before 10^{-43} seconds after the onset (referred to as the Planck period) is assumed to be the interval that contains the singularity at the inception of the big bang.

Physicists, based on past experience, expect that there was a time in the universe, early on and at very high energy just after the big bang, when gravity was united with all the other forces of nature and itself had a quantum of energy, the "graviton" (akin to the photon of light). However, a complete mathematical theory of nature that successfully includes this idea and makes testable predictions continues to be elusive.

In the 1970s, the inability of physicists to unify the forces of nature and complete the unification of the standard model prompted a search for new pathways to make it possible for them to construct a unifying model of the universe.

New work has moved onto a different path—that which addresses higher dimensionality. The best mathematical tool we have now, the standard model, is only known to be reliable up to the energies so-far explored at the LHC. Therefore, future colliders will enter uncharted territory and we may make only incremental improvements to what we know—or, we may discover a whole new realm of physical laws, unanticipated given our past experience. String theory, most particularly extra-dimensional effects and supersymmetry, now appear to be major targets for experimentation. New information should soon shed light on the number of extra dimensions there might be (if any), and whether and how supersymmetery will figure into these new findings (if at all).

Newly conceived-of fundamental particles called strings emerged from research during the 1960s and demonstrated that higher dimensions ("hyperspace") and supersymmetery would be required if the behaviors and interactions of strings were to mimic the standard model. Strings would require extra dimensions (or some other means) to make it possible to unify the forces of nature with all of the particles and forces of the standard model. This unification would include the incorporation of the graviton into the standard model.

The extra-dimensional path emerged in a paper written in 1919 by the physicist-mathematician, Theodor Kaluza and was modified later by another physicist, Oskar Klein. The first showed that it was mathematically possible to unify Einstein's gravity field equations with Maxwell's equations of electromagnetic field theory by employing an extra dimension (a fifth dimension).

The approach begins by using Einstein's equation, but doing so in a universe that has a fifth dimension, one beyond those of the usual three for space and one for time. This approach demonstrated that when a fifth dimension is added, the Maxwell equations also result naturally from the mathematics. This fifth dimension, Klein further showed, can be compactified into a small circle (10^{-33}cm) that appears at every location point in the space and, due to its small size, avoid detection.

Here the word "compactify" is quite appropriate and to make this use clearer, one can recall a scene from the James Bond movie *Goldfinger* in which an automobile is placed into a machine that compactifies it into a much smaller block of metal. Mathematically speaking, Klein took Kaluza's fifth dimension and carried out the same process. He further showed that doing these mathematical operations would have no impact on the four-dimensional behavior and interactions that are measured by the instruments of the time.

Debate on Kaluza theory proceeded and extra-dimensional behavior was tabled for years owing to the intensive work in the new quantum mechanics of the day. It was revived later, during the 1970s, as physicists, seeing that they had reached another stumbling block in their quest to unify gravity with the three other forces of nature, embraced the extra-dimensional pathway of string theory.

String theory had matured by a significant degree, offering a clear opportunity to unify gravity with the other forces of nature and the known particles. Four-dimensional string theory is a work in progress as its mathematics must evolve to maturity, so we continue to entertain multiple dimensions as a mathematical vehicle.

Extra-dimensional string theory unifies the equations of the macro- and micro-worlds into a single quantum gravity framework. The term framework, rather than complete theory, is used here because string theory mathematics sets up a situation wherein unification can occur, but it is not yet understood just what it means to have the ten dimensions demanded by the theory. The extra-dimensional framework allows physicists to study string phenomena and develop the laws that govern their behavior, while more conventional approaches continue to be pursued.

It is hoped by numbers of physicists that experiments can be performed with the LHC to assess the existence of extra-dimensional behaviors of strings. If experiments cannot be devised, that will be a signal that four-dimensional theories may have merit. Experiments will have to be done to establish concrete attributes of string theory, which, for now, are but a mathematical construct—another of the shadows on the wall.

Hyperspace is a major departure from the three space and one time dimensions familiar to our everyday life. If there are extra dimensions, how are they hidden from us? Does their hiding have consequences we can observe? How many such dimensions are there? If there are none beyond the four, can string theory be made to work in just four dimensions? These are major questions that we will explore as we move along.

The quantum theory, while setting the stage for continuing discussion of the universe, may fall short for the reader as it does not presently add much to the description of the universe. It is known to be a very reliable description of the subatomic particles that make up the atom, and all of the cousins of those particles, but atomic matter (and those known cousins) are not the bulk of what makes up the universe, as we shall see later. We are not yet certain that quantum theory is the correct and most complete description of nature on its smallest scales.

The "Authority" of Scientific Observation

Scientific observation is the arbiter of theory because it tells us whether these theories are reflective of reality. Observation has built the base for our understanding of the reality of the universe. Observation has been an active methodology employed since the beginnings of the explorations of mankind. Scientific observation builds reality into mathematical equations.

Scientists have performed many experiments and made observations of the real world that have verified some ideas about reality while disproving others. Those new discoveries and constructs were driven by data that made it possible to express behaviors mathematically. The question arises whether this new mathematics presents a restricted view of the universe. Twentieth century science supported and reinforced Newton's laws with little ambiguity. Successful spacecraft missions and the APOLLO mission to the moon have completely proven that Newton's laws of motion work on worlds beyond our own.

Einstein's theories have been put to the test in many forms and have been shown to be sound theories. Einstein's predictions of the existence of many unexpected phenomena have proven to be correct.

Careful measurement of the light in the universe shows that the universe is at least 13.8 billion years old and that there must have been a "big bang" or some other single event that was its beginning. Satellite findings reveal the remnants of a big bang that offers strong evidence to confirm that such a beginning did occur and is supported by ripples in the space-time fabric that are caused by clusters of matter and energy in arrangements formed during the beginning of this event.

Careful observation of the expansions of the universe and the light left over from the big bang reveals stranger cosmos than even Einstein's imaginings could have dreamed. As we will find, there is an unseen form of matter whose gravitational hand guides the formation of structure in the universe. This is "dark matter." The expansions of the cosmos appears to be increasing in pace. This is attributed to another kind of energy of space—"dark energy"—whose origin is not yet well understood. These are *major* challenges to a complete understanding of nature, since together these dark components appear to account for 95% of the stuff in the universe.

History teaches that just when one thinks one has conquered the unknown universe, one finds that only the surface of what needs to be known has been-scratched. In the chapters that follow we will continue to explore. The universe will slowly reveal more of itself under this pursuit. The reality we see in the heavens becomes less obscured as the shadows are lit by knowledge.

## Chapter 3

A View from the Shadows

If ever a science has danced around in the shadows, it was quantum theory in the early 1900s when particle physics, as we know it now, came into being. The heliocentric view of the universe had been firmly in place for centuries, and nature was beginning to be seen from a different perspective owing to advances in technology. Cyclotrons and other particle accelerators would soon reveal new characteristics of particles even smaller than the nucleus, and, more shockingly, produce new particles that had not been predicted. Improvements to the telescope opened the heavens to greater scrutiny than was ever before imagined.

Quantum mechanics, the mathematical description of the motion and interactions of subatomic particles, began in the early 20th century and would become key to understanding nature. Quantum mechanics prompted a new theory of the particle world, new mathematics, and an evolution of experimental apparatus.

Albert Einstein spearheaded a different conceptual leap forward by updating the description of gravity and defining the space-time characteristic of the universe. A century-long development of tools enhanced humanity's ability to move its understanding forward. Satellite and sensory technology developed during the last fifty years has been instrumental in observing and improving our understanding of cosmological behavior. Developments such as semi-conductors and the laser continue to evolve. Nanotechnology (dealing with objects and dimensions smaller than one billionth of a meter), first discussed in the middle of the twentieth century, has spurred further discovery.

The 20th century provided increasingly convincing scientific proof that the big bang theory is an accurate description of the birth of our universe by way of significant data collected by the COBE (Cosmic Background Explorer), WMAP (Wilkinson Microwave Anisotropy Project), and Planck satellites. They have provided observation of subtle patterns of light from the big bang that have been turned into information on the primordial characteristics of the universe and portray a detailed history of its evolution. In the optical realm, the

Hubble telescope's high resolution has shown us vivid close-up pictures of the universe that are to be surpassed by the James Webb Space Telescope (JWST) that is scheduled for launch in 2018.

New technologies, analytical tools, and facilities are influencing modern theories that originated with the observations and discoveries of the early 1900s, providing us better explanations of particle and celestial behavior. These studies have combined to improve our understanding of the standard model of particles and interactions, which now incorporates relativistic behavior.

The destiny of the universe is hardly a scientific certainty. For example, if the present circumstance continues unabated, the continuing expansion of the universe will affect life. As the planets, galaxies, and other objects continue to recede from the earth, the surrounding medium will become colder. Should this process continue indefinitely, the universe will become so cold that eventually stars and planets cannot form, or, if the expansion somehow reverses itself, it will compress and become so hot that the universe will return to being a soup of subatomic particles before a final collapse to a singularity.

It has been known since Newton that other bodies, particularly large ones like, say, Jupiter, influence the orbital paths of surrounding planets and all other objects in space-time due to gravitational effects. Knowing something about one thing causes us to question another.

For example, a question that presently stumps physicists is, how does the gravitational field fit into the scheme of things? Why is it very weak compared to the electromagnetic force and the nuclear forces? It is 10^{-38} times weaker than the strong nuclear force. Called the "hierarchy problem," this question has been the impetus for detailed research in many areas (including string theory) that seek to understand these force fields so that they can be unified with the laws that govern the behaviors of particles and the macro-universe.

We previously described the Bohr model of the atom, that depicts electrons orbiting a central nucleus as planets orbit the sun in well-defined paths, but this was not how the atom was subsequently observed to actually behave. The development of a more mature quantum theory in the wake of the Bohr model revealed that the properties of the electron (mass, charge, and spin) all play a role in the structure of the atom, including how one imagines it to look. The atom is not a planetary system with electrons in well-defined places in space, but more a blur of electron *probability*, with each electron having a definite energy but no singular and defined location in space. Uncertainty plays a pivotal role in the quantum world, as we have hinted. Newton's clockwork universe has little confirmation in this miniature cosmos.

Physicists know that the atom is divisible. Even though it *is* divisible, the atom can, for most purposes in ordinary life, and even in chemistry, still be thought of as a fundamental particle of matter because it influences the behavior of larger structures; molecules, cells, etc. This is perhaps one of the most puzzling attributes of the mathematics of quantum theory. It allows physicists to think of these minute parts of nature as either a cloud of probabilities or a particle. We now know that protons and neutrons, the basic constituents of the nucleus, can be described as combinations of other components (sub-atomic particles) called quarks.

The standard model, to be detailed later, is considered to be the culmination of quantum thinking. The physics community believes that, to be pursuable, any successor quantum theory must include the standard model. The standard model accurately describes the observation of particle interaction with the forces of nature in real-world terms. Thus, the results found in the standard model must be reproducible by any greater following theory.

In comparison to the science of the small, the universe has been increasingly better understood by large-scale studies of phenomena such as black holes and dark matter. Investigating observable physical objects such as galaxies (and speculative ones like worm holes), and a large array of celestial objects has enabled physicists to understand the cosmology of the universe while simultaneously advancing their understanding of the micro-universe.

Particle physics and astronomy, as two perspectives on radically different distance scales, have revealed both truths and mysteries in our universe. As we will see later, string theory seeks to unite these two disparate scales into a singular framework. However string theory remains elegant mathematics having no discernable or unique consequence on the observable cosmos, while the study of the sub-atomic and the study of the cosmological, each enjoy a level of precision observability that makes them daily useful tools.

Newtonian Physics

Newtonian physics has been the most instrumental form of study to date in advancing the science needed to understand the universe. Isaac Newton undertook the scientific approach of observation, theoretical description, mathematical analysis, and experimentation as his method for rigorous research in science. Newton's accomplishments and the equations and laws he defined continue to be valid today (within everyday domains) for most physical phenomena that occur on Earth.

His laws of classical mechanics are simple:

First Law: A body in uniform motion (at a constant speed and direction) tends to remain in motion at that speed and in that direction unless acted upon by an unbalanced force. (The law of Inertia)

Second Law: The relationship between an object's mass, m, its acceleration, a, and the applied force, F, is defined by the relation $F = m \times a$. Acceleration and force are vectors having both magnitude and direction (as indicated by their symbols being displayed in bold italic type); for this force vector the direction of the force is the same as the direction of the acceleration.

Third Law: For every action there is an equal and opposite reaction.

The first law recognizes Galileo's concept of inertia. It states that every body has a property that makes it tend to resist a change in motion or a change in direction. The smaller the body, the more easily it can change direction, but a large body, like a huge ship, tends to resist changing its direction. All bodies have inertia given by their mass, which causes them to resist changes in motion. What we call mass, a measure of the substance of matter, is also a measure of resistance to changes in motion.

The second law of motion is the most powerful of the three laws, because it allows one to make quantitative calculations describing the dynamics of a problem: how a body's speed, position, and direction change when forces are applied to it. This calculation is then used to describe how the path of an object is affected by the applied force. For example, a baseball moves in a prescribed path that is determined by the pitcher's arm and gravity.

The second law also marks one of the key moments in mathematics and science . . . the birth of differential calculus. Although it is not widely recognized by non-scientists, the second law is actually a statement about calculus! It marked the first time that a mathematical development called differential equations appeared in print. The subsequent work of two other giants of physics, James Clerk Maxwell and Albert Einstein, would not have been possible without this development by Newton.

The third law describes a situation with which we are all familiar: what happens when one steps from a boat onto a dock. Stepping from the boat causes the boat to move in the opposite direction. Rocket propulsion is another example.

Newton went on to define the laws by which objects in the universe are attracted to each other under the influence of the force of gravity as shown by the diagram in Figure 3.1. The two masses shown are attracted by a force attributable to their mass.

The symbol "F_{12}" indicates the force on mass number one caused by mass number two; "F_{21}" indicates the force on mass number two caused by mass number one. Before Newton, no one knew that these two forces must be equal but pointed in opposite directions. In the diagram, the force does not have to be caused by gravity. Newton taught via his third law that it is to apply to all

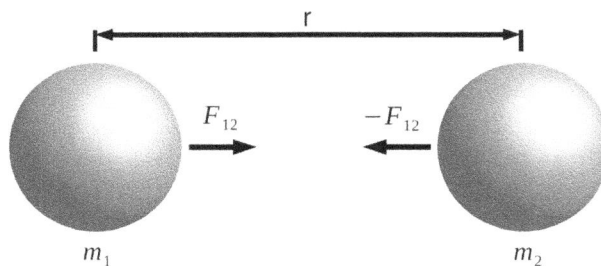

$$r$$

$$F_{12} \qquad -F_{12}$$

$$m_1 \qquad\qquad\qquad m_2$$

Figure 3.1   Newton's Law of Gravitation

forces, but, as a special case, it can also be applied to gravity. Here he went even farther because he determined how large the forces must be in terms of the size of the masses and their separation.

Law of Gravity: Two bodies are attracted to each other by gravity, a force that is proportional to the product of their masses and inversely proportional to the square of the distance between them.

When expressed as an equation it appears as:

$$F_G = G\,[(m_1 m_2)/r^2]$$

Note that m_1 and m_2 are the two masses, and G is the gravitational constant measured by Newton, given below in metric units

$$G = 6.674 \times 10^{-8}\,cm^3/gm\text{-}sec^2 \text{ (metric units)}$$

James Clerk Maxwell (1831–1879), lived and worked in the time between the lives of Newton and Einstein. Though not as widely known by today's public, he is regarded among physicists as an equal to the other two. His legacy is remarkable across a number of areas of physics. For example, he was the first person to understand how the human eye perceives color and made profound contributions to the concepts of heat and entropy, but his most widely felt impact on today's world are the laws that govern the behavior of electricity and magnets. These laws are called "Maxwell's Equations" and they are four in number. They did for electricity and magnets what Newton's laws did for forces, gravity, and motion. Einstein consciously styled himself after Maxwell and was successful in doing for space and time what his predecessors did in other realms, thus moving some of our understanding out of the shadows.

A Man and a Revolution in Physics

Albert Einstein received his doctorate degree in physics in 1905 from the Swiss Federal Polytechnic School in Zurich. He was little known in scientific circles at that time, working as an obscure technical assistant in the patent office. In 1905, the same year in which he received his Ph.D., Einstein published five major science papers. Three of those papers were of great import to our subject; the first on special relativity, the second on the photoelectric effect, and the third in the field of statistical mechanics, elaborating on the work of Ludwig Boltzmann (the Boltzmann constant), relating the pressure and volume of a gas to the number of its molecules and its absolute temperature. Einstein's most famous paper, on general relativity, which defines gravitational influence on the world, the concept of the space-time fabric, and the fourth dimension of time, was published in 1915.

Einstein was well-versed in the ideas of his day, but his distinguishing characteristic as a young physicist was a willingness to set aside grounded assumptions in favor of describing nature as it is, not as people would wish it to be. For instance, Maxwell's equations of electromagnetism described light as a wave moving at a fixed speed through empty space, but the Newtonian view of a wave (a mechanical distortion of a medium) demanded a medium in which the wave (light) moved. This medium was called the "aether." The brilliant physicist, Hendrik Lorentz, (1853–1928) developed a complete mathematical description of the compression of physical bodies as they moved through the aether, in an attempt to resolve some of the paradoxes between the Newtonian and Maxwellian views of physical bodies in motion.

The properties of the aether were well-predicted, and it was sought, its existence to be demonstrated with experiments. These experiments, the most famous by Albert Michelson (1852–1931) and Edward Morley, (1838–1923) failed to detect the aether. Einstein embraced this "null result," accepting that the speed of light, which appeared wholly constant regardless of motion in the Michelseon-Morley experiment, was a universal constant. In doing this, he abandoned the Newtonian notion of time as being absolute and fixed for all observers, which was a key assumption in Newton's original ideas. This preserved Newton's Laws of Mechanics and Gravitation while it sacrificed notions about time and space to provide "room" for new ideas based on the constancy of light.

Einstein recognized that Lorentz's mathematics, developed for describing motion in the aether, worked equally well at relating observations between two observers in motion relative to each other. This was to become known as the famous "Lorentz Transformation," the correct way for two observers to reconcile their independent and differing measurements of space and time by using the constancy of the speed of light. These realizations also led to another of his famous equations relating energy and mass, a recognition that mass is just another form of energy, such as motion or heat.

As a consequence, Einstein discovered the phenomenon of time dilation. Time is analogous to length and mass. The measurements of both of these depend on the frame of reference in which the measurement is taken. That is, both change as the speed of an object changes. When an observer is in motion, measurements of length and duration as they pertain to other objects depend on the speed of the frame where the measurements are being performed; clocks and rulers are affected commensurate with the movement of the frame of reference within which the object is moving.

Relativity

Einstein's theory of relativity stirred up a major controversy in the scientific community. For the first time in three centuries the motion of bodies was being viewed quite differently from that of Newton's understanding. The concepts presented in Einstein's papers on special and general theories of relativity revolutionized existing beliefs about the behavior of the universe:

1. Optical rays from a star passing near a large body (the sun, for example) would deflect (bend) under the gravitational pull of the large body by twice as much as was predicted by the Newtonian understanding of light and motion. Newton wrote in 1704, in his treatise on optics, a prediction that light, too, would be affected by gravity—a consequence of gravity exerting a field of uniform acceleration in a local region of space, independent of the mass of the body, even a massless one!
2. As a vehicle's speed approaches that of light, its clock will slow down (time dilation).
3. As a vehicle's speed is increased, its length, and that of all objects inside it, will contract (get shorter).

To understand these strange effects consider the first assertion. Einstein realized, after his detailed study of the movement of light through space-time and the way that mass can bend space-time, that the degree of deflection of light would not only be non-zero, it would be twice the degree of that predicted when Newton's notion of space, time, and gravity was used. This was confirmed by astronomer Arthur Eddington, along with his colleague Andrew Crommelin, during the solar eclipse of 1919.

One point to remember is that time dilation happens at all speeds, not just those near to the speed of light. This was tested in 1971 when four ultra-accurate atomic clocks were taken aboard a commercial jet. It was found that clocks that travel indeed do record the passage of time at a different rate than do ones at rest. The results agreed with Einstein's prediction.

At the ordinary speeds experienced by most humans, the changes are miniscule and far too small to notice. What an observer at these speeds perceives as a car passes in the opposite direction, is that it passes at a speed that is the sum of his car's speed and that of the other car ($v_1 + v_2$). If their speeds are both thirty miles per hour, then, the other car is observed to pass at a speed that is the sum of their speeds; sixty miles per hour. However, when the two cars are travelling at the speed of light, the other car passes the observer at exactly the speed of light.

There is no simple summing of speeds at the speed of light. This is due to the aforementioned contraction of space and the dilation of time, which conspire as one speeds up, to maintain the constancy of the speed of light. As one approaches the speed of light, the effect becomes more and more noticeable. The key factor that describes the degree of contraction or dilation is shown below—the Lorentz gamma factor.

As one accelerates, space outside one's own frame compresses in the direction of motion by γ. Also, time outside one's own frame appears slowed by the same degree, γ, so, while space intervals are shorter, time is passing more slowly, canceling these effects on light's observed speed. These characteristics have actually been observed. The first of the three assertions (deflection) was first observed at an eclipse of the sun in 1919. The second of these (clock slow-down, time dilation) was first demonstrated during the 1950s when a satellite carrying an atomic clock was synchronized with one on the ground before launch and, during its orbit, was observed to slow down, as predicted. Today atomic clocks in each of the GPS satellites (there are 24 GPS satellites) are corrected for time dilation by the Lorentz factor

$$\gamma = [1 / (1-v^2/c^2)^{1/2}]$$

Where γ (Gamma) = Lorentz transformation at the speed, v, in space
v = Speed of the vehicle
c = The speed of light

As another example, let a body at rest have a mass denoted by the symbol m_0. To compute its "relativistic mass" at a speed v, use the Lorentz factor shown in the equation below. Relativistic mass increases with speed since the energy of motion has an equivalence to mass:

$$m = m_0 [1/ (1-v^2/c^2]^{1/2}$$

where the symbols m_0 and v represent vectors. As we noted earlier, the first person to apply these mathematical ideas was Lorentz, which explains why, to this day, Einstein's work is often spoken of in terms of "Lorentz transformations."

The equation shows that inertia increases as the velocity, v, of the vehicle approaches the speed of light, c. At the speed of light, inertia becomes infinite. In collider experiments, it is observed that as one accelerates an electron, for example, stronger and stronger electric fields are needed to make the same incremental next-step-up in speed; the inertia of the electron is seen to increase as its speed increases. This theory should not be confused with the Lorentz equation.

There are many strange happenings that have been both predicted and observed. Inertia increases, distances become shorter, and so on. One of the most interesting strange happenings, as we shall soon see, is that Newton's law of gravity is not quite right.

The Equivalence Principle

Albert Einstein sat at his desk as a patent examiner in Switzerland and wondered about many things. On a particular day, after thinking about what would happen to a workman falling off a roof, he had an inspiration. He called it "the happiest thought of my life." He would later recast this idea in terms of "what if I had been in an elevator with no windows?" (like the stick-person in Figure 3.2). What would happen to the man when the elevator accelerated in the downward direction? He concluded that, due to a cancellation of forces, the workman could float freely in the space, weightless. The occupant's feet could

A man in an elevator is accelerating downward.

There are no windows.

He cannot see outside.

He is floating above the floor.

He cannot tell whether he is under the influence of gravity or is accelerating because he is being pulled by a force.

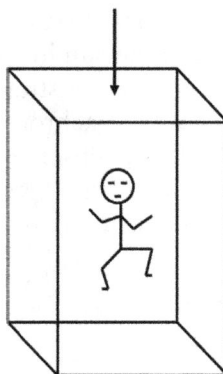

Figure 3.2 The Equivalence Principle: Downward force can either be gravity or an externally applied force.

come off the floor of the elevator and he would simply float there as it continued its descent. Einstein then went on to ask himself what explanation could the workman use to explain this phenomenon?

He reasoned that this could be caused by one of two things:

1. A gravitational force, like the force of gravity on the earth (or any large body) could cause the elevator to fall. As this occurred, the man would rise up from the floor of the elevator causing him to float as though there was no gravity.
2. A sudden external force (such as a rocket booster pushing downward on the roof of the elevator cab) could do the same thing, even if the elevator were in a place with no gravitational fields.

It occurred to Einstein that the elevator passenger could not tell which of these forces was the cause. It made no difference which type of force provided the influence, the result would be the same.

Einstein then wondered what might happen if a beam of light were to traverse the elevator horizontally as the elevator was moving downward at such an acceleration as to have the workman inside be weightless. Consider the situation of the elevator being pushed down by a rocket on its roof. The photons in the light beam, (entering the elevator cab free from external forces—including the rocket) would continue to travel through space free from external forces. Therefore, although the elevator would move down in space, the light (the photons) would follow a straight path through space.

To the workman, however, the beam of light would *appear* to strike the opposite wall at a location higher than the point at which it entered the elevator. Einstein reasoned that because both explanations for the workman's weightlessness are equivalent and indistinguishable, it must be true that in a gravitational field, the same deflection of light must occur! Acceleration due to gravity and acceleration due to any other kind of force are physically indistinguishable.

These thought experiments led Einstein to conclude that Newton's concept of gravity, space, and time—which keeps the three separate—misses the key point: that they are interlocked. That it is space and time *together* that work to explain gravity's ability to "act at a distance" that so puzzled Newton.

According to Einstein's general theory of relativity, bodies in the universe do not respond in exact accordance with the Newtonian equation of gravity. While, in most cases, Newton's equations are adequate, Einstein reasoned that time is a fourth dimension that must be taken into consideration. He perceived a gravitational field structure in which all matter is immersed in a space-time fabric like that shown in Figure 3.3. Let us use this analogy to create a word-picture about space-time and gravity.

This "fabric," when large bodies such as the sun, are immersed in it, warps. The large body causes a depression. Smaller bodies, like Earth, ride along the edge of the depression wanting to move closer to the sun. However, the circular motion of the smaller body causes a centrifugal force that pulls the body outward, canceling the force that seeks to move it inward toward the sun.

As large bodies move within the space-time fabric they cause it to warp, modifying its shape in response to their movements. The space-time fabric, therefore, is constantly in motion as large bodies move about in the fabric. In such case, smaller bodies like Earth appear to be attracted to the larger body as they "fall" into the warped depressions. This effect is what accounts for Newton's equation of gravitational attraction.

Warping produces the effect that planets and other objects are pulled toward each other. This, in turn, cause smaller objects to "roll around" in the warped space-time fabric as the larger bodies move within it and the warped fabric moves with it. This action gives the appearance of gravitational pull. When this occurs, the fabric influences the behaviors of the planets and other smaller objects to make it appear as though they are being pulled toward each other, but they actually are being drawn to the large-mass object near them due to the depression in the fabric of space-time that the large object produces.

Figure 3.3 Space-time Fabric Motion Acts Like Gravitational Attraction

If the large ball in our picture were alternately pushed downward and pulled upward rapidly, this would cause ripples on the fabric just like the ripples on the surface of a pond. We will discuss this in more detail later, but we here tease the discussion of gravitational waves by noting that the Laser Gravitational Wave Observatory (LIGO) and its scientific collaboration of hundreds of physicists and engineers reported during 2016, two successful observations of these waves, the first definitive detection of its kind. Such ripples are predicted by Einstein's theory of general relativity. It took a century from Einstein's completion of the general theory to be able to detect them. This finding of gravitational waves is now the beating heart of a wide array of studies, including experiments like the Planck satellite and telescopes like BICEP, KECK, and SPIDER that seek evidence of the imprint of gravitational waves on the cosmic microwave background, the light left over from the big bang.

A Quick Tour of Space-time

In this brief section we will show the reader some of the mathematical beauty and complexity that is general relativity. We provide some of its core mathematical equations, not expecting the reader to then want to become a general relativist, but as a means to note and appreciate how challenging this was, even for Einstein, to go from abstract mathematics to a concrete explanation for space, time, energy, and matter.

Einstein studied Bernhard Riemann's (1826–1866) work that created the field of non-Euclidian geometry (the geometry of non-flat surfaces) as a mathematical topic. He was able to apply Riemann's geometry to general relativity by asserting that the curvature of space-time is what creates the gravitational effects experienced by all bodies. Einstein applied Riemann's metric tensor to his problem, which became the basis for his field equation.

Einstein's Field Equation (EFE), defined by the elegant tensor equation (the equation shown below), is a set of ten equations that describe the fundamental force of gravitation that results from the curving of space-time. The equation contains, not numbers, but objects called "tensors." You can think of a tensor as a generalized number. A single, simple number can be used to describe something like, say, temperature. (It's 72°F in this room.) Single numbers are not always useful, though.

For instance, what if a friend asks you directions to a party? You could say, "Just go five miles and you're there." What useless instructions! You need in this case to give them numbers with direction. Such numbers are called "vectors," and are a slightly more general kind of number. A better answer to the question is, "Go three miles east and four miles north and you'll get there." The total straight-line distance is, indeed, five miles, but this vector information is far more useful. A tensor, then, is a collection of numbers that can describe many more dimensions, not just east-west or north-south, but also up-down, forward or backward in time, and so forth. Tensors are at the heart of general relativity, and its central tensor equation is given below.

$$R_{ab} - \tfrac{1}{2}Rg_{ab} \pm \Lambda g_{ab} = (8\pi G/c^4)T_{ab}$$

This equation contains three tensors, R_{ab}, g_{ab}, T_{ab}, each having ten elements and the gravity constant, G, that Newton defined. Einstein's field equations are all about describing how space-time curves in response to the presence of matter and energy, and how matter and energy are moved by the curvature of space-time. The symbol R_{ab} is known as the Ricci tensor, named after mathematician Gregorio Ricci-Curbastro. It describes how a volume of sphere in a specific space-time differs from its Euclidean (flat) counterpart. Thus, it is a measure of curvature in a volume. In general relativity, this tells us how matter disperses or collects due to the shape of space-time. The symbol, R, is the scalar curvature—a number defined at each point in space-time that tells us about the shape of space-time at that location. It multiplies the symbol g_{ab}, which is known as the "metric tensor." This tensor is the meter-stick of the universe, telling us about how to define distances, areas, volumes, and so on in space-time. Then there is Λ, which is known as "the cosmological constant." It describes the degree to which empty space is actually empty. Should empty space really contain an irreducible energy content—a basic level of energy at all points in the universe—that, too, affects the shape of space-time. Finally, on the right-hand side of the equation, we have T_{ab}, the "stress-energy tensor." This tensor describes the energy and momentum content of a region of space-time, so. if there is mass or light present in space, it is represented by this term in the equation. We should also recognize our old friend, "c," the constant speed of light.

The EFE, a tensor equation, relates this set of tensor equations to four unknowns that must be found to solve a given problem. These equations are analogs to the equations that Maxwell wrote to describe how electric and magnetic fields are related to electrical charges and currents.

This is written (see top of next page) using abstract index notation, and is shown for the purpose of illustrating to the reader the complicated, yet elegant, nature of Einstein's work. The metric tensor developed by Riemann was a key building block in Einstein's derivation of the equation. Using Riemann's metric tensor, the quantity g_{ab} becomes a set of four equations in four variables shown on the next page in matrix form (all four equations' coefficients make up the metric tensor array of numbers or terms). The Riemann tensor is not restricted as to dimensionality (it can have more than four dimensions, at least in terms of its mathematics) and string theory makes good use of that fact, as we will see. Tensors can be used to describe behaviors that have many dimensions. The Riemann metric tensor has coefficients (terms) that describe a multi-dimensional geometric surface at a point on a spherical surface, in this case a four-dimensional surface.

$$g_{ab} \quad = \quad \begin{pmatrix} g_{11} & g_{12} & g_{13} & g_{14} \\ g_{21} & g_{22} & g_{23} & g_{24} \\ g_{31} & g_{32} & g_{33} & g_{34} \\ g_{41} & g_{42} & g_{43} & g_{44} \end{pmatrix}$$

Each element of the array, g_{11}, g_{12}...etc., defines a geometrical condition in space-time that causes curvature of the surface it represents; there are ten numbers in each element that describe this condition, a. This curvature induces the forces that act in space-time.

The EFE effectively contains ten equations describing curved space-time (as expressed using the stress-energy tensor). The tensor equation addresses only four dimensions, x, y, z, (spatial) and t (time). However, tensors are not limited in the number of dimensions that can be described. The EFE are used to determine the curvature of space-time resulting from the mass and energy contained in the universe. That is, it determines the behavior of the metric tensor in space-time for a given arrangement of stress-energy in space-time. The EFE was first published in 1915.

Matter and Curvature

Even if you skipped the previous section, note that the Einstein field equations (EFEs) are generic. They describe curving space-time using any distance scale to include matter and energy on that same scale. You can consider a volume the size of the solar system, or one the size of an entire universe. In fact, this generality allows one to use EFEs to describe the whole of any universe, including our own. This contributes to what makes general relativity so exciting and additive to Newton's laws of mechanics and gravitation—that EFEs have the ability to describe everything because they contain everything. When Einstein first developed his equation, he showed that the universe could either be expanding or contracting, depending on the type of curvature it had. At the time, he thought that the universe should be static, that is, that the matter-energy density in the universe was a fixed amount, which says that the universe should behave in a static manner in space-time. He inserted a cosmological constant (lambda) into the equation to cancel out the expansion term.

However, Edwin Hubble's 1929 astronomical observations showed a slight red shift in the light received from the most distant galaxies *in any direction*. This red shift indicates that *all* bodies are receding from Earth in *all* directions, causing Hubble to define a new law stating that the universe is expanding at a constant rate; that is, that all objects in space are racing away from each other in all directions at speeds proportional to their distance from Earth. When he found what Hubble had observed, Einstein removed the constant from his equation although, as we shall see, its use has been revived in current-day physics.

We show in Figure 3.4 the shapes of possible topologies of the universe illustrating how matter density in the universe affects topology. Three topologies are depicted, depending on whether a curvature is positive, negative, or zero. Figure 3.5 shows a series of curves that indicate what the average distance between galaxies would be, plotted against time-since-the-big-bang. The curve created between two measurements in time would describe whether there is acceleration or deceleration.

The shapes in Figure 3.4 and the curves in Figure 3.5 assume possible values of matter density ratio, Ω_0, a ratio of critical density to the estimate of matter density assessed for the three values shown. Critical density is defined as a value of matter equal to ten hydrogen atoms per cubic meter. This amounts to finding one hydrogen atom within the volume of three basketballs. WMAP satellite data has determined that Ω_0 is very nearly unity. When Ω_0 is unity, space-time is a flat surface and the angles of a triangle (shown in each of the figures) add up to 180 degrees; there is no curvature in this case. When Ω_0 is greater than unity, the curvature is a sphere and the angles of a triangle add up to greater than 180 degrees; and when it is less than unity, the curvature is a hyperbola, and the sum of the angles is less than 180 degrees.

The effect on the structure of the universe is shown in Figure 3.5, where we see that the content of the universe affects the average distance between galaxies as a function of time. Looking at the curve labeled "accelerating," it shows where the universe seems presently to be, while Ω_0 has been measured to be very close to 1, it's crucial to know for certain whether it is exactly, less than, or greater than 1. Knowing this will tell us the ongoing fate of the universe—whether it slows to a stop in its expansion, expands forever,

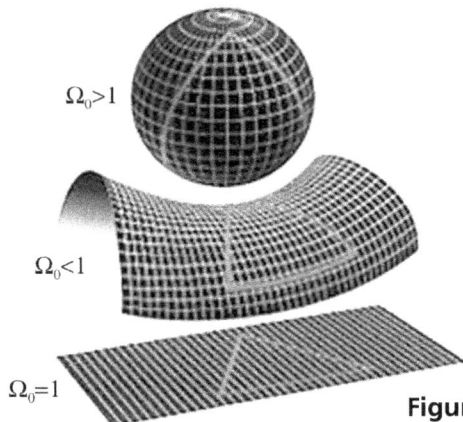

$\Omega_0>1$

$\Omega_0<1$

$\Omega_0=1$

Figure 3.4   Possible Shapes of the Universe

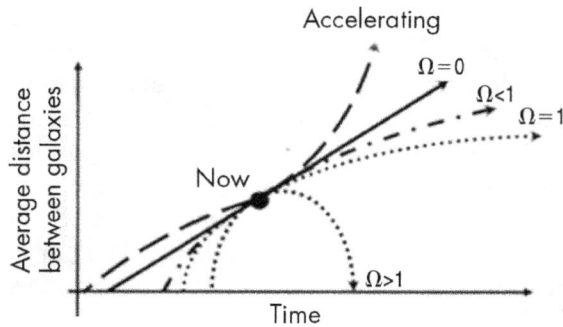

Figure 3.5 The effect of the content of the universe on its structure.

or collapses back onto itself. At the moment, we observe that the universe's expansion is accelerating.

When Lemaitre and Friedmann considered solutions to Einstein's field equations, they did so guided by the observations by Hubble of an expanding universe. Many other physicists, inspired by various other models for the cosmos, have done the same, and in doing so have developed a constellation of solutions to the equations of general relativity.It is possible today to use powerful computers to solve these equations for many complex scenarios, including what seems to be the present state of the universe, a universe whose expansion rate is rising, filled with the familiar matter of atoms as well as unfamiliar dark matter.

WMAP

Quite by accident, two Bell Laboratory researchers, Arno Penzias and Robert Wilson, while working on a satellite tracking antenna, detected noise from their antenna that they determined to be background microwave energy. This noise was attributed to a signal constant in amplitude and received from all directions.

Discussions with colleagues (and especially with physicist Robert H. Dicke at nearby Princeton University) suggested that the noise could possibly be a remnant of the big bang. Evaluating this noise turns out to be like taking the temperature of the universe in every direction using a radio antenna as though it were a thermometer. That is, the noise was seen everywhere in the heavens and its intensity at different locations was equivalent to that which physicists calculated would exist as a remnant of the big bang at that location in space-time.

Because this radio noise has a representative temperature, one can use the current temperature to infer how long the universe has been cooling. Based on this, it's possible to establish an age for the universe, enabling modern measurements to very precisely determine the time since the big bang. A sequence of experiments from the Cosmic Background Explorer (COBE) in the 1990s, to WMAP in the early 2000s, to Planck in the present decade, have refined this measurement using independent instruments and draw the same conclusions. The universe is, overall, quite geometrically flat, quite old (13.8 billion years), and only partly filled with the normal matter of everyday life.

Expansion

The radio noise in the universe, along with other observations (such as the relative amounts of hydrogen and helium in the universe) allowed physicists and astronomers to deduce that the cosmos began in a big bang. This is not a literal "bang," but an incredible explosion in the production and expansion of space-time accompanied by a tremendous release of energy. The fingerprints of this are all over the modern cosmos.

Penzias and Wilson had sufficient data to conclude that the temperature of the universe was just about three degrees above absolute zero, the coldest anything can be, and the temperature at which all atomic motion stops. Here are the broad conclusions from this work:

- A more accurate date for the beginning of star formation— 200 million years after the big bang
- There is evidence suggesting that the universe will expand forever

As this process advanced from the big bang, temperature, energy, and matter changed rapidly from a nearly uniform primordial plasma to one where there were noticeable hot and cold spots. The average density was higher in the hot spots than in the cold ones. This caused matter to be drawn in by the force of gravity. These were the seeds of cosmic structures that ultimately evolved into galaxies and even larger structures. At these beginnings it was a condensation-like process wherein there appeared the basic constituents of matter, and, finally, atoms. In the first billionths of a second of time, the four fundamental forces, strong, weak, electromagnetic, and gravity parted ways.

Physicists theorize that gravity was characterized as a soup of gravitons (the graviton is the still-to-be-discovered carrier of the gravity force) in the makeup of the primeval matter of the early big bang. As the cooling and expansion continued, temperature and energy density decreased. Assuming that it ever was a quantum mechanical force, gravity ceased to be one almost immediately, taking on its more general relativistic character quite quickly. Sometime after that, but still within a billionth of a second after the big bang, the building blocks of all matter, quarks and leptons, appeared. While we will discuss them more later, no one knows the origin of these particles. Eventually, the universe cooled enough for quarks to clump into protons and neutrons, the building blocks of atomic nuclei. Further cooling led to protons and neutrons combining, leading to the fashioning of the nuclei of atoms like hydrogen, heavy hydrogen, and helium—all of this happening within the first minutes after the big bang.

The universe was still quite hot at this point and nuclei and electrons (a kind of lepton) could not stick to each other to form stable atoms. This happened later, after about 380,000 years. It was at that point that the light we have come to identify as the cosmic microwave background was finally free to stream through the cosmos. The story of the universe is one of cooling and clumping.

Here are some more conclusions drawn from the satellite missions discussed earlier, as well as other independent astronomical measurements:

- The universe is 13.799 ± 0.021 billion years old.
- Its composition is 5% atoms and neutrinos, 26% unknown dark matter, and 69% dark energy (which acts to accelerate expansion).
- The universe geometry, based on the Einstein's equations, is confirmed to a very high degree of precision to be flat.
- There are tight new constraints on the burst of expansion in the first trillionth of a second.

This last point is very important. The universe began its existence as a space many orders of magnitude smaller than an atom. In order for it to grow into the magnificent structure it is today, it underwent a period of expansion called "the period of inflation."

This phase was posited by Alan Guth in 1981. According to this theory, within a tiny fraction of a second after the big bang, the universe underwent an incredibly rapid expansion. This explains why the CMB seems to be very homogeneous. (Figure 3.6) Regions of quantum-mechanics that dominate space were once able to exchange energy with each other, thus becoming similar to one another, are stretching apart, never to contact each other again.

After the Big Bang

In 1920, the work of Arthur Eddington led to the recognition that stars are powered by nuclear reactions caused by fusion of atomic nuclei under gravitational

Figure 3.6 WMAP cosmic microwave temperature of universe background

attraction. The formation of elements that started during the big bang with the creation of hydrogen and helium (and some other very light elements) continued in a straight-ahead march to heavier elements in the hearts of the stars that coalesced into existence millions of years after the big bang. What the big bang had not accomplished, gravity and radiation over time, would.

The formation of those initial light-weight elements is called "Big Bang Nucleosynthesis" (BBN), the process of temperature causing fusion of the primordial nuclear materials. This concept was provided support by the work of Hans Bethe (1906–2005), Carl Friedrich von Weizsäcker (1912–2007), and Fred Hoyle (1915–2001) who used this idea to determine how lighter elements are created in the nuclear environments of stars. Once this was known one could mathematically work out how each element was created and therefore its theoretical abundance in the universe. Consequently, the nuclear processes associated with each element are characterized by the temperature as evolution ensues following the big bang. If one has an idea of how the universe was created and subsequently cooled, there is the chance to understand why the lighter elements appear in the proportions in which they are found in nature. Understanding BBN and its related "supernova nucleosynthesis" (i.e., the creation of heavier elements by supernova explosions) permits understanding the distribution levels of all of the elements in the universe and the temperature needed to form each.

At approximately 6,000,000 degrees Kelvin, the fusion of hydrogen and helium takes place, thereby producing heavier elements. The hearts of stars allow for elements up to iron to be produced.

We show in Figure 3.7 an illustration of the history of the universe, including the expansion of space-time over the life of the cosmos. While the speed of light

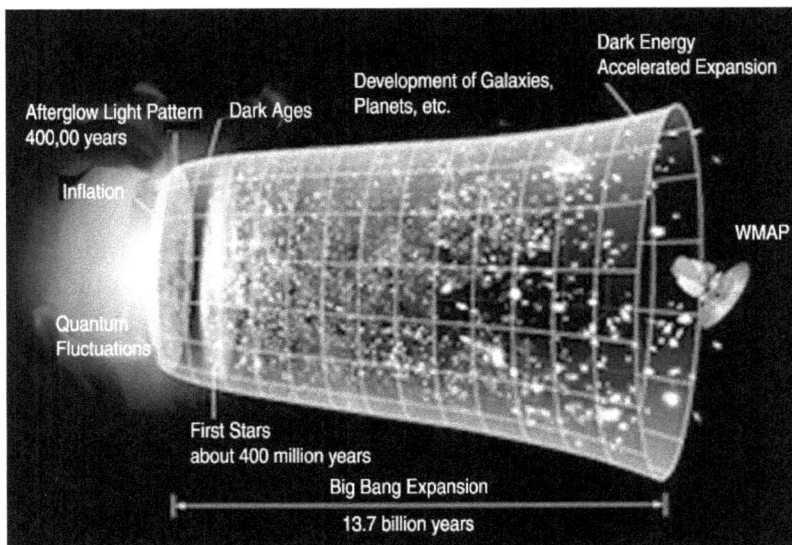

Figure 3.7
The Universe.
Forward Evolution
After the Big Bang

limits what we can see (light takes time to travel from distant objects to us), as we wait for light to reach us from the most distant objects space-time continues to expand, drawing those distant objects even farther from us. The oldest galaxy we have ever observed emitted the light we see now about 13.4 billion years ago, but while we waited for its light to arrive, that galaxy moved away. This means that things we see now are actually further than they appear to be, defining the "observable universe," a sphere roughly ninety-three billion light-years across containing trillions of galaxies, each galaxy containing hundreds of billions or more stars.

Restating what has earlier been discussed, the big bang caused a rapid inflationary expansion of primordial matter and energy which rapidly cooled down as inflationary growth reached greater proportion. This early inflationary period lasted until about 10^{-32} seconds after the big bang.

Being in an inflationary phase does not mean that the amount of matter in the universe is increasing. It simply means that space and the matter within it is moving apart. But where did that matter come from? This is a key question that still lies at the frontier of physics. We will continue to explore this.

The universe settled into a steadier expansion during which cooling was sufficient to permit various combinations of matter to aggregate and, in turn, form atomic matter. The initial atoms, hydrogen and helium, formed, creating stars. This was followed by the formation of new galaxies, each having billions of stars as gravitational attraction pulled these stars into groups. The expansion rate slowed until reaching a point about five billion years ago when matter, spread thin by the continued expansion of the cosmos, could no longer slow the expansion rate of the cosmos. This initiated a second inflationary period that astronomers observe today.

After 13.8 billion years, astronomers observe that the universe has many galaxies, stars, planets, and nebulae—new celestial objects. (The first human ancestors appeared on Earth about two million years ago.) The expansion continues presently and astronomers confirm that stars are born, they mature, and they die in a cycle of matter. By the late 20th century, physicists began to think that they had conquered the understanding of the whole of the universe—that their job was nearly complete.

The early excitement about a total understanding of the cosmos was, as it often is in the history of science, thwarted by better observations of the constituents of the universe. We will see later how our seeming mastery of matter and forces has been diminished by the observation of dark matter in the cosmos, and our assumed mastery of the energy of what we call "empty" space is challenged by the accelerated expansion of the universe. While we have a great understanding of the laws of the very big (general relativity) and the laws of the very small (quantum mechanics), there is frustration in making them work together to describe the total universe.

The difficulty in gravitational unification with other forces of nature has been a serious stumbling block in making this mastery true. Theoretical physicists have pondered the issue of the hierarchy problem (why three of the four fundamental forces are much stronger than gravity) for more than sixty years. During the 1970s a group of physicists broke away from mainstream thinking to pursue other ways to fit the gravitational force into the mix. This effort continues to move forward.

Building on those old ideas, there is currently an ongoing effort to fit gravity together with the standard model using string theory. As a means of achieving unification, string theorists have introduced the possibility that there are additional dimensions beyond the four we know. Even if these extra dimensions do not actually exist, their presence makes the mathematics easier.

We have explored the theory of the very great—from the bending of light around stars to the birth of the entire cosmos. We've seen how ten equations, the Einstein field equations, are capable of describing the entire universe and charting its fate according to the density of the matter and energy they describe that is contained in the universe. In exploring the cosmos to better understand the players in our universe, we have discoveries anomalies—the universe's expansion appears to be increasing in speed with time, and the majority of the matter in the universe is of a form unknown to us. But, before we continue to tackle those mysteries, let us shrink our thinking down from a whole cosmos to the small spaces occupied by its building blocks. We will first escape into the quantum realm and explore what is known about the smallest things in the universe. From there, we will chart a course forward, first mapping the familiar territory of our cave before questing its frontier—closer to the edges of the shadows we need to better describe.

## Chapter 4

The Quantum World

By the late 1800s, the behavior of the macro-universe had been well verified to be consistent with Newton's equations. The findings of electromagnetism, heat, light, optics, and the many observations that telescopes revealed were almost all in complete agreement with well-documented physics predictions and equations. By then, physicists had accepted that the universe was composed of atoms (at the smallest scales), comets, planets, stars, nebulae, and many other objects (at the largest scales), and that they operated in accordance with Newton's clockwork processes. Most everyday problems were understood; most questions encountered in the study of nature were solved with the equations of physics. There was growing unanimity of belief among scientists that there was little more that remained to be resolved. Engineers were developing new applications, processes, and products at an ever-increasing rate based on these theories. This demonstrated that science was no longer a matter only for philosophers, but a practical way to increase the quality of human life. Even so, the subject of physics was not considered to be very important in the 1800s. This would change!

Nearly instantaneous communications over huge distances is taken for granted today. Of course, mobile telephoning did not spring forth from nothingness. It was the result of much individual input and in a sense began with four equations. By 1873, James Clerk Maxwell had written equations that describe everything ever observed about electrical and magnetic phenomena. His equations suggested that, were one to arrange magnets, batteries, and wires in the right way, bundles of electromagnetic energy would fly off the wires!

In 1886, Heinrich Hertz built a device to measure those bundles of energy as they, indeed, traveled away from their wires. Many others such as Guglielmo Marconi, Nikola Tesla, Nathan Stubblefield, and even Thomas Edison, pursued physics experiments to create practical radios.

In the late 1800s new knowledge about the atom was revealed every day by chemists creating new compounds. Physicists knew little then about atoms, and among some there was even doubt as to whether they existed. However, questions about the structure and behavior of matter were many.

Physicists were about to open up a field of science that would establish a conceptual framework that would be used to explore and uncover new knowledge about the universe. Let's focus on the quantum realm.

Elementary particles were not generally known to physicists of that time. Chemists were the ones who studied the world of the small. They had to understand elemental behavior if they were to make new products from chemical compounds. Medical science of that period relied on chemistry for medicines and drugs.

Coming upon this scene, an electrochemist, George Johnstone Stoney, began unknowingly to investigate the realm of the atom. In 1891, based on his study of the behavior of gases, he conceived of a "fundamental unit quantity of electricity" that he named the "electron" after discarding the name "electrine." It was that very object that J.J. Thompson would later find in his laboratory in 1897. Stoney also cast his research eye widely on physics issues from planetary dynamics to the theory of gases. Indeed, the Planck mass, the maximum unit of mass, which will play a major role in later chapters, was already in Stoney's mind, although he considered it to be a unit about ten times smaller. Stoney had thereby become the first person in history to conceive of something smaller than an atom, although the size of the atom was unknown at that time.

With no knowledge of an atom's behavior, physicists could not understand how the universe was formed, how the planets and stars got where they are, and how these things interact with the gravitational force. There was much data missing from the description of atoms as posited by chemists because chemists focused on the chemical interaction of elements rather than on their physical substance and the explanation for their behavior.

The Search for the Atom

The existence of the atom as a substantive unit of matter was established by a long process of observation without clear explanation. For instance, working on the masses and other properties of the chemical elements, the chemist Dmitrii Mendeleev (1834–1907) organized the first "periodic table of the elements," a way of looking at matter that made it clear that elements have a pattern of organization about them.

Physicists like Wilhelm Roentgen (1845–1923) and Marie Curie (1867–1934) made fundamental observations about the instability of matter—its ability to spontaneously emit energy that we call "radioactivity." However, at the time they made their observations, they did not yet understand the subatomic processes that contributed to cause this, nor the hidden world of new forces that these behaviors revealed. Albert Einstein, in his "miracle year" of 1905, published an explanation of "Brownian motion," the jittering of dust motes in a drop of water, providing definitive proof that atoms were real. It was atoms, he reasoned, zipping about in random directions next to the dust mote, that sometimes buffet it from the left and sometimes from the right, causing the mote to jitter around. Each of these scientists, and many others, provided pieces of a puzzle, but it would take decades of work for the picture to emerge from those pieces. Let us look at the how-so of this occurrence, starting with the nature of the electron.

Joseph John (JJ) Thomson (1856–1940) was a physics researcher who wanted to understand atomic behavior. He began in 1896 to experiment with what were then called cathode ray phenomena.

Electricity, the flow of charge through solid or liquid material was a well-established and well-understood phenomenon by the time that Thompson began his research. What was curious about cathode rays was that they represented some kind of electrical phenomenon that could traverse even empty space. The cathode ray phenomenon could be produced by placing two metal electrodes inside a high-vacuum chamber, a chamber almost completely devoid of matter. A high voltage was placed at one end (the "anode") and the other metal electrode, the "cathode," was then seen to emit a kind of ray that traveled the empty space between the electrodes and came to the anode.

Thomson exposed these rays to many trials. He noted what happened when one allowed the rays to travel through electric or magnetic fields, observing accelerations and deflections that indicated that they were electrically charged. By carefully measuring these effects, he concluded that they possessed negative electric charge and had a mass about a thousand times smaller than that of a hydrogen ion. These were not atoms, they were something else.

It was George Fitzgerald, nephew of George Stoney, who suggested the name "electron," resurrecting his uncle's ideas about a smallest unit of electrical flow. The name stuck, and to this day we still know this first subatomic particle by this name.

Electrons are also the first particle discovered that represents one of the two known classes of matter. These are quarks and leptons. Electrons are leptons, although their membership as a related class of subatomic particles would not become clear for many decades after Thomson's ground-breaking work.

Thomson realized that he had isolated electrons—small, negatively charged particles, a part of the atom that had been released from the cathode due to the large voltage that had been applied to it. For his work in cathode ray phenomena, Thompson received the Nobel Prize in physics in 1906.

Electrons were thought to move only along wires carrying electric current when under the pressure of voltage from batteries. Chemists had discovered that atoms contain electrons and those electrons appeared to freely exchange place with other electrons in other atoms farther along a wire in a common direction called current.

The Call to Precision

Before we continue with the story of quantum mechanics, let's pause to reflect on the important of the precise understanding of nature. In the late 1940s and early 1950s, Richard Feynman (1918–1988), Julian Schwinger (1918–1994), and

Sin-Itiro Tomonaga (1906–1979), working independently, developed a method for solving the complex mathematics of QED (quantum electrodynamics—how light and matter interact). Feynman approached the problem by developing Feynman diagrams, graphical analogs of the mathematical equations that describe a particle's behavior, making it easier to analyze and solve the equations that describe particle interaction. This technique simplified the complex mathematical rigor needed to analyze particle behavior. Feynman shared a 1965 Nobel Prize with Sin-Itiro Tomonaga and Julian Schwinger.

Quantum electrodynamics extended the understanding that chemists began from their research into chemical reactions. It would now be used by physicists to investigate particle behavior under various influences such as electric fields, magnetic fields, and relativity. It is used to evaluate the characteristics of particles such as their spin, angular and linear momenta, and other behaviors caused by the influence of the forces of nature. QED is simultaneously the best theoretically understood and the most stringently tested-for-accuracy theory in all of physics and in all of history.

As one indication of this, it predicts that electrons possess a property called the "electron anomalous magnetic moment," which means that, in certain ways, electrons behave like magnets. This property of QED is described using a quantity called "g-2" that can be calculated using hundreds and hundreds of Feynman's graphs, calculus manipulations, and computer results, but it is also measurable in the laboratory. Using both approaches, it yields answers that agree with each other to better than one part out of a trillion. There is no other number in all of science where such uniform agreement between theory and experiment has been demonstrated!

So why should one care about such close accuracy? The answer speaks to the dual roles of science in society. One of these roles is to enable the discovery and development of new technologies. As Theodore von Karman (1881–1963), scientist and co-founder of the Jet Propulsion Laboratory once said, "Scientists discover the world that exists; engineers create the world that never was." If scientists do not do their work accurately, then neither can engineers nor technologists reliably develop new products in the realm of nanoscale engineering and its applications . . . think smartphone and tablet computers.

A second role of science is to provide humanity with the means to accurately extend what it knows about our universe. If one wishes to have an accurate understanding of events near to the time of the big bang, for example, one must have the most accurate possible understanding of how things work at the smallest scales. Because, as we have learned, what happened during the big bang continues to affect the universe today and has future implications.

This points to something that is not widely appreciated. Science is a unity, not a disparate collection of belief systems. One cannot with intellectual rigor and honesty accept some parts of it, while simultaneously rejecting other parts. Accuracy in all things drives all of science. Science is able to make predictions only on the basis of accurate measurement compared against the most rigorous logic and mathematics.

Quantum Theory Moves Forward

At the beginning of the 20th century, theoretical physicists, being mostly satisfied that they understood the behavior of the universe as a clockwork process, and believing their quest for understanding nature was almost complete, modified this position as quantum mechanical behavior exhibited itself. With Planck's quantized view of blackbody energy, Einstein's discovery of the quantized photon, Dirac's findings evolving to show that Schrödinger's wave equation could be extended to include Einstein's rules of space-time relativity, and the discovery of the electron, the first subatomic particle, the stage was set for further research into the makeup of the atom and its particles.

There were many mysteries about the atomic realm. Energy, radiating from atoms acted upon by electrical forces, emerges in the form of light, but not all colors (frequencies) of light were emitted by atoms. This phenomenon, "the atomic spectra," would not be understood until a more rigorous model of the atom emerged from Thomson's early model. Further, what differentiated elements from each other? Why is hydrogen, a highly reactive gas, while helium, the next-heaviest element, an inert gas? And, the even more complex question, why are gravity and electromagnetism, the two forces known at the time, so different in character from each other? These issues were unclear at the beginning of the 20th century.

Gravity and electromagnetism are both infinite in range. Separate two masses by greater and greater distances, and one only diminishes their mutual attraction, never quite driving it to zero attraction unless they are an infinite distance apart (an unphysical prospect in a finite universe). Similarly, separate two electrons from each other and their mutual repulsion, due to the sameness of their electric charges, diminishes, but does not vanish except at infinite separation. Yet, the strength of electromagnetism is so much greater than is the strength of gravity on the same distance scale.

For example, a coin-sized permanent magnet can pick up a paper clip even though the paper clip is acted upon by the gravitational force *of every atom that is the Earth!* From this it is clear that gravity and electromagnetism are wildly disparate forces. Why?

The 20th century would bring new questions as physicists explored the mysteries left over from the 19th century. In an era in which physicists probed deeper into the structure of matter than at any time in the past, they learned that the clockwork universe of Newton, with its certainty about the future given the precise knowledge of the past, would have to be abandoned for something far more like the roll of a pair of dice. From this non-deterministic realm of atom and sub-atom would emerge one of the most precise and accurate theories of nature ever devised. To understand how this happened, we must begin with a simple question: what is the atom? Here emerged new mysteries, and from the solution of these mysteries, new understanding about the universe. Unlocking the atom opened our minds to knowing why the stars burn bright in the emptiness of space, how to build new devices like the transistors that are at the heart of every electronic computer, and how to look inside the human body without making a single incision.

The Early Model of the Atom

Scientists, with chemists in the lead, reasoned that atoms contain electrons and protons that were thought to be held together by electromagnetic forces. Later, after the discovery of the neutron, physicists began to wonder about the structure of the nucleus; that protons and neutrons were held together by something about which they knew very little. They surmised that the structure of the atom was determined by fundamental forces in nature. These forces were later defined as the strong nuclear force, (binding protons and neutrons together in the atom's nucleus—as a residual effect of an even stronger force that would later be known as "quantum chromodynamics" or QCD), the electromagnetic force, the weak nuclear force (that influences electrons and particle decay), and the gravitational force.

The strong nuclear force works in the nucleus, keeping protons and neutrons in a stable relationship. The strong nuclear force is the "glue" that holds protons and neutrons together in the nucleus of the atom.

The weak nuclear force influences particle decay. The best example of this is the neutron. If a neutron is placed outside the environment of the nucleus, it will decay (transform) into a proton, electron, and another particle called a neutrino.

Within protons, neutrons and similar nuclear particles, the gluon is the strong-force carrier particle (comparable to the photon that serves as the carrier of electromagnetic force), that imparts unusual behavioral characteristics to quarks (a type of sub-nuclear particle).

The strong force, when quarks are close together, is incredibly strong, overcoming the repulsive electrical forces experienced by grouping positive like-charged quarks closely together. Within this environment, the strong nuclear force acts strangely, behaving like a rubber band or taffy, "slapping" the quarks tightly together and binding them strongly until they are extremely close to each other, requiring very large energies to separate them. When the quarks are extremely close, the force binding them behaves like a rubber band—the tension is removed and relaxed, allowing the quarks to behave almost as if they are free—i.e., not subject to any forces.

The notion that atoms are a fundamental, indivisible particle was long over in the 1930s when it was realized that even the nucleus of an atom could be split. Electrons and protons could be released under bombardment by other particles. Atoms were found to give up matter in the form of beta rays (decaying) as the work of Marie Curie indicated during her experiments with radium and other substances. The theory that the atom was indivisible was disproven by work at Los Alamos and elsewhere showing that splitting the atom would release tremendous energy, just as Einstein had predicted.

The instability of the nucleus of the atom, in light of the strong force that binds protons to protons, protons to neutrons, and neutrons to neutrons, is particularly curious. Radioactive nuclei decay with differing half-lives (the time required for a particle's original radioactivity to fall to half of its intensity). Radium, for example, decays very rapidly—its half-life is short. All known elements (even elements like gold) have unstable isotopes (radioactive forms of the element). In consideration of this, an entirely new science related to weak nuclear force interactions was discovered. It is sometimes called "weak interaction physics."

Early illustrations of the atom looked like the one shown in Figure 4.1, the Rutherford model. This model shows the atom to have a nucleus containing protons and neutrons tightly bound together by the strong nuclear force. Electrons orbit the nucleus like little planets around the sun. The size of the atom was determined to be about 10^{-10} meters (up to a few factors of 10 depending on the number of electrons orbiting the nucleus).

It was thought early on that electrons followed Newtonian rules about the laws of inertia and the electromagnetic force—that they are repelled by other electrons and attracted to the positively charged protons of the nucleus. The "Copenhagen interpretation" (developed between 1924 and 1927 by Niels Bohr and Werner Heisenberg) showed that only the *probabilities* of electron properties such as positional momentum could be determined, while charge can be determined with certainty.

It's Waves, Waves, Waves

In order to grasp the Copenhagen interpretation, which continues to be in use today to understand and explain measurement in the realm of the atom and

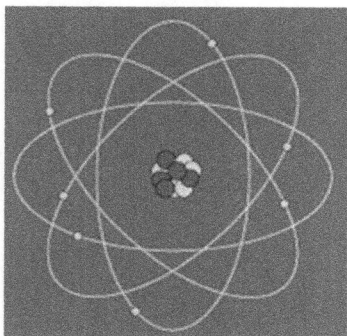

Figure 4.1 The Rutherford Model of the Atom

Figure 4.2   Water wave

the subatomic, we need to understand the key idea of the quantum realm: it is *waves, waves, waves.* If one had to sum up quantum physics in a single sentence, it might be: *"It's waves all the way!"*

Let's review waves. In the physical Newtonian context, waves are distortions of a medium that travel through that medium. Consider a question about the water waves shown in Figure 4.2: *Where is a wave?*

This is not a simple question, although it might seem that at first. You might point to the crest of the wave and say, "there it is!" You might also point at a trough and say, "no, the wave is there." There is no single correct answer to this question. You could make an argument that the wave is an extended phenomenon, naving no single location in space—that there is a little bit of it everywhere and that they are not well defined or localized in space or in time.

We can employ useful mathematical quantities to summarize a wave without having to state its exact physical location in space. For instance, waves have *wavelength* (denoted by λ, the lowercase Greek letter "lambda"), the distance between any two identical parts of the wave; for example, the distance between two crests or two troughs of the wave. They also have *frequency (f)*, the number of times per second that the same part of a wave passes a given point; for example, a wave with a frequency of 5 per second, or 5 Hertz (written "Hz," the unit of frequency), is a wave that crests 5 times in one second.

Let us now look at wave behavior in the quantum realm, beginning with light. The exact nature of light was a scientific debate that stretched across centuries. Arguments for the *particle* nature of light were advanced by such luminaries as Isaac Newton, who argued that light's ability to bend around obstacles was evidence of its particle identity. Others conducted experiments on light to find that light beams *interfere* with each other, creating bright and dark patterns where they intersect. This was a familiar behavior, because waves do this.

Think about sound waves created by two speakers in a room. In some places in the room the volume seems louder because waves are cumulative when two crests (or two troughs) arrive from the separated speakers at the same place at the same moment (simultaneously). In other places in the room where a trough arrives at the same time as a crest, the volume will appear less as the waves cancel (interfere with) each other. This interaction of waves is called phase. When waves arrive synchronously, they are "in phase." When crests and troughs arrive before or after each other, they are "out of phase."

Phase is described mathematically as a circle. When two crests arrive in phase they are at 360 degrees of phase (or zero, too). When the peak of a crest arrives simultaneously with the bottom of a trough, they are said to be at 180 degrees of phase (or they can be said to be 180 degrees out of phase). One can carry phase identification to different increments of a circle depending on what part of a wave meets its counterpart at what point in time.

The debate about particles and waves raged until Maxwell's equations seemed to definitively describe light as an electromagnetic wave. But, then, in 1905, Einstein showed that the photoelectric effect could only be explained by a particle hypothesis for light. So which is it?

Since there was evidence for both behaviors, a view emerged: perhaps small particles, like those that compose light, exhibit wave behaviors under some circumstances and particle behaviors under others. The is "wave–particle duality." The defining principle that determines which behavior happens is based on the wavelength (frequency) of the phenomenon (such as light).

Consider a wave that encounters an obstacle. If the wavelength is much shorter than the size of the obstacle (its thickness), the interactions will seem very particle-like and can even be described by particle mathematics. To understand this, think about a once-popular toy; the vortex cannon. This is a small air cannon that, when fired, produces a donut-shaped ring of compressed air several centimeters in thickness. The effect of being struck by this ring is quite stunning. It feels as though someone has hit you with a large rubber ball. Yet, you're merely being hit by a wave of air! This is an example of a wave phenomenon that feels particle-like because the thickness of the wave (its wavelength) is shorter than you.

Wave behavior, on the other hand, manifests when the wavelength is much greater than that of the thickness of the obstacle. Consider the air from the vortex cannon again. If one increases its wavelength, exposing the human body to the same total energy, but spread across a greater thickness than that of the human body, one will feel instead a gentle increase in air pressure and then a gentle decline. That's very wave-like.

A gray area emerges when the wave and the obstacle have comparable sizes. Both behaviors can manifest. This is the sense in which duality is meant, but it is all explicable using fundamental wave behavior. Wave-particle duality is a concept that is made to appear strange or confusing, but it's not. Once you grasp this concept, you are nearly ready for the Copenhagen interpretation.

The last piece of this puzzle has to do with matter. Photons are carriers of a force field, the electromagnetic force. They are observed to manifest particle-like and wave-like properties, depending on the situation. If light can do this, why not matter? What's so special about light?

Louis de Broglie, whom we mentioned earlier, made just this conjecture. In his Ph.D. thesis in 1924, he postulated that since photons can manifest wave-particle duality, so could electrons. He then predicted "matter waves"—that matter particles also have wave properties, but because their wavelengths are typically so short, we don't notice this and their behavior seems like the scattering of billiard balls rather than the coalescing activities of waves.

This was a risky conjecture. There was, as yet, no experimental evidence for his idea. There was only the work on the particle nature of light, and how particle behavior manifests when wavelengths are sufficiently short. So de Broglie democratically applied such thinking to matter, specifically, the electron. When wave-light behavior was discovered in electrons just a few years later, it's no surprise that de Broglie's insights garnered him a Nobel Prize in Physics. This was a stunning feature of nature. Electrons (and presumably all other matter particle phenomena) are waves, too! Since that time, the wave nature of matter has been revealed by observation again and again.

de Broglie showed how one would relate particle properties—such as the definite momentum and definite energy of a particle—to wave properties, such as wavelength and frequency. These relationships survive today and are a cornerstone of quantum theory. For instance, if one knows the momentum, p, of a particle (in Newtonian physics, the product of mass and velocity), then one can determine the wavelength of that particle:

$$\lambda = h/p$$

where h is Planck's constant. If one knows the energy (E) of the particle and wishes to find its frequency, there is an equally simple calculation:

$$f = E/h.$$

(keeping in mind that E, here, is related to mass and momentum via Einstein's famous formula). The summary properties of wave behavior are exactly matched to the precise properties of particle behavior.

Are you a wave? Do you exhibit duality? To answer that, you have to calculate your wavelength. At standard walking speeds, 3 mph (1.3 m/s), and factoring in your mass, your wavelength is approximately 10^{-36}m! For scale comparison, the size of the nucleus of an atom is about 10^{-15}m. It's no wonder we don't notice the wave nature of the things we encounter! Compared to the scale of daily things, our wavelengths are very tiny. Only our particle nature is evident in everyday life. This explains why Newton never included this in his ideas—he couldn't have observed it.

What about the wave properties of an electron? With a mass of just 9.11 x 10^{-31} kg—far, far smaller than a human's mass—we can already see that things are going to be more interesting because $E=mc^2$ and mass, here being small, can result in lower frequencies and thus larger matter wavelengths. Electrons in atoms move around the nucleus at speeds of about a hundredth that of light. If we put that into the equations, we find that the wavelength of an electron, when orbiting a central atomic nucleus, is about 10^{-10} m. This is about the same size as an atom. No wonder humans began to notice that Newtonian physics breaks down as we study the atom! Electrons in the atomic realm have matter wavelengths that are comparably sized to the atom in which they exist—they will manifest both their wave and particle properties.

There is one last concept to take away before we revisit the model of the atom. As an electron speeds up, its wavelength relative to its wave property decreases. If you want to make a better microscope, use fast-moving matter particles. For a microscope, wavelength matters. As example, because we use the human eye to view light in a standard laboratory microscope, and the human eye can see only visible light, this limits our precision to the wavelengths of violet light (the shortest wavelength light that we can see)—about 400 nm, or 400 billionths of a meter. This permits us to see, with the eye, animal and plant cells and even the organelles inside these structures, but the membrane of an animal or a plant cell is not visible. A cell's membrane is just a few nanometers thick, smaller than the wavelength of violet light, and thus invisible to our eyes. To see a cell's membrane, one has to use shorter wavelengths of light—x-rays, for example.

If you want to see the atom, however, or see deep inside an atom, even x-rays are not enough. You need to go to gamma rays. However, gamma rays are difficult to control and to work with, and they have the nasty ability (given how much energy they contain) of tearing atoms apart. No, if you want to see inside an atom you need to discover a more gentle surgeon.

You have seen that the everyday electron has a wavelength small enough to resolve atoms. An electron with a much gentler energy level of just a few tens or hundreds of electron volts can easily resolve an atom without tearing it apart. The wave nature of matter, revealed through countless observations since the 1920s, has proven a most useful tool for making a better microscope. We will utilize this principle again in the later discussion of particle accelerators.

Revisiting the Model of the Atom

The Copenhagen interpretation revealed a much different behavior from that of the older Rutherford concept. It shows that the place around a nucleus at which an electron might be found will be at one place at one time and a different place at a different time. Plotting these positons over time, electrons would appear as if they were clouds surrounding the atom's nucleus—that electrons do not move in well-defined classical (circular) Newtonian orbits.

This began a search for a robust, more accurate theory of atomic structure. The notion that electrons were arranged in deterministic orbits was fully dispelled when the cyclotron and more advanced particle accelerators showed that electrons *actually do appear as clouds* around the nucleus of the atom. This was observed in collider experiments conducted during the 1960s and 1970s. Electrons around the atomic nucleus distribute themselves in accord with the probabilities described by Schrödinger's wave equation. This was based on finding them probabilistically distributed around the nucleus as shown in Figure 4.3.

So far, we have talked about the movement and position of electrons around a central atomic nucleus as a defining characteristic of the atom. We've seen how the model of the atom needed refinement as more observations were made of atomic behavior. One behavior, spin, is essential to the structure of the atom, although its occurence was only directly observed beginning with experiments in 1922, more than two decades after Thomson found the electron. Spin is a property of matter that manifests like the energy associated with a rotating object, thus its name. As we noted earlier, however, as applied to the atomic and subatomic realm, spin is an analogy for what is going on. Electrons don't actually spin (if they did, the universe would be in trouble as electrons would rotate at speeds faster than light). They merely possess internal energy akin to that possessed by a macroscopic spinning wheel.

Quantum spin is essential to our existence. The probabilities of the locations of electrons arrange themselves around the nucleus in shapes according to the amount of energy they carry. These clouds take the shapes of the wavefunction proposed by Schrödinger's wave equation. In any atom it has been observed that only two electrons are permitted to have the lowest possible energy allowed by Schrödinger's equation.

The lowest energy level has but one orbital configuration in which electrons can be placed, and this is known as a the "s" orbital (Table 4.1). Two electrons can be placed in such an orbital, and no more; to try to do so violates a core tenet of the quantum realm for electrons: thou shalt not have two electrons with identical quantum descriptions (i.e., the same location in space and orientation of their spin). However, if one goes up to the next allowed energy level of an atom, one finds not only the "s" orbital, but another configuration called the "p" orbital; this second orbital can hold 6 more electrons. The second energy

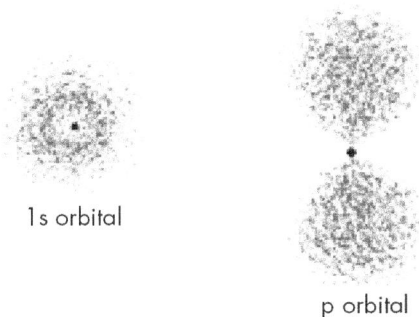

1s orbital

p orbital   **Figure 4.3** Examples of s- and p-Type Orbitals

Sublevel	# of Orbitals	Maximum Number of Electrons
S	1	2
p	3	6
d	5	10
f	7	14

Table 4.1   Orbital Electron Capacity of the Four Named Sublevels

level of an atom can be host to up to 8 electrons, but no more. At the next highest energy level we find an additional available orbital configuration, the "d" orbital, which can hold an additional 10 electrons, allowing for up to 18 electrons to be in this third energy state.

The atomic solutions to the Schrödinger wave equation, including spin, exactly predicts all of this and, in turn, it results in the periodic nature of the periodic table of the elements that Dmitri Mendeleev constructed. The regularity of chemical elements is governed by the quantum realm. This defines chemistry, and without such chemical properties there could not exist the properties that control the emergence of even the simplest biological systems; not to say more complicated systems, including we humans.

So far, we have concentrated on the electrons. Let's turn our attention to the nucleus of the atom. The nucleus is the smallest volume of the atom. How small is it? Let's think about some relative distance scales to get a feeling for the compactness of the atomic nucleus.

The distance required to complete a first down in football is ten yards—the team must move the ball from, say, the 30-yard marker to the 40-yard marker. The distance from New York City to Los Angeles is 2,475 miles. The ratio of these two numbers is about the same as the ratio of the nucleus of an atom to the total size of that atom. Let's imagine being far enough out in space to see the lights of both cities. From that height it would be very difficult to see even a *very* powerful search light being moved the ten yards from the 30-yard marker to the 40-yard marker! A very powerful telescope would be needed. Without it, much will be missed; but, with it, one should expect to find surprises.

When research on the atom began, the forces of nature were believed to consist only of gravity and the electromagnetic force. Those two forces had been known to exist for at least two centuries. From experiments performed in chemistry and physics, it was learned that two additional forces bind the nucleus and electrons together; the strong nuclear force controls the protons and neutrons, and the weak nuclear force controls the decay process, while the electromagnetic force acts on electrons. Gravity, of course, applies to the behavior of all particles. These four forces and their properties are described in Table 4.2. which shows their relative strengths and the theories that embody them. Note that the strength of each force is specified relative to that of the gravitational force. Prior to the advent of superstring theory, gravity was not consistent with quantum theory, but the other three were.

Note from the last column that the two nuclear forces (strong and weak) have small ranges of effectiveness that do not exceed 10^{-15} meters. The electromagnetic force and the gravity force have a range of effectiveness that is infinite. The behaviors of these forces are defined in the column labeled "Long Distance Behavior." For example, the gravity force diminishes as the square of the distance between two objects. "Current Theory" shows which mathematical framework applies to a given force.

Quantum electrodynamics (QED) governs electromagnetic behavior through the photon, whereas quantum chromodynamics (QCD) governs the behavior of the strong nuclear force through the gluon. The weak nuclear force is governed by the electroweak theory.

The QCD force has eight gluons that act *only* within protons, neutrons, and hadrons. Gluons are the force carriers that attract quarks. Gluons and quarks have what is called "color" charge. This charge is not actually a color, but is more like an electrical charge. There are three different types of color charge, allowing each to balance the other, rather like a triangle with three equal sides. When color charge is balanced it is as if one is looking at a color-neutral object. This can be thought of as much like combining red, blue, and green light to result in white light.

We've mentioned the importance of spin in the structure of the atom, but spin goes far deeper than just for how electrons arrange themselves in the orbits

Table 4.2 Forces of Nature and Their Actions

Interaction	Current Theory	Force Carrier	Relative Strength	Long-distance Behavior	Range (m)
Strong Nuclear	Quantum Chromodynamics (QCD)	gluons	10^{38}	1	10^{-15}
Electromagnetic	Quantum electrodynamics (QED)	photons	10^{36}	$\dfrac{1}{r^2}$	infinite
Weak	Electroweak Theory	W and Z bosons	10^{25}	$\dfrac{c^{-m_{W,Z}\,r}}{r}$	10^{-18}
Gravitation	General Relativity (GR)	gravitons (not yet discovered)	1	$\dfrac{1}{r^2}$	infinite

around a central atomic nucleus. The electron carries internal spin angular momentum, and does so in a specific unit: 1/2 ℏ (said as "h-bar"), which is an amount of energy that is about a trillionth of a trillionth of a trillionth of the energy of rotation that a single car wheel possesses when driving at 60 mph. This is the fundamental internal unit of energy that, along with its mass, defines an electron.

What makes a photon different from an electron? The photon possesses zero mass, and that certainly distinguishes it from an electron, whose mass is about 2000 times smaller than that of a hydrogen atom. But what also distinguishes an electron from a photon is the spin of the photon, which is ℏ—twice that of the electron. In fact, physicists have regularly observed that all particles of matter, like electrons and the quarks inside of protons and neutrons, possess half-units of Planck's constant in angular momentum, while all force-carrying particles (like photons, gluons, etc.) possess integer units of Planck's constant. This creates the two classes of fundamental particles mentioned earlier, bosons (those with integer multiples of Planck's constant for spin) and fermions (those with half-integer multiples of Planck's constant for spin).

As we shall see later, the exact structure of the forces of nature was revealed over decades of work by hundreds of physicists, some of whom we will meet, but the basic ideas are these: forces are carried between matter by particles from the boson class, and matter is composed of particles from the fermion class. The nucleus of the atom is home to both the strong and weak nuclear force, the former creates stability and certainty in the nucleus and the latter creates instability and change in the nucleus. Electromagnetism plays a small role in the nucleus, but outside of the nucleus, it rules, creating the structure of the atom with all of its allowed electron orbits and energies. While gravity, in principle, acts in all spaces in the atom, its effects are negligibly weak, weaker than even the weak nuclear force by a huge amount. It is not until we have vast collections of atoms near other vast collections of atoms that the force of gravity becomes noticeable. It is worth noting that understanding this structure, which took decades, was so fundamentally important that it garnered dozens of Nobel Prizes for those centrally involved in the key discoveries and insights that led the way from the atomic to the subatomic, and, finally, into the sub-nuclear realm.

Particle and Force Interactions

Before we look again at some of the ways in which the quantum realm differs from the familiar macroscopic realm, let us meet the building blocks of matter and the force-carrying particles that bind them. We'll learn more about some of these later, but here, let's simply take a tour of the family photo, a photo described by the most successful theory of nature ever devised, the standard model.

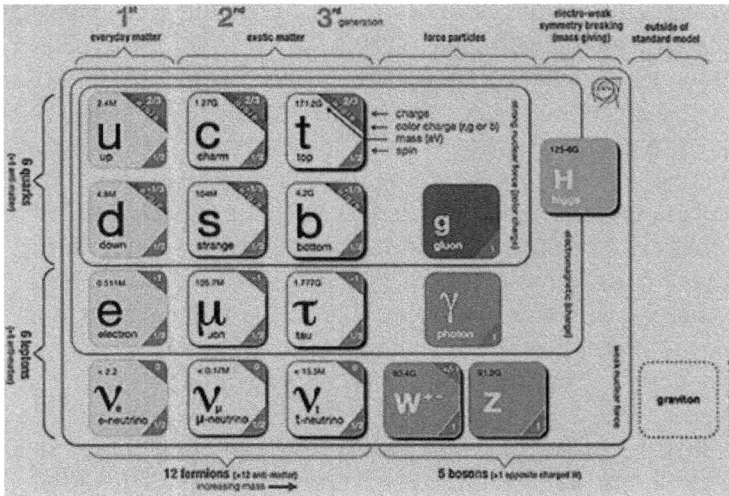

Figure 4.4 The Standard Model of Particles

Figure 4.4 will help to guide our tour. Matter is constructed from two families of particles, quarks and leptons. Quarks are the building blocks of atomic nuclei, but there are more quarks in nature than are needed to build atomic nuclei. We do not yet know why this is so, but it may be a clue to anticipating some deeper truth about the universe that we will continue to explore. All quarks have mass, but not the same mass. Again, we do not know why this is so, but physicists will explore this too.

Up and down quarks form the first "generation" of matter. The term "generation" is used here to identify characteristics of quarks that group them with similar particles in the standard model—similar in how strongly the weak interaction connects pairs of quarks. Up and down quarks are found inside all protons and neutrons as well as some other, more fleeting, subatomic particles. They are also the lightest (having least mass) among all quarks.

The second generation of quarks contains the strange and charm quarks, and the third generation are the top and bottom quarks—these are the heaviest.

The second family of matter are leptons, and they, too, are arranged in generations. The lightest generation contains the electron, familiar building block of the atom, and its partner electron-neutrino, a nearly massless zero-charge particle that is produced readily in some kinds of nuclear decay. The electron conveys the expectation for atomic stability, while the electron-neutrino is a suspect of nuclear *in*stability. The second generation of leptons contains the muon and the muon-neutrino, and the third generation, the tau and the tau-neutrino. We'll meet these leptons again, later.

One of the enduring mysteries of nature is why are there two kinds of matter, quarks and leptons, and why does each kind come in generations, and why there are three generations? Another mystery is their masses; the top quark is so heavy, yet the lightest neutrino will have a mass that is on the order of a trillion times smaller than the top quark!

Before we look at forces and how we imagine they function on the quantum level, let's pause and consider the fields of force that surround particles with charge. As we think about this, and draw in some of the workings of the quantum picture of nature, there will be revealed a strange phenomenon in the space around a particle, a phenomenon with real consequences for the universe.

Physicists found that electrons, protons, and neutrons are arranged in fixed patterns by fields created by force-carrying bosons such that their electric charges and other such properties are balanced within the atom. The nucleus contains positively charged protons and uncharged neutrons. Negatively charged electrons are found outside of the nucleus in accordance with their probabilistic distribution.

The expectation one might form of a charged particle surrounded by a field of force could be a simple one, but the quantum world is more complex than is expected. For instance, if one were to zoom in—way in—on the empty space surrounding an electron, one would find that that emptiness is not real. According to quantum mechanics, because there is no absolute certainty about position when measuring momentum, and no absolute certainty about momentum when measuring position, one must abandon certainty about the energy content of that space. That is, one cannot be sure that it is empty, and contains no particles at all.

This is a consequence of something called the "Heisenberg Uncertainty Principle," a concept first illuminated by the physicist Werner Heisenberg. It has caused much confusion in public discourse on physics. Let us work to dispel some of that.

The principle states that there are inseparable pairs of properties, such as position and momentum, in which certainty about one implies uncertainty about the other. You can design an experiment that is very precise about measuring the momentum of a subatomic particle, but you will find that, as you dial up the precision on momentum, you lose control of the notion of where the particle is in space. This seems odd, but let's apply an analogy.

Imagine that you are playing a game in which you are blindfolded and must locate where there are other people in a room. Those people are in motion. The only way you are allowed to find them is to throw a bowling ball, hope that it strikes one of the people, and then study how the ball has been deflected as a means to determine where the person had been located when struck. This is an unlikely game to play, but it's very like what must be done in the quantum realm each time we wish to locate something in space. "Seeing" occurs when one kind of particle bounces off of another and the scatter is then detected with some kind of instrument (e.g., the human eye, a camera, a particle detector). We must shoot something at an electron in an atom if we want to locate it. But, just as occurs with the atom, the game requires that we must sacrifice something to gain our knowledge.

We can evaluate the course of the thrown ball and, knowing its original trajectory, figure out where the person was when the ball hit, but if we also want to know how fast the person was moving when struck, we have sacrificed the knowledge of that because, upon being struck by the ball, the person's velocity was changed when hit by a heavy, fast-moving bowling ball.

This analogy, gives one a "feel" for the consequences arising out of the Heisenberg Uncertainty Principle. It has limits. While the bowling ball and a person are solid things with well-defined locations in space, electrons and photons are waves with poorly defined locations in space. Waves doing wave-like things comes at a cost in certainty.

So it should come as no surprise that, as we zoom in on the space around an electron we find that the expected empty space is actually filled with particles such as pairs of electrons and anti-electrons (positrons) that pop into existence and, as suddenly, go out of existence. These are called "virtual particles." They are a consequence of the uncertainty principle. This may sound fanciful, but it has real consequences.

For instance, the strength of the electric field around an electron is diminished by the presence of these charged particles appearing and disappearing. The closer we probe to an electron, the closer to the "bare strength" of its field we get. In fact, we have observed this! Also, should one want to compute properties of the electron, such as its g-2 number, one must take these virtual particles into account. Only by doing so does one obtain mathematical accuracy in the measurement of the number.

We have considered matter and the structure of the "empty" space around it that affects matter's properties and those of the fields that surround the particles. Let us now consider the forces that define those fields.

Each of the four fundamental forces are now known to have at least one force carrier associated with it as shown in Table 4.3. That force carrier provides the attractive or repulsive force for an interaction. These force carriers, in exchanging between two particles, create interactions at a distance between them. In this exchange they behave in a manner similar to the bowling ball scenario described earlier. When one person throws a bowling ball at another, there recoils trajectory change from the force of the ball. (For every action, there is an equal and opposite reaction.)

Table 4.3   Forces of Nature and Their Force Carriers

Force Type	Force Carriers
Strong Force	gluons (eight)
Electromagnetic Force	photon
Weak Force	$W^{\pm}$ and Z
Gravity	graviton

However, in the world of the small, if the charges on the two agents are opposite to each other, the recoil in that case would not be a recoil, but an attraction that would pull each toward the other. This interaction is the same for all of the exchange particles shown in the table. The conclusion that can be drawn from this is that one must determine the force carrier actions and reactions (their magnitude and direction) in order to establish the total effect of the interactions.

In 1971, a key development that forwarded the unification of the three forces of the standard model (exclusive of gravity) was based on research into the Yang-Mills field theory. This work, showing that the weak force is wholly consistent with quantum theory, changed the landscape of theoretical physics. The research was conducted by Dutch graduate student, Gerard 't Hooft, who was in his twenties at the time. Today 't Hooft is a professor of theoretical physics at Utrecht University. Later, investigations that t' Hooft performed independently were carried out by Leonard Susskind, resulting in the holographic principle of string theory. We will learn more about this later.

The Yang-Mills Field

Physicists believe that the weak and strong forces are caused by the exchange of a quantum of energy in the context of the mathematics of the Yang-Mills field. Widely recognized as having been discovered by C. N. Yang and his student, R. L. Mills (1927–1999) in 1954, the Yang-Mills field is a generalization of the electromagnetic field described a century earlier by Maxwell, except that the Yang-Mills field has many more components and can carry an electric charge (a photon carries no electric charge).

Often in the history of physics, there appear curiosities about such discoveries. Here there are little-recognized hints that Klein (the same one whose work relates to that of Kaluza) and a student of Salam's (Ronald Shaw) also found the mathematics of the Yang-Mills fields independently. The photon and the electromagnetic force are involved with the Yang-Mills field description. For weak interactions, the quantum force carriers are the W± and Z particles. The gluons are also described by this same mathematics of Yang-Mills Theory.

The quantum carrier for the strong force is the gluon, the taffy-like particle described earlier that holds quarks together in all hadronic matter (i.e., objects similar to protons and neutrons). The Yang-Mills field is an over-arching framework that combines the characteristics of QED (quantum electrodynamics) and QCD (quantum chromodynamics) into a single mathematical structure.

Initially, a perplexing issue of the Yang-Mills field W± and Z particles was that it appeared that when all possible fluctuations are taken into account via Feynman diagrams, the mathematics could not be used to make predictions. Physicists say that such equations are not "renormalizable."

It was more than twenty years before this was resolved and the theory could be used to solve typical quantum mechanics problems. 't Hooft's Ph.D. dissertation proved the final argument that the Yang-Mills field is renormalizable. It was this Nobel Prize-winning work of t' Hooft that opened the way for the equations of Yang and Mills to be used to make measurable predictions.

Symmetry

To help identify the particles and their behavior, physicists used the tools that were available in classical physics. One of these tools, very important in isolating particle behavior during interaction, is the physicists' knowledge of symmetry. Symmetry is important because it helps to simplify the mathematics. Anyone can think of a kind of symmetry that helps to identify things.

As an example, geometrical symmetry is useful in identifying objects. A football is quickly distinguished from a baseball, tennis ball, or golf ball because its shape is less symmetrical than the others. Geometrical symmetry allows an observer to separate different shapes inside an opaque bag of objects. If the bag contains objects that are rectangular, spherical, and triangular, feeling them will enable one to separate them without seeing them. It is the mathematical analog of this process that physicists use to solve problems.

Geometrical symmetry is observed in snowflakes. (See Figure 4.5a) This kind of geometrical symmetry is called rotational symmetry. The snow flake is a six-sided figure that, when rotated one-sixth of a turn, looks exactly as it did before it was rotated. Many particles have rotational symmetry. What physicists realized in the 20th century was that all particles that possess charge possess rotation-like properties that are indissolubly linked to the charge.

When one views an object from another vantage point, the object is said to be rotationally symmetrical if that object looks the same as it does when viewed from the earlier vantage point. If we replace the snowflake by a circle, then the image would have rotational symmetry no matter what angle is used for the rotation. But the reader may object that this is only true in two planes. If considered in three planes, is it still rotationally symmetrical? If the snowflake or circle is looked at edge-on and then turned 180 degrees to the flipped edge, is that considered to be rotationally symmetrical? The answer is "no," but if one wishes to have rotation invariance in three planes, one should replace the circle by a sphere!

This discussion shows that different shapes or situations can have differing amounts of symmetry. In fact, a mathematician named Élie Joseph Cartan during the 1920s and 1930s deduced every possible such symmetry that generalizes rotations! Although the physicists at that time did not realize it, his insights into symmetry would play an enormous role in understanding forces and force-carrying particles. His work is a mathematical basis for the work of Yang and Mills.

Another kind of symmetry, mirror symmetry, is illustrated in Figure 4.5(b). The left page of the book shows the left-hand image of a butterfly while the right page shows its mirror image. No simple rotation can achieve this effect, making it distinct from rotation. Mirror symmetry is what is known as a "discrete symmetry."

A mathematical theory can be said to have symmetry if a change in perspective is made but there is no change in the mathematical result. The mathematics of symmetry is called group theory. It is thought of as a mathematical formalism that describes symmetries that exist in the world.

In the 1930s, when scientists sought to understand the behavior of protons and neutrons with regard to the strong force in the atom's nucleus, they learned that the label proton or neutron was irrelevant so long as one considered only the strong nuclear force. The strong force cannot tell the difference between two kinds of quarks—all it detects is color charge, and quarks don't carry different degrees of color charge (for instance, an up quark is not twice as charged than is the down quark). As a result, the strong force is blind to quark types, and binds them democratically. If one had eyes that could see only gluons, a proton would look identical to a neutron! This is a kind of symmetry. Under the electromagnetic interaction, however, protons and neutrons look *very* different; photons will interact with a proton, owing to its positive electric charge, but not with the neutron, which carries a net zero electric charge. Protons and neutrons are not symmetric under the electromagnetic interaction.

Using experimental results described in terms of mathematics, and subtracting out the effects of electromagnetic forces using mathematics, it was seen that protons attract protons, protons attract neutrons, and neutrons attract neutrons. They found that the predictions derived from this purely mathematical and theoretical viewpoint, without including electromagnetism, gave answers that were remarkably close to what was seen in the laboratory!

How close? Well, the ratio of the actual mass of a proton to the actual mass of a neutron is 938:939. This number differs from 1 by about one part in one thousand. That is how close their mathematical predictions agreed with observation. From this it was concluded that the strong force has (approximate) proton/neutron symmetry. This is like concluding that the picture of the snowflake would not change by rotating it a one-sixth turn.

Symmetry simplifies the physicists' view of our universe by suggesting what structure should be sought to control the form of the equation they choose to use to describe the behavior of fundamental physics.

There is an important attribute of symmetry that the reader should know, discovered by Emmy Noether in Germany during the late 1880s and early 1900s. She is the discoverer of Noether's theorem: that every continuous symmetry (a symmetry like that associated with a circle, not a snowflake) that is also "local," has a conservation law.

Rotational symmetry
(a)

Mirror symmetry
(b)

Figure 4.5 Examples of Rotational and Mirror Symmetries

A conservation law states that if a process begins with a quantity that undergoes many changes, the original quantity will not have changed at the end of the process. This is similar to the law of the conservation of energy.

The most widely known conservation law states that energy can neither be created nor destroyed. If a process begins with a given energy level it may change its form, but at the end of the process it will have the same total energy with which it began.

There are many kinds of symmetries. For example, in translational symmetry you can do an experiment in one room and perform the identical experiment next door or down the street. An experiment done in your basement can be brought to a friend's house and the result will be the same. Alternatively, if there are two identical experimental devices, and the same experiment is performed on either, the results should be same regardless of the distance between them. That this distance can be made to vary from small to large without "jumping over" any distances is the meaning of continuous.

In her work, Amalie Emmy Noether (who is generally referred to as Emmy and whom we'll revisit later) shows that there is a mathematical result that implies that translational symmetry leads to the conservation of momentum.

"Every symmetry yields a conserved quantity" is an incorrect interpretation of the Noether theorem that often appears in the literature. The more accurate statement is, every *local* symmetry yields a conserved quantity. (So, dear reader, you might expect that we now have a brief discussion of the meaning of "local" in this context. Unfortunately, its precise meaning would require the use of calculus—there really are some things in nature that require the use of calculus to be understood—and would therefore be well beyond the use of algebra to show that there is no change in the observational consequences for the symmetry, thereby going beyond our promise to not use intricate levels of mathematics in this book.)

Another characteristic of local symmetry is that symmetry can be spontaneously broken. The Higgs mechanism has this property. Force-carrying particles that have mass cannot be mathematically described except by beginning with equations that have no mass—but, ab initio, include the role of the Higgs boson to induce mass for the force-carriers! We will learn more about this later.

An illustration of spontaneous symmetry breaking has been given by physicist Michio Kaku as follows: A top spinning at high speed can be made to stand vertically. Thus, when it is viewed by someone from above, the top chooses no particular horizontal direction along the table as being special. As it slows down (due to friction) it will eventually fall over. The important thing to note is that when it does fall, one direction along the horizontal direction is chosen as special because the top, lying on its side, points along it.

This is exactly like the Higgs mechanism, the established mechanism by which the known subatomic particles acquire mass (with, perhaps, the exception of neutrinos, something we will discuss later).

*****

This chapter has focused on the physical theories that describe the understanding that physicists have developed about how the universe works at the quantum level.

While doing this, physicists learned that symmetry plays an important role in manipulating the complex mathematics that describes how the universe works. Understanding symmetry's role helped them to deal with the complexities of those mathematics. Even in the realm of the large, as in Einstein's description of the universe called general relativity, it turns out that symmetry seems also to play an outsized role. It appears that the "Laws of the Cosmos" are expressed broadly by symmetry. The simplest observation of theoretical physicists is that for every force that has ever been discovered in the realm of the extremely small or the extremely large, there is always found a mathematical symmetry relating to its existence.

## Chapter 5

Finding Reality

The quest for a particle theory began in the mid-1940s with experimentation by groups of physicists who sought a new approach to the comprehension of matter from a quantum viewpoint. They reasoned that the fundamental particles in the atom might be found were they to first develop an approach to the quantum theory of particles focusing on the nucleus and its interaction with the strong force. This would focus these researchers on the theory of the strong force and its interactions with the neutron and proton. That theory, arrived at after many dead ends and decades-long frustration, is now called quantum chromodynamics (QCD). Paralleling these developments, collider experiments are effective in examining interactions, corroborating analyses, and confirming mathematical predictions that describe the behavior of fundamental particles and forces when studied under a variety of circumstances.

A great challenge for theoretical physicists is to not get lost in the elegance of their mathematics. By this it is meant that however assuring in predictive results they may be, the purpose of the mathematics is to make predictions that are falsifiable (concepts that are able to be proven false). Many different approaches to this challenge have been pursued by many scientists over time.

In this chapter we look at the quest to make sense of the subatomic world by studying it in ever-greater detail with powerful instruments: the particle accelerator and particle collider. Recall that all of the constituents of the universe known so far have a wave nature underlying their behaviors. This idea was introduced by the physicist Louis de Broglie (1892–1987) and verified with experimental measurements.

As one accelerates a particle to higher and higher speeds, the corresponding wavelength of the particle decreases. Just as using shorter wavelengths of light allows one to build a microscope that probes smaller and smaller structures, using highly accelerated particles—even those other than light—allows one to probe incredibly small structures. Modern accelerators probe the subnuclear realm, digging deeper and deeper down to shorter and shorter distance scales, and correspondingly larger and larger energies (like those that were present close to the beginning of time). Particle accelerators and colliders are both a microscope and a time machine, taking us deeper into the nooks and crannies of space while cranking up the energy, taking us back to moments in time that were far higher in energy, and thus far earlier in the history of the cosmos. Recall the big bang hypothesis, that is confirmed with much observational evidence and tells us that the universe was in a state of very high energy during very early times. Correspondingly, producing a state of very high energy in a laboratory environment is akin to recreating the universe at those very early times.

We will begin with a look at particle acceleration, and then discuss what has been revealed by generation upon generation of advancing accelerator technology. We will look at a few cases along the historical timeline that illustrate how a new technology, advanced by the discoveries of the past, can usher in an era of discoveries. These may have no clear pattern at first, but the work of theoretically minded physicists, searching for patterns, can close in on the best explanation—the one that survives the next test, to go on, perhaps, to predict new and as-yet-unknown features of nature. We will learn something about the structure of nature, underneath its plentiful zoo of particles, to find that new, more fundamental building blocks were lurking there the whole time.

This chapter will set the stage for a look at one of the very latest discoveries found within particle accelerators and colliders: the Higgs boson. This curious player in the life of the cosmos has an even larger role to play later in this book, but let's see now how this particle, "born on the Fourth of July," was discovered after decades of creative thinking and the difficult labor of building one-of-a-kind, frontier instrumentation.

Particle Acceleration

Historically, experimentation in cyclotrons and accelerators (which become "colliders" when the accelerated particles are smashed into a target) were the principal methods by which to discover mathematical equations that accurately capture reality in the behavior of subatomic forces and particles within a controlled environment. Experiments can be done to examine the behavior of particles and forces (and their characteristics and properties) under various situations. In the early days, the methodologies were primitive, but technology advanced quickly as the questions came into focus. At first, it was extremely difficult to predict the outcome of collisions on targets. No attempt was made to predict the results of particle collisions, as there was no clear mathematical foundation upon which predictions could be made. New particles came pouring out of colliders as we began to learn about, see, and sift, better and better, the products of collisions.

Since those early days, we have developed a firmer understanding of what to expect when particles collide. It is now possible to first provide a theoretical analysis to establish whether experimentation is likely to be fruitful before making a large investment in new colliders and detectors. That said, the unexpected is at the core of scientific progress. If we could predict everything with complete certainty, science would no longer be experimental. Planning is an intricate process, an intertwining of theoretical expectation with the thrill of the unknown. An example of this was demonstrated through confirmation of the existence of the Higgs boson. The mathematics that predicted the existence of the Higgs particle were presented more than forty years before it was possible to make an experimental confirmation. However, those very mathematics were used to finally guide the successful search to its discovery!

The primary role of particle accelerators and colliders is to manipulate Einstein's famous relationship between energy and mass, $E = mc^2$, in order to release the particle's energy by collision and convert it into mass and motion in the resulting explosion. Colliders are one of the purest expressions of Einstein's famous equation, demonstrating with each collision the relationship between energy, motion, and mass. Let us take a look at a few historical advances in particle acceleration and collision technology, to set the stage for the modern machine: the Large Hadron Collider (LHC).

The first particle accelerators were the forebears of the cathode ray tube in television sets. One example of a very early particle accelerator is the "Crooke's Tube," invented by William Crookes (1832–1919) and others in the period 1869–1875. It consists of a partially evacuated glass tube containing two metal electrodes, one at each end of the tube. By applying a high voltage, an electric field is established in the tube that rips electrons from one metal electrode, that are attracted toward the other. The electrons are compelled to leave their mother metal by the influence of the external electric force field. This is the essence of all modern particle accelerators. The core idea of a Crooke's Tube—tearing apart atoms using a strong electric field and compelling the electrons or ions to then accelerate within the field—is found in every research accelerator and every accelerator-based cancer treatment center in the world.

The Crooke's Tube is an example of one of the two major types of particle accelerator, a *linear* accelerator (written as LINAC). It takes its name from the fact that particles are accelerated along a straight line with as little bending of their paths as possible. While electric fields cause acceleration along the line, one can introduce magnetic fields at angles to the path of linear acceleration and compel the particles to focus and defocus. Magnetic fields are to electrically charged particles as glass lenses are to light. Both serve the role of focusing (brightening) or defocusing (dimming)—either gathering together or spreading apart the beam of particles that pass through them.

Accelerators do not accelerate one particle at a time, but, instead, bunches of particles, each of them carrying the same electric charge and therefore wanting to repel away from the others. These bunches have to be coaxed to stay together. Magnetic fields do the trick nicely. As a result, whether you are using a LINAC or another kind of accelerator, some amount of deflection of charged particles is necessary. A LINAC minimizes this by keeping the bunches moving on a mostly straight path, bending them only to keep the bunches tightly focused.

Linear accelerators have many advantages, such as there being little energy loss in having to constantly use energy to? deflect particles, but there are also disadvantages because one needs a longer and longer straight line (great pieces of geography as a railroad would need) to achieve more and more acceleration. The other type of accelerator sacrifices the low energy loss of the linear accelerator by trading into a *circular* accelerator, or "cyclotron," requiring less land usage.

The cyclotron was invented by Ernest Lawrence (1901–1958) in 1929. It has a circular cavity, as seen in his first patent application shown in Figure 5.1. Surrounded by magnets, a particle stream is introduced into the cavity of the device and accelerated by an oscillating alternating current voltage across a cathode–anode pair. The magnets surrounding the cavity bend the particle stream's path, causing it to move in a spiral trajectory as the stream travels from the cathode to the anode. As the stream of particles, bent by the magnets into the cavity's path, circles the cyclotron, it's sped up by pulsed external electric field energy that increases its speed. At first, as in a linear accelerator, the energy and speed of the particles is quite low. As the stream continues within that confined area more acceleration along the path of motion is achieved by continually passing through the same external electric fields, further increasing their speed.

This same principle is at work when you watch a child swinging on a swing. If she kicks her feet at just the right moment, she will cause the swing to go higher and higher. This kicking motion must be synchronized to a specific point in the arc of the swing to get the maximal effect; kicking at other times can lessen the transfer of energy from the body to the swing, or even reduce the motion of the swing. This same principle is at work in the synchrotron. The electrical field used to speed up the electrically charged particles is applied at key moments along the circular path. This is done by pulsing the electrical field when the protons are at that specific point along the circular path and then

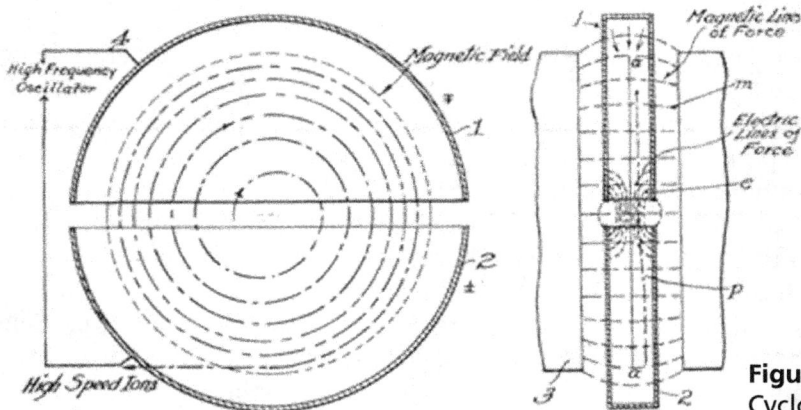

Figure 5.1 The Cyclotron, Circa 1929

awaiting their return to do it again. This is performed repeatedly at that point and, like the child on the swing kicking to get higher and higher, the protons reach higher and higher speeds. This synchronized application of the electrical field is the reason that the device is called a synchrotron.

Of course, running the particles over and over again in a circle has one disadvantage: they would never hit anything! If you want to use the beam, you need to smash it into something. The original cyclotron designs were such that as the speed of the particles increased, they had a harder and harder time being held in a circular motion by the external magnetic force fields. Eventually, they moved so fast that they escaped the accelerator; this was designed to happen at one specific place. The beam exited the accelerator and struck a target, achieving the collisions vital to making use of the beam.

For an accelerator, one of the key benchmarks of its performance is the energy with which particles exit the acceleration process and enter the collision process. The particle's energy level is expressed in terms of the voltage level that is needed to accelerate the particle to that speed. Based on Einstein's full equation that relates energy, motion, and mass, the higher the mass of a particle, the higher is the energy required to move/accelerate it to a high speed within the device. The other key benchmark for a collider is its intensity—how many particles per second, per unit area, it is able to accelerate to the desired energy. More particles mean more collisions, which means more opportunity to discover something, even if it's rare in nature. More energy means more access to new particles that might be so heavy (which require more energy to cause them to appear) than earlier colliders that would not have seen them. The twin requirements of energy and intensity drive the design of better and better accelerators.

During World War II, the cyclotron was first used to better understand particle behavior, particularly the proton. This timing was not an accident. The increased performance of radar, due to the demands of the war, had led to the rapid development of a device called the "klystron." Their use was adapted to delivering ever-increased power to accelerators like the cyclotron. They are particle-accelerating devices that propel a charged particle stream (e.g. one made from electrons or protons) to a speed sufficient to "knock out" subatomic particles from the nucleus of the atom. This provides the experimental physicist the opportunity to see what might be rattling around inside the nucleus or even the electron itself.

Physicists wanted to know how the atom responds to such things as charge, gravity, and magnetism. They would need some kind of collider to cause atomic particles to collide with other atomic particles under very high energies. Until the invention of the first linear accelerator that delivered more than 200,000 volts (by J. D. Cockcroft and E.T.S. Walton in 1932), cyclotrons were limited as to how much energy they could attain to continue to increase the final energy they could achieve.

Marrying a linear accelerator to a cyclotron, to kick-off the acceleration process, which can take the pre-accelerated beam and accelerate it the rest of the way to the target energy, allowed for more and more powerful accelerators to be developed. One could not delve very far into strong and weak force behavior until the energy levels of experimentation were increased. This allowed exploration deep inside the nucleus of the atom, a region of extreme interest wherein the discovery and cataloging of the atomic nucleus, its constituents, and the many ways nuclei emit energy and fall apart could be studied.

The early method of using particle accelerators was to divert the beam from the accelerator and slam it into a target at rest in the laboratory. One famous example of this approach was to cause a single proton stream to collide with a target material such as gold to determine what would be found in the particle debris. This limited the pace with which new energy regimes could be reached with these tools. During the design of the atomic bomb, a new era of speed in pushing the boundaries of collider energy was reached. The idea behind a particle collider is to make one beam of accelerated particles collide head-on with a second, independent beam of accelerated particles. By doing this, instead of striking a target at rest in the laboratory frame, the energy of collision is greatly magnified.

Modern particle accelerators and (colliders) can be understood by referring to Figure 5.2. The high-altitude photograph shows two rings at the Fermilab collider complex. The "Tevatron," the ring to the rear, operates at energy levels of almost 1 TeV. Each ring is four miles in circumference. The system begins particle stream acceleration in the forward ring, transitioning it to the Tevatron ring when its energy has reached a value appropriate for transition. Positioned around the rear Tevatron ring are stations where experiments are performed on the debris. Each experiment looks for results anticipated from the collision and for particle behavior that is expected to occur at specific energy levels. Each experiment is a particle detector, described below, designed to sift through the debris from the collisions and make sense of what happened when the beams

Figure 5.2 The Tevatron (Rear Ring)

interacted in the accelerator. Magnets are positioned around the circle to manage the behavior of the particle stream's orientation and direction. and to bend and focus the accelerated particles, keeping them in the ring.

Let's revisit Einstein's famous realization that $E = mc^2$. With this equation, we can not only begin to understand the energy units used by particle physicists to describe the performance of accelerators, but why it's so important to constantly push the boundaries of accelerators upward in energy. First, let's consider the units of energy: electron-Volts. This is the amount of energy gained by one electron moving through a one-volt potential difference. (Voltage potential is similar to water pressure.) If you make an electron travel between the terminals of one of those little, rectangular 9-volt (9V) batteries, you have accelerated that electron up to an energy of 9 electron-Volts (eV). If you instead use the electric field present in a standard 480V high-voltage panel (emblazoned, perhaps, with a sticker that says "DANGER! Arc flash hazard!"—i.e., Don't mess with me!!), you can get one electron up to a total energy of 480eV. Now the energy is getting big and it's convenient to switch to a bigger unit. 1000eV is called a "kilo-electron-Volt" and written "keV." So 480eV would be equivalent to 0.48keV, or about half a keV.

Keep cranking up the strength of the electric field and you can switch to even larger units—mega-electron-Volts (millions of eV, written MeV), giga-eV (billions, or GeV), and tera-eV (trillions, or TeV). Early cyclotrons achieved energies in the MeV range.

Let us now consider the usefulness of cranking the energy up to such high levels. According to Einstein's equation, $E = mc^2$, mass and energy are equivalent. Thus, the mass required to produce subatomic particles specifies the energy level at which a collider must operate. One can express mass in units related to the eV using this equation. If we rewrite Einstein's equation to solve for mass, we find that mass (m) can be expressed as $m = E/c^2$. For instance, the proton, in these units, has a mass of about 1 GeV/c^2. This is a convenient benchmark for expressing masses.

So, to produce a proton, you have to muster an energy of 1 GeV in a particle accelerator. (In practice, thanks to various laws of conservation in nature, one is forced to make at least two particles to satisfy nature. To make at least one proton means you must get above its mass in energy so that other particles can also be made.) In comparison to the proton, which is made from two up quarks (there are six kinds of quarks; up, down, strange, charm, bottom, and top) and a down quark (and a whole mess of gluons), the top quark has a mass that is about 172 times that of the proton. This is why the Tevatron at Fermilab was required to discover it; only the Tevatron had enough energy in its proton/anti-proton collisions, at a total energy of 1.96 GeV (1.96 billion eV!), to make top quarks for the first time in human history. The top quark was discovered in the early 1990s.

When one designs a new particle collider, there are two major factors that come into consideration: what energy is needed and what intensity is needed to achieve the physics goals of the experiments? For instance, if a new but rarely produced heavy subatomic particle is expected to be produced, it may not be sufficient to simply fire up *just* enough energy. If such a particle is only produced in one in every 100 billion collisions, you had better be able to make as many collisions happen per second as is reasonable, given technology and cost. The Large Hadron Collider was designed to achieve both of these things: to be a very high energy collider (in order to produce the top quark and the Higgs, which are the frontier matter and force particles) and to be a very intense collider (because making a Higgs particle is a rare phenomenon—otherwise, it would have already been discovered at the Tevatron). The LHC began operations at an energy that was half of its design energy—3.5 TeV per beam, for a maximum possible of 7 TeV of energy per proton-proton collision. Since beginning operations in 2009 and 2010, the LHC has ramped up its energy toward its design goal, first stepping from 7 TeV to 8 TeV, and then stepping to 13 TeV. The design goal is 14 TeV.

The LHC is not just a high-energy machine, but a high-intensity machine. It now routinely delivers well over 1 billion proton-proton collisions each second. Since, at the energies that the LHC currently operates, the probability of a single proton-proton interaction making at least 1 Higgs particle is a miniscule 1-in-a-billion, the LHC currently produces about 1 Higgs boson during every second of operation. This is why, to study the Higgs boson in detail, the LHC must run not only for months and months each year, but for years and years in order to produce enough Higgs bosons for physicists to understand the details of their behavior.

Two powerful accelerators, the Large Hadron Collider (LHC) at CERN, and the Fermilab Tevatron are examples of multi-TeV devices. Without these powerful accelerators it would have been impossible to understand interactions among the particles sufficiently well as to observe new ones that enable atomic structure and the standard model to be constructed.

In collider physics, all the easy stuff, for the most part, has already been discovered—because that stuff has been produced so readily in earlier particle collisions. For each singly interesting Higgs boson event, there are about a billion uninteresting proton-proton collisions that must be rejected in some clever way in order to remove the "background noise" without also squelching the interesting event.

Here we find a crucial intersection of theoretical physics, computational techniques like simulating nature, and the experience of experimental scientists with their instruments. To design a search procedure for that rare gem of a particle, one must carefully craft a strategy to remove the "noise" of the decades of easily discovered particles that preceded the present work.

Let's take a closer look at the LHC before moving on to the next crucial tool—particle detectors—and learn about some of the ways physicists sift through data for those precious particles that may not have ever before been observed.

The Innards of the LHC

You can learn a lot about a tree by studying its rings. The rings of a tree result from the yearly cycles of its growth and slumber. If you look closely at the rings of a single tree, you can even learn about the kinds of struggles it has gone through over its history. The flattening of the rings on one side on just one or two rings might imply that it was struck on that side, perhaps by a hard object. The thickness of the rings can tell you about which were good or lean years for the tree, based on temperature and other environmental phenomena. Much like the rings of a tree, we can learn a lot about a circular particle accelerator by studying its rings.

To feed the accelerator, protons are liberated from hydrogen gas by ionizing the gas. The protons are then injected first into a LINAC. The LINAC accelerates the protons up to 50 MeV. This is not a lot of energy compared to the final energy achievable by the LHC, and brings the protons only to about 5% of the speed of light—the speed of light being the fastest they could ever conceivably travel. This LINAC, known specifically as "LINAC 2," was not built for the LHC; it was constructed in 1978 to provide proton beams to much earlier generations of circular accelerators. This is one of the older "tree rings" of the LHC accelerator complex, designed to intensify proton beams over an earlier LINAC built at CERN.

The LINAC injects these protons into the Proton Synchrotron Booster (PSB), which further accelerates the protons up to 1.4GeV, bringing them to about 83% of the speed of light. Neither was the PSB built specifically for the LHC; it was constructed in 1972 to provide a "boost" in energy to protons emanating from the LINAC before they entered another circular accelerator, the Proton Synchrotron, to be discussed in a moment. Like the LINAC, the PSB is another tree ring in the great complex of machines that feed the LHC. It's an older ring, informing us about an earlier era in proton beam physics.

From the PSB, the protons are injected into a larger circular accelerator, the Proton Synchrotron (PS), which continues the acceleration to 26GeV (99.935% of the speed of light). The PS was also not built just for the LHC. It was its own major accelerator project constructed in the late 1950s and first operated in 1959. The PS is extremely versatile and can accelerate more than just protons. The PS was used to provide protons to earlier projects at CERN, including the first "hadron collider" in the world, the Intersecting Storage Rings (ISR), that ran from 1971 to 1984. It also provided particle beams for a bubble chamber experiment called "Gargamelle," delivering not protons, but a secondarily produced neutrino stream. Gargamelle is famous for detecting subatomic particle interactions that had to be mediated by an electrically neutral particle, but not the photon—it had to be some other neutral, force-carrying particle—one of the foundations of the great discoveries in the early 1980s of the force-carrying particles of the weak nuclear force. We'll talk about the Gargamelle again later.

Note that in each instance of the improvement of the accelerator stages listed above, the leap in energy has been big—the first one was from 50MeV to 1.4GeV, a factor of 28 in energy—but this results in an increasingly modest gain in the speed of the protons (taking them only from 5% to 83% of the speed of light, a gain of just 17—smaller than the gain in energy). The leaps in energy continue, but the gains in speed become ever more modest. This is Einstein's special relativity in action. You can continue to put in more and more energy, making the collision potential of the particles huge, but each major leap in energy results in more and more paltry gains in speed. Particles with mass cannot reach the speed of light. This fundamental behavior of nature is evident in the workings of the pre-accelerator ring systems of the LHC.

From the PS, the protons are injected into the Super Proton Synchrotron (SPS), where they are accelerated to 450GeV (putting them within a tiny fraction—one millionth—of the speed of light, or about 1,500 mph below the maximum possible speed permitted by nature). Again, this accelerator is a "tree ring" in the great CERN accelerator complex leading to entry to the LHC. The SPS was designed to be a frontier energy proton collider, and was the first circular collider to smash protons into anti-protons. The SPS was the machine that produced the collisions for the famous CERN experiments UA1 and UA2, that resulted in the discovery of the W and Z bosons—the carriers of the weak nuclear force. We'll talk about them again, too.

We have reached the end of the history of CERN and about the particle physics that can be told with these ever-larger accelerators. We have finally reached the LHC. The protons delivered from the SPS are accelerated up to 6.5TeV in the main ring of the LHC, which brings them to a speed just 7 mph below the speed of light (99.99999999% of this fantastic speed). They are moving faster than humans have ever before accelerated objects.

There are many independent experiments that operate on the LHC—the beams and the accelerator serve many physics programs, each one operated by hundreds or thousands of scientists and engineers. Now that we have surveyed the ideas and some of the technology and achievements in particle accelerators, let's look at the other side of the problem: once you have successfully made particles collide, how do you identify the debris so you can learn something from the collisions?

Detecting Particle Collisions

If particle accelerators and colliders are meant to provide illumination of the subatomic world, what is it that captures the information occurring in those collisions? For this, one needs particle detectors. Detectors are the complement of colliders. Particle accelerators and colliders are highly complex instruments requiring a devoted, independent team of scientists and engineers to develop, upgrade, operate, and maintain them. Detectors are similarly complex, requiring their own team of scientists and engineers to design, develop, upgrade, operate, and maintain them. Let's look at some basic ideas involved in detection, and some past technologies used to achieve this process. Then we'll look at a modern particle detector and the ways in which these devices push the bounds of present technology.

The basic method of particle detection is this: in front of the particles escaping the point of collision, place layers of material whose job it is to cause the escaping particles to lose energy; capture that energy loss using instrumentation and store the information about the lost energy; develop ways of relating the lost energy back to the kinds of particles that could have lost them, and from this information build a descriptive picture of the debris created by the particle collider.

Much like a forensic scientist analyzing the scene of a multicar accident, the scientist was not there when the accident occurred so there is no first-hand knowledge of the exact details of the accident—which car began it, how it happened, and what was the path of destruction as each car struck the other. The forensic scientist uses the aftermath of the collision to piece together these details, if imperfectly: which car struck first (perhaps based on the one having the most damage, indicating it might have been moving the fastest), and the sequence of car strikes (based on the angles and amounts of damage); the path of destruction, tracking the origin of the accident; and then further clues found at the scene about what might have caused the accident (for instance, debris on the road, or some kind of mechanical failure, or driver error).

Just as in accident scene investigation, experimental physicists, using data from a particle detector, are using their past experience with known particles to interpret the energy losses in material, make best guesses about which particles lost and left that energy, and then track back to find where all the particles came from, thence learning what might have initiated them in the first place. Decades of experience then allows cross-checking their assumptions and inferences, making as sure as they can that they have not misled themselves with wrong inferences.

Let us look at some historical techniques (that have gone through a kind of renaissance in the modern era in looking for dark matter—about which we'll talk more later) followed by a look at more modern techniques for doing the same job, even in the face of far more particles produced at a far faster pace.

A classic technology for seeing traces of subatomic particles as they pass through material is a "cloud chamber" or a "bubble chamber". The ideas are similar in both cases. Cloud chambers contain mists of various liquids, typically water or alcohol while bubble chambers use, not a mist, but a very cold and highly compressed liquid, such as liquid hydrogen.

In a cloud chamber, the mist is caused by super-saturation of a liquid (e.g. 99% pure isopropyl alcohol) in a very cold environment such as dry ice. When a charged particle passes through a cloud (or bubble) chamber, it disturbs the mist (or liquid), allowing one to observe its path as it traverses the chamber. Some cloud chambers are better suited to observe electrons and others are better suited to observe protons (or other charged objects such as those that produce beta rays). The choices one can make in the chemical content of the vapor determines the type of particle that will be observable.

The invention of the cloud chamber is credited to a physicist from Scotland named Charles Thomas Rees Wilson (1869–1959). The technology and scale of these chambers was improved over decades, and by the time of the explosion in theoretical and experimental work in quantum mechanics that took place in the very early 1900s, the cloud chamber was a tool-of-choice for probing deeply into the subatomic world. Cloud chambers allow chemists and physicists to study nuclear radiation, the energetic particles ejected from atomic nuclei when instabilities are induced by the nuclear forces. The cloud chamber was central in many particle discoveries, including the positron, the anti-matter counterpart of the electron, that was predicted to exist in 1928 by Paul Dirac (1902–1984) and found by Carl Anderson (1905–1991) in 1932 and the muon, the heavier cousin of the electron, found in 1936 (also by Anderson).

After the invention of the cyclotron in 1929, cloud chambers were used in conjunction with the cyclotron to observe the paths taken by particles under the influence of nature's fundamental forces. While the positron and muon were discovered in cosmic rays, the marriage of the cloud chamber to the particle collider allowed for the discovery of even rarer and harder-to-detect particles. For instance, while the pion—the proposed carrier of the inter-nuclear force binding proton to proton, proton to neutron, and neutron to neutron—was discovered using cosmic rays and a photographic film-based detector technology, the combination of particle accelerator and cloud chamber allowed for the discovery of a strange cousin of the pion—the kaon—in 1947. As the technology for accelerating, colliding, and detecting subatomic particles advanced, the moving frontier revealed more and more subatomic particles. An entire "zoo" of strange creatures, from kaons to lambdas to sigmas, poured forth from the colliders, found by ever-evolving detector technologies.

A modern particle detector combines many technologies, layered in well-engineered structures, to serve the needs of the physics goals of the experiments. For a modern collider-based detector, a goal is to be able to detect multiple kinds of particles, as many particles are produced in each collision. For instance, in the ATLAS detector at the LHC, the design of the detector was driven by the goal of observing the Higgs boson (should it exist at all), studying in more detail the top quark (presently the heaviest-known fundamental particle in nature), while searching for a wide-ranging suite of phenomena that are not described by the standard model, but could represent a more fundamental theory of nature that includes the standard model. The ATLAS detector, like its counterpart, the CMS detector, needed to be able to see the particles that were expected to be produced if theoretical frameworks like supersymmetry or extra dimensions of space (more about these later) play a role in the proton-proton collisions in the LHC.

The ATLAS detector is eight stories tall. It's half-a-football field in length, weighing in at about 7,700 tons. It is, in fact, a huge digital camera, capable of taking 40,000,000 pictures per second! Each picture is a digital collection of "signatures" of particle interactions (such as electrical pulses read out by different detector subsystems) from each of the subcomponents of the detector, totaling more than 100 million channels—this is the equivalent of a 100-megapixel photograph. (The iPhone 7 produces a 12-megapixel photograph.) The detector is cylindrical in shape. Protons collide at the center of the cylinder, its shape designed to photographically capture the debris from those collisions.

The innermost layers of the ATLAS detector are intended to detect particles with electric charge; the materials that make up these inner layers—silicon and gas, instrumented with electronics—were chosen to highlight the behaviors of charged particles passing through them. In addition to charged particle detection, the device has an energy measurement system—the "calorimeters." The inner-most calorimeter is designed to detect the interactions of electrons and photons, while the outer-most calorimeter is designed to detect the interactions of particles like pions and protons.

Some particles pass through the ATLAS detector leaving little or no information behind as they do so. For instance, the muon, the heavier cousin of the electron, interacts little with the ATLAS detector on its way away from the collisions that produce them. The farthest layers of the ATLAS detector are designed to catch energy from muons as they exit the detector and continue their travel, right through the earth. The freedom of muons, having escaped the AT-LAS detector, is bittersweet in their frame of reference; they only live about 2 millionths of a second. So, no sooner than they escape ATLAS to begin a great journey, they fall apart (due to the quantum mechanics of matter), converting their energy into other particles (electrons and neutrinos).

The neutrino is another particle that is often produced in proton-proton collisions, but it is even *more* elusive than the muon. A neutrino is capable of traveling through a stack of lead bricks an entire light-year in length and *might*, in that time, *might,* encounter one collision with an atom! Explaining this phenomenon, unlike electrons, muons, and taus, which can also experience electromagnetic interactions (or quarks that can experience the strong interaction), neutrinos experience only the weak interaction, This makes neutrinos the most elusive known form of matter. It has nothing to do with size, all fundamental particles are equally sized "points." It has everything to do with forces. The fewer forces interacted with, the less noticeable is that particle. Further, neutrinos passing through the ATLAS detector have an easier time eluding detection, because ATLAS does not contain much lead to slow them down.

Neutrinos are detected, not by seeing them, but by noting in a given proton-proton collision that an amount of energy is missing from a careful accounting of the collision of the protons. We detect, not the neutrino itself, but in the bookkeeping of the experiment, the absence of its energy. This is a precision numbers game!

This is the way physicists see and evaluate the debris from particle collisions. Some physicists use large arrays of detectors on the earth's surface to observe the debris from cosmic ray collisions in the atmosphere. Some physicists use ultra-cold liquids and solids to detect the rare interactions of particles in matter as they hunt for tiny corners of nature where physics outside of the standard model might first manifest. Some physicists put particle detectors high above the earth's atmosphere to watch the cosmos, using not only light, but all of the messengering-methods that the cosmos flings at our planet.

Such one-of-a-kind digital cameras found in detectors are cutting-edge, built by a close team of physicists and engineers and controlled by advanced custom-built computer software. Students often participate in these experiments, constructing hardware, writing software, testing the technology to make sure it operates as designed, finding new uses for the same instrument, and analyzing the piles of data produced by the detectors. For experimental particle physicists, the accelerators and detectors are the crucibles in which Ph.D.s are forged.

How do we make sense of what we see in these experiments? A particle detector presents a blinding pile of subatomic particles. How does one see patterns? Finding the patterns in the data, the patterns that tell us about the fundamental laws of nature, is the most difficult aspect of these experiments. Let's take a look at one historical moment during which the patterns of detected particles led to a much greater understanding of the universe.

Particle Analysis and the Finding of Strange

The boom in particle discoveries in the 1940s, '50s, and '60s left physicists with many puzzles. What was the pattern, if any, that related these particles to each other? Was there something fundamental beneath the seemingly endless stream of new states of matter that were being uncovered in photographic emulsion and cloud and bubble chamber detectors? Earlier, we mentioned that a particle called the kaon was discovered in 1947. The kaon was one of a strange class of newly discovered particles. It was heavy, about four times the mass of the pion (making it about half the mass of a proton). Other particles in this class were objects known as "hyperons," examples of which are the lambda (Λ) and the sigma (Σ). It wasn't the mass itself that made them strange; it was how oddly long-lived they were, given their heavier mass. Naively, it was expected that the heavier the particle, the faster it would decay. The other thing that made them odd was that they were easily and copiously produced in particle accelerator experiments. A particle readily produced would suggest one just as ready to decay to something else; a door easily opened is easily closed again.

The kaons and the hyperons bucked this expectation, being easily produced but slow to decay. The tracks—the trajectories of electrically charged particles in the detector medium—consisted of either straight paths, paths that were bent due to magnetic and electrical forces, or very curled paths that were often the signature of decaying particles.

The image in Figure 5.3 is produced by first using a collider to create a beam of particles that are aimed at a target. The bubble chamber is then placed on the side opposite to the incoming beam so that after collision with the target, the resulting debris has to travel through the bubble chamber. As it traverses the chamber, the spray of particles produces the tracks shown in the image.

Occasionally, however, tracks were found with strange upside-down "V" formations (see Figure 5.3). These were first seen using detectors that looked at cosmic ray interactions, and they were later found, as well, using detectors at particle accelerators.

The "Vs," as they came to be known, were a signature that led to the discovery of kaons and hyperons. Using careful measurements of distances in the pictures, and using the geometric information about the bending of tracks and the thickness of those tracks, physicists could convert photographic information about these chains of particle decay into information about the energy of the particles. Most interestingly, one could apply the conservation of energy and momentum and apply Einstein's famous formula to arrive at the mass of decaying particles. These "Vs" arose from a new class of particles, the hyperons, that were heavy but very long-lived. This, as we noted earlier, was strange.

Murray Gell-Mann and another brilliant physicist, Kazuhiko Nishijima (1926–2009), independently studied these V particles and postulated that, to explain their heavy masses but paradoxically long lives, one could invoke the

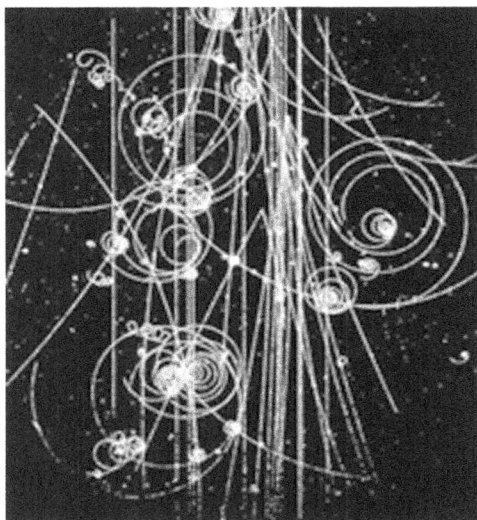

Figure 5.3 Typical Output Tracks From a Collider/Bubble Chamber Experiment

existence of a *new* quantum number. Spin is a quantum number. The spin angular momentum of a particle tells us its overall particle class: fermion (half-integer spin) or boson (full-integer spin). Electric charge is another quantum number. Quantum numbers label states that matter can take, and there is almost always some kind of law of conservation associated with a given quantum number, or with some combination of them. Gell-Mann and Nishijima postulated that the reason that Vs were so long-lived, despite being so heavy, was that nature was being forced to conserve some previously undetected quantum number that had not been seen before by humans. They called this quantum number "strangeness," in honor of the odd behavior of the Vs, and constructed a formula based on the data from particle accelerators and detectors to explain why the Vs were so long-lived. Nature was being forced to respect a conservation law, and that conservation law extended the life of a particle a bit longer than would have been naively expected based on the mass and past experience.

This is an example of how one can go from data that doesn't make sense, to a hypothesis that explains existing data while predicting something new that needs to be tested. In this case, the hypothesis explained the data (why these strange heavy particles were long-lived) while postulating a new thing in nature—strangeness—that is a quantum number. It should have other consequences. Also, it begged for an explanation—what was the origin of the strangeness quantum number?

The work of Gell-Mann, Nishijima, and many others in this turbulent era of particle discovery and the difficulty that theoretical physicists had in making sense of the data was a highly productive period for physics. We had gone from atoms, made from protons, neutrons, and electrons, to a zoo of other particles (the muon and the kaon) that seemed to have *nothing* to do with the atom. Sure, they could be produced by smashing things into atoms, but that was just $E=mc^2$ in action. What was the *need* in nature for these particles? What was the *explanation* for their existence in the universe?

Fundamentally, when particle physicists talk about "particles" they are speaking in a short-hand way for "fundamental, point-like objects with no internal structure." At low enough energies, even states of matter made from smaller bits—think "atoms"—can behave as if they are just tiny billiard balls banging off of one another in a cosmic game of snooker or pool. But crank up the energy—shrink the de Broglie wavelenths at which you probe nature—and you might find that the things you thought were ideal, point-like objects are really made from something else.

While there were many competing hypotheses that attempted to explain the zoo of new particles that were discovered in the '30s, '40s, and '50s, realize that *only one of these ideas could turn out to be true* (assuming that a correct one had been proposed).

One of these competing hypotheses had to do with what it meant to be a "point-like particle"—and whether all of these newly discovered particles, and even some of the old standards, like the proton or neutron, were actually point-like at all. If protons and kaons and hyperons were, perhaps, made from something more fundamental—more point-like and much smaller—perhaps this could explain the incredible number of particles in nature that had nothing to do with making atoms. After all, this idea had previously worked in making sense of atoms. The periodic table of the elements may categorize the elements in nature, and even let you predict the existence of new and undiscovered elements, but the reason for the periodic nature of this table is not explained by the table itself. That had to wait until the discovery of the electron, and the proton, and eventually the neutron. Atoms were made of smaller, more point-like things. Could subatomic particles be the same way? Might they be made of more fundamental, point-like things?

A Quark in the Dark

Let's take a look at another example of how data from particle accelerators and detectors, and the hard work of experimentalists and theorists, helped to explain the mystery of the particle zoo. In the 1950s, using previously confusing data from particle accelerator and particle detector experiments, Gell-Mann and Nishijima discovered the presence of a new quantum number—strangeness. By the early '60s, they had both successfully developed what they believed was a classification scheme that put the particle zoo on the same footing that the elements were on when they were organized into the periodic table by the Russian chemist, Dmitri Mendeleev (1834–1907). Gell-Mann called this scheme "The Eightfold Way," a reference to a practice of Buddhists (though Gell-Mann was not a Buddhist himself), because this first organizational chart of the particle zoo contained eight particles organized by their electric charge (positive, negative, and zero) and their strangeness quantum number (again, positive, negative, and zero). By organizing the particles in this way—specifically, the spin-1/2 kaons, pions, and eta meson—Gell-Mann, Nishijima, and, independently, the physicist Yuval Ne'eman (1925–2006) found a way not only to represent the particles, but to understand that they were parts of families related to each other through changes in quantum numbers like charge and strangeness.

Having successfully organized the spin-1/2 particles into such a chart, the work was repeated for the then-known spin-3/2 particles—the hyperons and other similar particles. Organized by strangeness and electric charge, these particles formed an inverted triangle shape having a curious feature: the point of the triangle *was missing*! All the then-known spin-3/2 particles formed the bases of the triangles, and a descending ladder of "floors" making up the rest of the triangle. (See Figure 5.4.) But the tip of the triangle at the bottom of the chart was missing. Gell-Mann and Ne'eman saw this, not as a problem for the idea, but as an opportunity—so they predicted in 1961 that there must be an as-

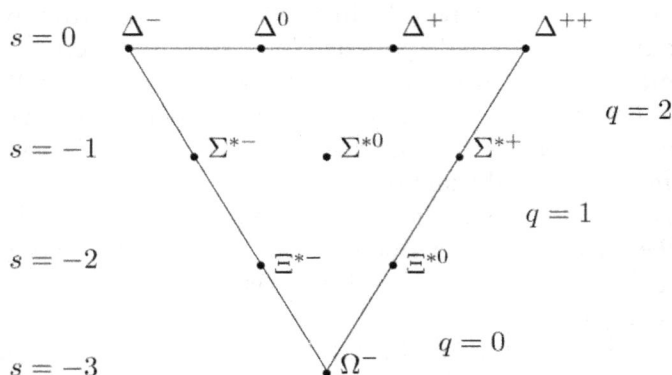

Figure 5.4 Baryons organized into an decuplet according to the Eightfold way
© Laurascudder via Creative Commons

yet-undiscovered particle having the right quantum numbers to fill in the blank in the chart. This was called the Omega-Minus particle (Ω^-), with negative charge and a strangeness quantum number, $s = -3$. Gell-Mann and Ne'eman even used the masses of the other states in the chart to predict the mass of this undiscovered particle: 1680 MeV.

In 1964, an experiment did discover this predicted particle. The experiment that found it was conducted in an 80-inch bubble chamber at the Brookhaven National Laboratory, a collaboration between physicists from the laboratory, Syracuse University, and the University of Rochester. They discovered a previously unseen new state of matter, with a mass of (1686 ± 12)MeV/c^2, with negative electric charge and a strangeness quantum number of $s = -3$, exactly as predicted by Gell-Mann and Ne'eman!

It is important to remember that the ideas of Gell-Mann and Ne'eman, although both elegant and predictive, did not have to be correct. Nature does not always reward such boldness. They turned out to be right, but as is the case in science, there are often very elegant and beautiful ideas that, when tested, fail the test and have to be discarded or modified. In this case, we merely have illustrated the power of organizing information so as to observe patterns that actually predict nature. This is the task of the theoretical physicist. Not all such labors are rewarded with evidence that supports the claims of a new idea, but here we have a case where the process yielded a discovery that further confirmed that this method of organizing nature led to something correct. There are some times when a labor such as this is rewarded.

One of the problems with such a reward is that it leads to more questions, almost guaranteeing more work for the theoretical physics community. Certainly, the "Eightfold Way" was the right way to organize particles by their measured information, but this did not answer the question *why? Why is this the way nature is organized? What is the underlying, unifying principle that results in the Eightfold Way?*

You begin to see in this historical example the ladder of questions and experiments that leads up to new knowledge about the universe. We might be painting too linear a picture of this process in this book, but in truth it is difficult (without a book of its own, of which a few already exist) to convey the messiness of science. Certainly, science as an idealized process, operates under guidelines—the scientific method—that each scientist tries to follow and improve upon as they go. But, when many scientists are working all at once on much the same confusing information, trying to make sense of it, there is a competing marketplace of ideas. In this case, many physicists studying the same data may come up with many explanations about why the data turned out the way it did. Not all of those ideas will be right; in fact, they may all be wrong. Experiment and the subsequent finding of evidence are the great arbiter of these ideas. If the ideas pass the test, they survive for another round. If not, they will probably be cast into the dustbin of history. This is how we make progress and come to a better understanding of the "truth" of the universe.

There were many competing ideas about organizing the particle zoo as to why the Eightfold Way was the right explanation. Historically, it turns out that Gell-Mann and another physicist, George Zweig, hit on the correct explanation. But, in truth, their explanation was not widely taken seriously because its implications were considered ludicrous in the light of our experience with the universe. Yet, not every idea that flies at odds with experience is proven incorrect; this example is illustrative of how, sometimes, ideas that challenge the orthodoxy of experience can win the battle to be subsequent come a correct description of nature.

By 1964, the year of the discovery of the Ω^-, Gell-Mann and Zweig had each sorted out their ideas about what explained the Eightfold Way. They named their explanation different things. Zweig called them "aces," while Gell-Mann called them "quarks," but the core of their ideas were essentially the same: that there must be a new class of more fundamental particle that, when combined in various ways, explains the patterns of spin, charge, and strangeness encoded in the Eightfold Way. Gell-Mann's naming convention was inspired by the fact that his scheme called for three new, fundamental subatomic particles and he wanted a word that, as he writes in his book *The Quark and the Jaguar*, needed to sound like "*kwork*" (he was uncommitted as to how it was spelled). When he stumbled on the word "quark" in James Joyce's book, *Finnegan's Wake*, he had his word.

Zweig preferred the term "aces," but the word "quark" caught on for these new building blocks of nature. Either scheme, Gell-Mann's or Zweig's, called for three of these building blocks. Two of them would not carry the quantum number called "strangeness," but the third one did. This worked to explain how the proton or the neutron did not have the strange behaviors of the kaon or hyperon (they contain no strange-carrying quarks), but other particles might (they contain one or more strange-carrying quarks). To explain the pattern of observed states of matter known in 1964, only three quarks were needed: up, down, and strange. (Up and down are named to refer to another property called "isospin," which had been identified earlier, explaining the close masses of the proton and neutron, and was part of the Gell-Mann—Nishijima formula.) A proton is made from two up quarks and one down; a neutron is made from two downs and one up. Notice the near identical composition of the proton and neutron, differing in only one kind of quark. This was a temptingly simple explanation for the similarity of the proton and neutron. An Ω^- particle, having $s = -3$, is made from three strange quarks.

One uncomfortable thing about this idea was that it required quarks to carry fractional electric charge. To explain the proton, the up quark had to be charged to only two-thirds of the elementary charge, written as +2/3. The down quark had to be charged to only one-third of the elementary charge, written –1/3. Adding the charges of two ups and a down quark yields the observed charge of the proton: +1 elementary charge. The idea of fractional charge was a sticking point for many physicists. The convention for electric charges having a sign was established by inventor and scientist Benjamin Franklin (1706–1790); the minimum unit of charge being established by the physicist Robert Millikan (1868–1953). Perhaps Millikan was just unlucky in that he found a unit of charge present in the electron (and the proton) but, given the low energy of his experiment (at room temperature), he couldn't have known about quarks, which are so much smaller than atoms, else he would have used other than an integer to account for charge and more readily establish a numerical convention to embrace quarks.

Another uncomfortable thing about quarks was that no one had ever seen them before. Physicists have seen protons and neutrons, and pions and kaons, and other states of matter made from quarks, but *why had they never seen a quark on its own*? This suggests that some new principle was at work—something operating in nature to hide quarks from view by packing them together into states like pions and protons. So, by invoking quarks one must allow that there are other undiscovered features of nature. This idea created either crisis or opportunity. It turned out to be an opportunity.

To address these issues, Gell-Mann and physicist Harald Frisch introduced a new kind of charge that could be present in quarks, but not in other particles. Such a new charge would help resolve some of the puzzles mentioned above. Gell-Mann and Frisch proposed that this charge came in three kinds, unlike electric charge which comes in two kinds (positive and negative). The use of color names ("red," "green," and "blue") suited their needs, as it provided a familiar basis for humans to understand pictorially why quarks are not found by themselves in nature and why the properties of particles made from them, like pions, are observed to be what they are.

They proposed that all states in nature had to be colorless—that nature prefers the absence of net color charge in the universe. If a stray quark were ever to wander from its parent state, the color charge acts (in accordance with other quantum laws and conservation of energy) to restore colorless balance to the universe. This would explain why quarks are never seen alone in nature, and why no experiment had yet detected them. Protons, containing three quarks, would be explained to be colorless so long as one quark was red, another blue, and the third green. The combination of red, green and blue light makes white when viewed by the eye—and white can be seen as having no single color; i.e.,it is described to be colorless.

As is the case with particles like the electron, its anti-matter counterpart carries the opposite charge. So, to accommodate anti-matter, there must also be anti-red, anti-green, and anti-blue. A particle, like a pion, is composed of two quarks—a quark and an anti-quark—so a pion can contain either red/anti-red, green/anti-green, or blue/anti-blue quark pairs.

Gell-Mann and Frisch proposed color charge in 1972, and very soon after proposed the existence of an entire mathematical framework akin to the ones that describe electromagnetism and the weak nuclear interaction, based around the idea of this new charge. This became known as Quantum Chromodynamics (a nod to the color charge in its name), to parallel Quantum Electrodynamics, the successful quantum theory of the electromagnetic interaction. Quantum Chromodynamics (or QCD) would emerge as the most successful explanation for the particle zoo that was discovered in the 1940s, 50s, and 60s. Physicists in the early 1970s would motivate the existence of a fourth quark (charm), which was discovered in 1974. A fifth quark (bottom) would be discovered just a few years later, and the sixth quark would be postulated (top) but not discovered until the early 1990s. The 1970s, 1980s, and 1990s were a booming era for the confirmation of the mathematical ideas laid down in the 1960s and 1970s, based on the explosion of particles discoveries in those preceding decades. And all of this was ushered in by the technological innovations at the heart of particle accelerators and particle detectors. Data drove analysis, and analysis produced new ideas that drove new experiments and resulted in new data. This is the great and messy cycle of science.

Wrapping Up

In this chapter we have looked at the tools of the modern particle physicist: the particle accelerator and the particle detector. The accelerator serves as both microscope and time machine, probing deeper and deeper into the heart of matter, higher and higher in energy, and thus recreating more and more the conditions of the early universe. Using such tools revealed in the 1940s–'60s a zoo of particles. This zoo, which seemed to have nothing to do with the atoms that were then known to make all matter, cried for an explanatory framework, one that could unite all of these mysterious particles under one umbrella. This was found first in the "Eightfold Way," a system of organization for the zoo, and later in the more fundamental explanation of this system: quarks, color, and the theory of their interaction—Quantum Chromodynamics (QCD). We now have the basic ingredients we need in order to step into the story of the Higgs boson. Where did it come from? Why was it needed? What would be its properties? And, most important, how does one discover it if it even exists at all? You now have the basic tools to step into this fascinating set of questions.

## Chapter 6

What the Heck's the Higgs? —Part I

As the story goes, Peter Higgs was driving to the Institute for Advanced Study (IAS) at Princeton, very close to his destination, when he pulled off the road. He was a nervous wreck. He was about to present his recent work on the mass of fundamental force-carrying particles, that included a concept to determine how this mass could come into being. If it was the correct idea—and if he had done his mathematics correctly—this could explain why some forces have very great range (think light, which, undisturbed by matter, can travel forever without stopping) while other forces have a very short range (think the weak nuclear force, which, locked in the nucleus of the atom, was detected only when the odd behaviors of radioactive nuclear decay were detected). He was about to present his work at one of the most renowned institutions for theoretical physics in the world, a place that had been (until his death in 1955) the home of the great Albert Einstein. Freeman Dyson, another famous physicist, had invited him to give the seminar.

It was 1966. Higgs had published his idea two years earlier, in 1964. He was about to have this idea gutted by some of the smartest people in the world, or perhaps it would live to survive another gutting some other day. Such is the life of the scientist—the quest to be correct at the risk of being shown to be very, very wrong. It was no wonder Higgs had pulled to the side of the road.

He gave the seminar. The ideas he presented then, having taken root in 1964, would seep into the basic theoretical framework that was used to successfully describe nature during the next decade—the Standard Model of Particle Physics. All of the implications of his ideas would turn out to be true, as they were substantiated by experiments in the 1970s, '80s, and '90s. But the cornerstone of his idea, that there must be an unknown force-carrying particle in nature, by which, through its interactions, mass is generated, would not be proven correct until July 4, 2012—nearly 48 years after the idea was first presented. Such is the "life of waiting" endured by a scientist.

The Problem with Forces

The phrase "quantum field theory" is used in colleges to warn undergraduate physics majors of the challenges ahead on the road to becoming a physicist. Graduate students who take the course are planning to become theoretical physicists or experimental physicists, or they are masochists. But what is there in life that is worth doing that is always and ever easy? The quantum field theory is the key idea that has led to a super-precise understanding of the quantum realm—so precise that it has led us to one of the most well-known numbers in the universe—the magnetic moment of the electron.

But, although they are so good at what they do, quantum field theories had a rough start. Many theoretical ideas do so in their infancy. Quantum field theories developed out of quantum mechanics and special relativity. They represent an evolutionary step in the quantum idea. If the first step is to describe atomic and subatomic matter constituents via wave behavior, quantizing the values of properties that are associated with those constituents, the next step is to apply the same logic to the force fields that cause matter to affect matter. Often referred to as "second quantization" (although this is a quite outdated and an often-misunderstood term used in modern college courses), it led to a deep understanding of both matter and forces in the manner in which they are used to represent states of nature. While matter is fundamentally different from force (you cannot create or destroy matter, but in interacting via a force field, two matter particles will exchange quanta of the force, requiring force quanta to come into existence, transmit the force, and go out of existence at the end of the interaction) the same theoretical ideas are extended to both.

Marrying quantum mechanics to special relativity led directly to the prediction of anti-matter and showed how spin angular momentum as a basic feature of matter is naturally a part of any such theory, whether you want it to be there or not. Quantizing force fields led to a deep understanding of the electromagnetic force, uniting the ideas of fields of force and quanta of energy (photons, in the case of light) beneath a singular umbrella.

One of the authors of this book, SS, recalls an extremely profound moment in graduate school when his quantum mechanics professor, Dieter Zeppenfeld, at the end of the second semester of the course, showed how second quantization leads to mathematical operations that create and destroy field quanta (photons). If one then imagines an enclosed cavity containing an electromagnetic field, one can ask the question: what is the energy intensity spectrum of photon radiation that would be emitted from such a cavity? He showed that the field theory elegantly reproduces exactly the spectrum of radiation called "the blackbody curve," the very curve whose mysterious properties first kicked off the quantum revolution through the work of a then-young Max Planck (1858–1947). There are often those profound moments in physics when a discovery out of the past paves the way to deeper insight into the cosmos—intellectual left turns in the mathematics that are found out to be supremely true in the light of experiment. These yield, in turn, better explanations for the mysterious phenomenon that started the whole thing in the first place.

As you can imagine, quantum field theory was quite an achievement. It was extremely successful at describing interactions like those found in matter having an electric charge and the given electromagnetic field associated with that charge. The culmination of this was Quantum Electrodynamics (QED), a theory built upon the difficult theoretical work of many physicists including famous individuals like Paul Dirac (1902–1984), Enrico Fermi (1901–1954), Hans Bethe, Sin-Itiro Tomonaga (1906–1979), Julian Schwinger (1918–1994), Richard Feynman (1918–1988) and Freeman Dyson. Truly an achievement, but it did not achieve perfection—when it came to other known forces of nature, such as the nuclear forces, quantum field theories had a serious flaw: their mathematics always predicted that such forces would be infinite in range.

But the nuclear forces are not infinite. They are supremely constrained to distances the size of just a bit more than a "nucleon"— that is just a bit more than the size of a proton or a neutron. The reasons for that confinement had been inferred by physicist Hideki Yukawa (1907–1981): that force carriers of the nuclear forces have mass, or behave as if they have mass. This limits their range, due to the form of the interaction. Yukawa's work appeared in 1934, and made a definitive prediction: to explain the short range of the nuclear force that binds protons and neutrons in the nucleus, the mass of the force carrier would have to be about 100 MeV/c^2. This force-carrying particle was called a "meson," from the Greek for "intermediate." When the muon was discovered later, in 1936, it was at first mistaken for Yukawa's predicted meson—thus the original name for this particle, the "mu meson," (later shortened to muon, which stuck), but is really a misnomer, as muons play no direct role in nuclear interactions

The discovery of the pi meson (or "pion") in 1947 by Cecil Powell (1903–1969), César Lattes (1924–2005), and Giuseppe Occhialini (1907–1993), led to the confirmation of Yukawa›s proposition that the shortness of the nuclear force was connected to massive force-carrying particles. The idea of massive force-carrying particles seemed a reality that could not be avoided.

At that time, the photon was the only known force carrier, and it was massless. The pion, though, seemed to transmit a force, *and it had mass*. Therefore, massive force carriers are something quantum field theory would have to describe.

The problem was that the quantum field theories under development at the time—the ones that would culminate in QED for the electromagnetic interaction and electric charge—could not reproduce this key feature of the short-ranged and massive nuclear forces. Neither the mass, nor the range, seemed within grasp of these powerful mathematical ideas.

It is important to pause at this moment and consider something that is crucial to quantum field theory: symmetry. We've discussed symmetry before. The symmetries that are built into quantum field theories are not easy to imagine, unlike the rotational symmetry of a snowflake that appears the same under certain choices of rotational angle. The symmetries that occur in quantum field theories also have to do with changes introduced to the mathematics that describe matter particles and the force fields. The equations that describe these things remain invariant under those changes. These kinds of symmetries are known as "gauge symmetries." An analogy may help to begin to grasp this key idea.

Consider water. If one leaves it sitting out in a cup in a room for a long time, it reaches the same temperature as the surrounding environment. Hold your finger in the air for a moment just above the cup of water before dipping your finger into the water in the cup. It feels to be the same temperature as the air, right? (Assuming that you really allowed the water to sit out for long enough.) Temperature is a common concept that we learn about from early in our lives ("Here comes dinner. Don't touch the pan. It's hot!"), but very few of us can actually *define* temperature.

Physicists in the 1800s studied heat, and the energy that is associated with heat, and came to understand that what we call temperature is a measure of the energy of the moving atoms in a body of matter. For instance, when water is hotter than our skin (and thus we call it "warm" or "hot"), it's because the atoms in the water are jiggling and jostling at average speeds that are faster than do the atoms in our skin. The opposite is true when something feels cold—the atoms in the "cold" water are jostling more slowly than those in our skin. The skin and water exchange kinetic energy—the energy of motion—when they come together, and your speedy atoms bang into the slower-moving atoms of the water, causing them to speed up, bringing the energies of the two bodies slowly to an identical state of average motion. Such energy transfers are how you make cold things warmer and warm things colder.

What does all this have to do with quantum field theory? Specifically, what does this have to do with *symmetry*, and very specifically with the *gauge symmetry* present in quantum field theories of nature? The answer lies in the *temperature scale*—the system of numbers we associate with different heat energies. Most people in the United States are comfortable with the Fahrenheit scale of temperatures, while most people in other countries are familiar with the Celsius scale. How are people able to communicate information about temperature between the U.S. and, say, Canada, when they use totally different scales of temperature? The answer is *gauge symmetry*.

The key idea is this: water freezes at the same heat energy content, and water boils at the same heat energy content, independent of what scale of temperature a nation uses to describe that content. This is *gauge invariance*. The physical universe and its behavior is not affected by what scale you choose to describe those behaviors. Not only that, but there is a continuous transformation that allows you to relate the two scales, so it doesn't really matter which one you choose (there are good reasons to prefer the *Kelvin* scale over either Celsius or Fahrenheit, but that's a conversation for another book). The independence of physical phenomena from the choice of scale is a core idea of gauge symmetry; in fact, the name of this symmetry, gauge, comes from the idea that the absolute scale on a gauge (an instrument for assigning numbers to behaviors, like a speed gauge or a fuel gauge in a car) doesn't affect the physical phenomenon that the gauge describes—nature is invariant under the choice of your gauge's scale.

So, water boils at the same amount of heat energy regardless of the temperature scale you use. On the Fahrenheit scale, this happens at about 212 degrees F, while on the Celsius scale this happens at 100 degrees C. How does one relate these two scales? Here is the continuous transformation that allows you to do this:

$$T_F = (9/5)T_C + 32$$

You'll note that at some commonly known temperatures, like boiling for water or freezing for water, this formula returns the familiar numbers. On the Celsius scale, water freezes at 0°C, and we see that if you plug 0°C into the formula above you find that, in the Fahrenheit scale, this happens at 32°F— exactly as is known from experience with these scales. This is how you relate any temperature in one scale to the other. As a fun exercise, the engaged reader might try inverting this equation to come to the one that takes a Fahrenheit temperature and returns the equivalent temperature in Celsius.

The gauge symmetries that are found in quantum field theories are quite a bit more abstract than this, but you get the essential idea from this example. If the core equation of a quantum field theory is invariant under a change of gauge using certain properties of the players in the equation, then the theory is said to be gauge invariant. In fact, a delightful feature of the quantum field theories being developed in the 1950s and early 1960s was that they had gauge symmetries in abundance, making calculations quite a bit easier than the earlier more fractured quantum physics.

But the beauty of these theories turned out to be the key to their failure. After all, is the world around you symmetric? If you look eastward, does the world look exactly the same as it does it you look westward? If someone blindfolded you, spun you about in a room, and then removed the blindfold, would you be able to tell whether your final orientation is different from your initial one? This is almost certainly the case—most people don't have rooms that look the same in every compass direction.

Symmetries are broken all the time in the world around us. They are cherished in mathematics because they greatly simplify the labor in using the equations; but, in the case of quantum field theories, they were also incapable of addressing the very problem that physicists wanted to solve in the 1950s and 1960s: the problem of short-ranged forces.

Precious Symmetry

How did anyone every figure that out? It might seem obvious to the reader that, of course, mathematical theories that have so many symmetries might not be very realistic. But, remember that QED, which came from these early quantum field theories, was a spectacularly successful theory of the electromagnetic interaction. It's not a huge surprise, then, that it wasn't incredibly obvious to most physicists that the symmetries might be the very problem they were trying to abate. Symmetries are associated with beauty. They have a kind of aesthetic attractiveness—even the abstract symmetries present in quantum field theory. More importantly, a brilliant young mathematician, initially shunned in pre-World War I German academic society simply because she was a woman, had taught us that symmetries and conserved quantities in nature go hand-in-hand.

Her name was Amalie Emmy Noether (1882–1935), though she is known today simply as Emmy Noether. She is probably the most famous mathematician you've never heard of, but despite a short life filled with tremendous social upheaval she discovered one of the most important mathematical truths of the universe. Noether was born in Erlangen, Germany. Her father was a mathematician who taught at the University of Erlangen. While her performance on her required exams suggested she was more fit to study human languages, she elected instead to study the language of nature—mathematics—at the University of Erlangen. She completed her dissertation in 1907 and went on to teach at her alma mater for seven years *without pay*. Why?

Noether was tangled up in the culture of her day, a culture in which women were almost completely locked out of faculty positions in academia, not because of their level of qualification, but because of their gender. Nonetheless, her prowess in mathematics caught the attention of other brilliant mathematicians of her day, most notably the mathematicians David Hilbert (1862–1943) and Felix Klein (1849–1925). In 1915, just as Albert Einstein was about to complete his general theory of relativity (this coincidence has consequences for our story), Hilbert and Klein invited her to join them at the University of Göttingen, one of the most renowned institutions of its day for the study of mathematics. This invitation was extended over the objections of faculty from the language and history departments! It is notable that Hilbert extended the invitation knowing that there would be objections; he was intentionally challenging the academic and social norms of the time. Sadly, those objections had consequences, as Noether spent the next four years unable to work independently, lecturing under Hilbert's name—he got the credit for her work (whether he wanted it or not) while she received none.

After World War I, the German academic systems were "liberalized" and this made it much easier for women to receive consideration for positions. Noether was finally permitted to obtain her "habilitation"—a post-doctoral qualification that permits further promotion in the field (common to many European countries). This allowed her to attain the rank of *Privatdozentin*, which reflects that Noether possessed sufficient academic qualifications to teach specific subjects at the university level. Now, rather than Hilbert (or someone else) receiving the credit for her work, she could instruct students, do research, and take full credit for the outcome of her own labors.

It was during this period at Göttingen that Noether did some of her most remarkable work. In 1915, around the time that Noether arrived, Einstein had given a series of lectures at Göttingen. He would remark later that he was delighted that, through these lectures (and, one presumes, private conversations) that Hilbert and Klein seemed convinced that he was on the right track. In 1916, Einstein completed his theory by deriving, at last, the "field equations" relating energy and mass to space and time. He then published what is considered to be the seminal paper on the General Theory of Relativity. Hilbert, using mathematics that he had derived himself (the "variational principle"), published his own paper that same month, arriving at the same equations as had Einstein. This was a very exciting time! It was during this period, having joined Hilbert and Klein, that Noether became involved in the study of general relativity as a mathematical framework.

Despite the publication of the seminal work on the subject by Einstein and the confirmation of this work by others using independent methods, there were nonetheless unresolved problems about the theory. One of these problems was that it was known for decades that energy was not conserved in the way it had been understood to be mathematically described. The notion of conservation of energy had been treated in very specific ways based on experience, but there was no strong mathematical foundation supporting this key idea. Using those ways of describing energy conservation, it appeared that general relativity violated this cherished principle of nature. This had the potential to doom it as a reliable theory of nature. It was a difficult problem.

Noether became deeply involved in this issue. In letters, Felix Klein refers to her constant engagement in the subject. She scrutinized the mathematics of the general theory and hit on a revelation. Mathematicians and physicists had been holding *too narrow* a definition of the conservation of energy. Einstein's theory did conserve energy, but in a way that was broader and deeper than was appreciated by the notions of what it meant to do so at the time. Consider the following situation.

The earth goes around the sun. If we ignore all other planets and bodies in the solar system for a moment, this is a perfectly simple system that general relativity should have no problem in describing. However, if one studies this system under general relativity, using the understanding that early physicists and mathematicians had established through the field equations, it would seem that, over time, the earth would lose energy. If you start up the entire earth-sun system with some initial energy, there is no apparent place where the earth would be "wound down" in its motion—it is orbiting the sun in empty space, meaning that there is no friction—no resistance to motion. This is a "closed system," one in which energy should not be able to sneak away via sound, heat, or friction. Yet, general relativity predicts that the earth will slow down very, very gently over time. Does this doom the theory?

Noether did careful and painstaking work on this question and discovered that the problem came about in the definition of "the system." One cannot consider just the earth and the sun; one must also consider that space-time is part of the picture, too. Although the most interesting part of a play can be the actors, in general relativity one cannot neglect the *stage* – it, too, is a crucial actor in the cosmic play! As the earth warps space-time and moves through the warped space-time caused by the sun, energy is slowly radiated away into space-time in the form of tiny ripples in the fabric of the universe—gravitational waves. Energy is not being lost to nowhere. It merely changes form from the earth's energy of motion into the rippling of space-time as the earth moves about the sun. The earth and sun, in the grander scheme of things, are not that massive, so, for our solar system, this process takes a ridiculously long time.

There are far more massive bodies in the universe, such as neutron stars and black holes. When these orbit each other the amount of energy lost to gravitational radiation is not only huge, it's measurable. In fact, the Nobel Prize in Physics in 1993 was given to Russell Hulse and Joseph Taylor for their observation that binary pulsar PSR B1913+16 spins down at *precisely* the rate calculated through general relativity, when one accounts for the loss of energy of motion converted to then-unseen gravitational waves.

While Klein was the public face of this mathematical discovery, he gave immense credit to Noether for having the more general and fundamental ideas about what it means to conserve energy in such a theory. It was clear from Klein's own words that Noether was the one driving this effort. In the process of doing this, Noether was also laying the groundwork for the remarkable ideas that bear only her name to this day, Noether's Theorems.

Before Noether's work, conservation laws were a bit "magical"—they were not really on a firm mathematical footing. It was easy to write one down—*what you start with is what you end with*—but the *why* of this was a real mystery. Others had made conjectures and tried to put conservation laws onto some footing, but it was Noether's insight into the simple cause of conservation that completely set the framework in place for discussing these features of nature.

In Noether's famous paper, "Invarianten beliebiger Differentialausdrücke" ("Invariants of Arbitrary Differential Expressions"), she lays out what are now known as "Noether's Theorems." In one, she shows that the origin of conservation of energy is that the equations that describe the laws of motion are *invariant* under shifts in time. It doesn't matter what specific time one puts into Newton's Laws—one obtains the same mathematical form of the laws at any time you choose to start the calculation. This, in turn, forces energy to be conserved! Noether had found a deep connection between symmetry and conservation.

She discovered that the reason for conservation of momentum in such equations is that they are invariant under changes in spatial location. It is permitted to shift the entire universe one yard to the left. The laws of physics remain unchanged under such a shift! This, then, forces momentum to be conserved. There is a deep connection between symmetry—the invariance of a system of laws under a transformation—and a conserved quantity. These symmetries are called "continuous symmetries," because a big shift in spatial location can be represented by summing many tiny spatial shifts. Make the shifts small enough (infinitesimally small), and it is as if you were continually shifting around the coordinate system of the universe.

These ideas created a generalized footing, so that one need not speak only about concrete concepts like space and time. One can apply the same framework to abstract quantities, like those associated with the gauge symmetries of quantum field theories. One begins to see why it's so hard to conceive of taking a beautiful mathematical theory of nature, with its many potential continuous symmetries, and choose to actively shatter those symmetries. For every continuous symmetry, there is a conservation law associated with it. This is a powerful feature of any theory, and one would be unlikely to simply abandon it at the outset.

To close the tale of Emmy Noether, we must visit some of the darkest events in recent human history. Noether remained at Göttingen until the rise of the Nazi Party in Germany, when Adolf Hitler became the Chancellor of Germany in January of 1933. As the Nazi influence spread through the country, a purge of Jewish academics began. Noether's parents were Jewish, making her a target of this horrible discrimination. What being a woman in this period of time had not done, being of Jewish heritage would. She, like many other academics of her day, lost her position at Göttingen. For a brief time, she taught from her apartment. She sought to support her colleagues during this difficult period and remained upbeat and positive about the future, even after she left Germany that same year.

As events turned toward the abyss in Germany, colleagues in America and England tried to find ways to help their fellow German academics. Positions were created at U.S. institutions to attract academics from Germany to pull them out of harm's way. That is how Einstein came to the U.S., with a position at Princeton's Institute for Advanced Study. Noether was sought by both Bryn Mawr College in Pennsylvania and Somerville College at the University of Oxford in England. Noether relocated to Bryn Mawr in October of 1933.

After two years in the U.S., Noether learned that she had a tumor in her pelvis. During surgery, doctors discovered other problems, including a very large ovarian cyst and other tumors in her uterus. While she initially showed positive signs of recovery, she succumbed to a fever and died shortly after. Such a brilliant mathematician, such an interesting life was tragically cut short. It was a huge loss to both mathematics and physics.

The Breaking of Symmetry

The role of broken symmetries in nature was discovered by many physicists, some of whom we'll discuss here. Eventually, the idea of broken gauge symmetry would become a core concept built into the most successful theory of nature devised so far, the Standard Model of Particle Physics.

The kind of breaking of symmetry that matters to this discussion is known as "spontaneous symmetry breaking." There is no specific cause for the breaking (although identifying the cause is part of building a better theory of nature); rather, a symmetry present in a theory is simply and spontaneously no longer respected, and the system becomes asymmetric. A good analogy for this can be done with a simple, plastic coffee stirrer. (See Figure 6.1)

Take one of those short plastic coffee stirrers—the ones that are like tiny straws—and hold it between your thumb and your middle finger. Don't press on it too hard; hold it lightly enough that it doesn't bend, but firmly enough that it doesn't fall from your hand. Now turn your hand and look at the stirrer from different sides. If someone were to hold the stirrer very close to your eyes (to prevent you seeing what their hands are doing), and ask you to close them, to open them a moment later and state whether the stirrer had been rotated, you'd have a very hard time telling unless you could see how their hands had moved. It's very symmetrical—the stirrer, rotated about its long axis, looks very much the same from every angle.

Now, press gently and evenly with your thumb and forefinger on the ends of the stirrer—gentle pressure is all that is required. Eventually, under increasing gentle pressure, the stirrer will bow. One cannot predict in which direction the bowing will occur, but before bowing occurs, all directions of bowing are equally likely and valid. Eventually, enough pressure will cause tiny non-uniformities in the construction of the stirrer to bow preferentially in one direction and not another. This is a spontaneous breaking of symmetry. We cannot predict in which direction the bowing will happen, or exactly when it will occur, but when it occurs, it breaks the rotational symmetry of the original system. Now, if someone holds the bowed stirrer up to your face, has you close your eyes and then reopen them a moment later to tell if the stirrer had been rotated, you could absolutely tell whether it had, even without referencing their hand. The bowed side would now point in a new direction in space if the stirrer had been rotated by the holder.

This is spontaneous symmetry breaking, as a physical analogy. The relationship between mass, particles, and such broken symmetry was first recognized in the study of theories of particles *without* quantum spin – these are the so-called *Goldstone bosons*.

Jeffrey Goldstone was born in England and received his university education from Trinity College, Cambridge, including his Ph.D. He would help lay the foundation for the effects of broken symmetries, and was the first to learn mathematically that, when a symmetry is spontaneously broken in a theory, it implies the existence of new, massless particles associated with the symmetry breaking – the Goldstone bosons. In fact, the pion, the lowest-mass combination of a pair of quarks, was identified by the physicist Yoichiro Nambu (1921–2015) as the "pseudo-Goldstone boson" of QCD. Nambu was born in Tokyo, educated in physics through his Ph.D. in 1952, and made professor at Osaka City University by the time he was 29, in 1950, when he still held only a master's degree. He moved to Princeton's Institute for Advanced Study in 1952 to establish the mathematical groundwork for symmetry breaking in 1960. It was on this work that Goldstone further built his ideas, and, elegantly, it was Nambu who recognized that the pion was a kind of "Goldstone boson." The pion's mass is not zero, but it is much smaller than typical masses of matter made from quarks. There is an abstract form of symmetry, called "chiral symmetry" (taking its name from the idea of handedness—being left-handed or right-handed) that is spontaneously broken in the fundamental theory of QCD and thus results in the very low mass of the pion.

However, applying Nambu's and Goldstone's ideas to theories other than those having spinless force-carrying particles— like the quantum field theory describing electromagnetism or one describing the nuclear forces—was much harder. Goldstone himself commented on this work in 1961, that his simple models would need to be taken much further to make progress on the subject. Not only that, Goldstone bosons are always without mass. If there were massless, spin-zero particles zipping about in the universe, they would have been noticed. It just wasn't clear that this idea would have value going forward to the problem of force-carrier mass in quantum field theories.

In hindsight, the ideas we've just described were the foundations upon which the final pieces would be laid, but the next piece would not come from particle physics. It would come from a different sub-field of physics where ideas about broken symmetry were first recognized to be important in describing physical

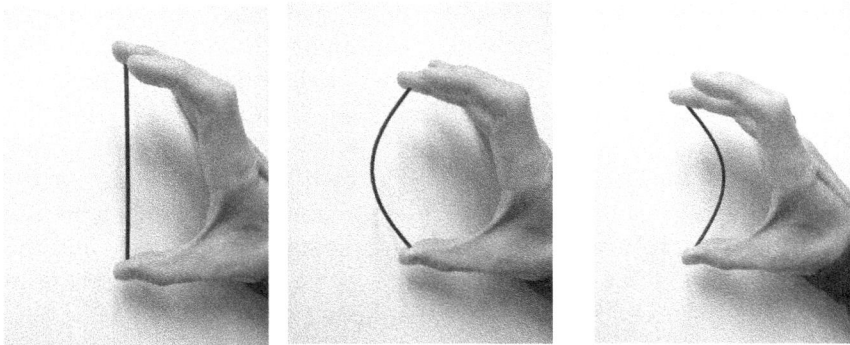

Figure 6.1   Spontaneous symmetry breaking

phenomena. Superconductivity in solid matter was that area where these ideas helped explain the behavior of a material. It wasn't until 1962 that another physicist, Philip Anderson (1923–), would demonstrate the connection between broken symmetry and massive states.

Anderson was working on the behavior of plasmas—highly ionized gases in which electromagnetism is such a strong force (unlike a liquid, solid, or gas, where electrons and protons are bound up in atoms and mostly cancel out each other's electric charges) that the calculations of its behavior, such as how waves travel in plasma, become much more complicated.

Understanding waves traveling in a plasma (by observing collective disturbances of the constituents of the plasma) is gained by applying "plasmon theory." Just as the waves of electromagnetism can be understood by discussing a quantum of the field, the *photon*; and just as vibrations in a solid can similarly be understood by describing a quantum of sound, the *phonon*, *plasmons* are the quantum of waves in a plasma.

It was known, when Anderson published one of his seminal papers in 1963, that below a certain frequency—the "plasma frequency"—certain kinds of waves could not propagate in a plasma. Above that frequency, however, all modes of waves could propagate. While studying this, Anderson had a revelation: this special frequency, the plasma frequency, acted like mass in a quantum field theory. While symmetry breaking should prefer massless force-carrying particles, the classical field theory of electromagnetism applied to plasmas contained a special frequency, below which only certain kinds of waves could travel (akin to massless particles in a quantum field theory) and above which all kinds of waves could travel (akin to particles with mass in a quantum field theory).

Anderson's insight spelled out the mathematics in the case of this classical field theory, showing that what happens in such a theory is that gauge invariance is still respected, as is the conservation of particles, but a mass-like phenomenon emerges because the vacuum of the theory—the lowest energy state of the system, typically containing no particles—gains structure: its own symmetry is broken while the symmetry of the overall theory is preserved.

Closing with a speculation, this paper, published in 1963, proposed that these ideas should apply equally to quantum field theory. Anderson himself did not immediately work with the challenge he had laid out in the closing of his paper—but, very quickly, others did.

In 1964, a series of three papers were published in *Physical Review Letters* (a premier physics journal), each with its own specific insights but all of them, together, laying out the mathematics of spontaneous symmetry breaking in the vacuum state, preservation of gauge symmetry in the theory, and the appearance of massive force-carrying particles. Now, with the hindsight of more than fifty years on this subject, they were a theoretical tour de force!

The first paper was by Robert Brout (1928–2011) and Francois Englert appearing in August of 1964. The paper made no reference to Anderson's 1963 paper, but very clearly laid out the mathematics of how to break vacuum symmetry and generate masses for force-carrying particles in a quantum field theory. The second paper, by Peter Higgs, published in October of 1964, was received by the journal on the very day that it published the Brout and Englert paper. The quickness with which this paper was received by the journal confirms the record of work by the physicists involved in these efforts: that each group was working independently on these ideas and sought to publish them as soon as they felt comfortable with the quality and readiness of their results.

Higgs's paper did cite Anderson's paper, arriving at similar conclusions on the subject to those found in the Brout and Englert paper, with one notable distinction: Higgs was the first of the authors to mention that, as a consequence of breaking the symmetry of the vacuum state, there is a left-over feature in the theory that would manifest as a new particle. Higgs reports that his paper was at first rejected for publication by a European journal. Upon receipt of this rejection, he saw the need to add additional text that heretofore had not contained hints of great future consequences. Indeed, one sentence mentions the potential appearance *of a new particle* as a consequence of the symmetry breaking process. He then submitted the paper to an American physics journal, *Physical Review Letters,* where it was accepted for publication. In the penultimate sentence of that paper, there appears that crucial sentence describing for the first time, the new particle that would later become famous as the "Higgs boson."

The final paper that year on this subject was by Gerald Guralnik, Carl Hagen, and Tom Kibble (1932–2016). It was published in November ,1964. Again, that it was received before the Higgs's paper appeared in print shows how much these six individuals were engaged in independent research, all of them culminating on similar time scales, and all of them in 1964! The Guralnik-Hagen-Kibble paper was, in many ways, the most complete of the three papers, describing the details of the new field that Higgs had discussed in his paper, that would later be known as the "Higgs field." They cited the Brout-Englert and the Higgs papers, each of which were independent steps along the way to the full solution provided by the three papers.

Within a year, based on the foundations of ideas put forward by Nambu, Goldstone, and others, these physicists had laid their own foundation for how breaking the symmetry of the vacuum state, while preserving the overall symmetry of the theory itself, allowed for mass to emerge from massless states.

Of course, there is always a "catch." Nothing this good comes without consequence. This catch, a "good" catch was a process left over from spontaneous symmetry breaking— it was the sentence that Higgs had added that contained the first inkling of this consequence— the prediction that there is another particle lurking out there in nature, a particle with very specific properties that must exist if this idea is a correct one.

The Standard Model of Particle Physics

In the history of science, with its slow progress stretched over decades or centuries, what happened next happened very quickly. By the end of the 1960s, particle physicists had the basic outlines of a fundamental theory of nature in hand. It contained many attractive features. It united previously disparate forces—the electromagnetic and nuclear forces—in a single framework that seemed even to predict that, as one cranks up the energy, these forces become indistinguishable from each other. This is called "unification"—when previously distinct aspects of nature are revealed to be different aspects of a singular idea, more fundamental than the aspects themselves. It predicted the behavior and nature of the force-carrying particles of the nuclear forces, especially the weak nuclear interaction. It held the ability to generate mass for force-carrying particles, allowing for them to be short-ranged. It preserved the massless photon. It was gloriously predictive. And it has, much to the aesthetic chagrin of many particle physicists, one of the blandest names in the history of science: the Standard Model of Particle Physics.

Do not let this featureless name fool you. The standard model, as it is shorthanded, is the single-most successful description of nature ever constructed by humankind. Its name belies its power, its scope, and its beauty. It is a great example of why it is unwise to judge a book by its cover, or the character of a person by their physical appearance. It's what's on the inside that counts!

It is worth taking yet one more historical step before proceeding to the most important predictions of the standard model. In this step, we will meet a few of the key players in the development of the model. The most famous— those who received the Nobel Prize in Physics for the development of what came to be known as the standard model—are Abdus Salam, Sheldon Glashow, and Steven Weinberg.

We almost met Salam and Weinberg a bit earlier, in the discussion of Goldstone's work. Now it's time to bring them out of the shadows and into the forefront of this journey. Abdus Salam (1926–1996) was born in the Punjab State in British India (which later, when British India was partitioned in 1947, was split into East and West Punjab, the part of Punjab where Salam was born becoming part of what is now Pakistan). He proved himself an outstanding scholar quite early on, scoring the highest marks ever recorded for the entrance examination into the Punjab University. While he originally pursued the study of literature, his interests were soon hooked by the language of nature—mathematics—and its ability to tell tales of the fabric of reality. Although he was under pressure to teach English, he committed to the study of mathematics and earned his bachelor's degree in 1944. After trying (and failing) to join the Indian civil service, he earned his master's degree in mathematics before earning scholarships to study in England. He went on to earn his Ph.D. from the Cavendish Laboratory at Cambridge University. His thesis, which dealt with fundamental work in quantum field theory, specifically QED, brought him to international fame in the physics community by the time it was published in 1951. While pursuing his Ph.D., he solved a problem that had stumped the great minds of Paul Dirac and Richard Feynman—the elimination of troublesome infinities in calculations within the theory of mesons (the messenger particles in the nucleus). Within six months of taking up the problem, he had a full solution. This brought him attention. While at Imperial College in London in 1961, Salam collaborated with Goldstone to prove the conjecture that Goldstone had earlier made about spontaneous symmetry breaking and the appearance of spin-zero particles as a consequence. This transformed Goldstone's conjecture into a full-fledged mathematical theory, referred to as "Goldstone's Theorem," discussed a bit earlier in this chapter.

Salam's work on quantum field theory, symmetry, and the properties of particles in these theories was carrying him toward a culmination of effort in 1968 that would end in the standard model. But also on that trajectory were Sheldon Glashow and Steven Weinberg. Let's meet them.

Sheldon Glashow was born halfway around the world from Salam, in New York City. His parents were immigrants of Jewish heritage who had come to the United States from Russia. He graduated from the Bronx High School of Science, in the very same class as Steven Weinberg, whom we'll meet in a moment! Glashow earned his bachelor's degree from Cornell University in 1954 and his Ph.D. in Physics from Harvard University in 1959. His advisor was the famous physicist Julian Schwinger (1918—1994), whose ideas about symmetry, mass, and force-carrying particles would later inspire Philip Anderson's work on plasmons and symmetry breaking in 1962–1963. Glashow would hold professorships first at the University of California Berkeley (1962–1966) and then Harvard University (1966–).

Glashow had some notable accomplishments in the 1960s. In conjunction with James Bjorken, they predicted that a fourth quark was needed. In 1964, when there was only evidence for three quarks in nature (up, down, and strange), the prediction, a part of the effort to make sense of the "particle zoo," resulted from the decades of particle accelerator and particle detector experiments. The actual discovery of this fourth quark would wait for a decade after its prediction,

The other highly noted accomplishment that Glashow had during that decade was to correctly propose the group of transformations that define the symmetry of the electromagnetic and weak force unification program. Schwinger had already developed models in which electromagnetism and the weak nuclear force are united in a single framework, but Glashow's work added a crucial element to these models: a new particle that transmitted "neutral current interactions" in the nucleus. This particle would come to be known as the Z boson.

What are "neutral currents?" This concept is most easily understood by considering an electromagnetic analogy. Let's say we send two electrons speeding toward one another, but not head-on. They would miss each other by just a little, were it not for their mutual electromagnetic fields. Because they possess electric charge, they excite quanta of the electromagnetic field—photons—and exchange them. In doing so, they interact, and because they are same-charged, they will repel each other. The electrons will scatter away from each other, but will leave their original electric charges *unchanged*. This is a neutral current interaction—the photon, which has no electric charge (is neutral), is exchanged between them as a flow (a current), transmitting the electromagnetic force (repulsion, in this case) between the two electrons, whose electric charges remain intact.

Glashow's work, extending the ideas of Schwinger, led to the prediction that there were such neutral currents in nuclear interactions, specifically those of the weak interaction. It was already known that there were charged currents—currents involving a force-carrying particle with its own electric charge, either positive or negative. But no one had observed neutral currents—this was a definitive prediction of this idea. There must, then, be an associated, new, massive force-carrying particle that transmits this interaction, one that had previously not been detected.

It is time to meet our final player in this part of the story: Steven Weinberg (1933–). Born in New York City, attending the same high school during the same years as Glashow, Weinberg would also go on to earn his bachelor's degree from Cornell University in 1954. He completed his graduate education at Princeton University, earning his Ph.D. in physics in 1957. He earned his first professorship at the University of California-Berkeley in 1960, becoming a lecturer at Harvard in 1966 and serving as a visiting professor at MIT for one year, in 1967. It was during that year at MIT that he completed his own model of unification of the electromagnetic and weak interactions, paralleling the work of Salam and Glashow. The masses of the force-carrying particles were achieved via the spontaneous symmetry breaking mechanism outlined by Brout, Englert, Higgs, Guralnik, Hagen, and Kibble. It possessed the very same symmetry that was proposed by Glashow in 1961. It, too, predicted the existence of a new heavy force-carrying particle—the Z boson—that transmitted a neutral current interaction in the nucleus. The paper in which this work was published has the very unassuming title, "A Model of Leptons." Again, that simplicity belies the beauty of what Weinberg would describe in his mathematical work.

1968 is the year during which all three lines of this work met. What was distinctive about what Salam, Glashow, and Weinberg accomplished was that they had applied consideration to something which need not have had anything to do with the weak interaction, and, in doing so, not only successfully described the weak interaction but also united it with electromagnetism. That core idea was gauge symmetry. It worked spectacularly well for QED, the quantum theory of electromagnetism, but past is not always prologue. Just because an idea works for one aspect of nature is no guarantee that it will work for another.

Salam, Glashow, and Weinberg pressed this concept, building on the discoveries of Nambu, Schwinger, Goldstone, Brout, Englert, Higgs, Guralnik, Hagen, Kibble, and countless others to develop a complete theory that united electromagnetism and the weak interaction under a single mathematical umbrella—gauge symmetry (and the spontaneous breaking of gauge symmetry).

Their work made specific predictions. As noted earlier, they predicted the existence of neutral currents in the weak interaction, whose explanation would be the Z boson. They predicted that the charged currents of the weak interaction would be explained by a similar massive force-carrying particle, the W boson. The photon, the trusty carrier of the electromagnetic interaction, was of course part of this model. A consequence of using spontaneous symmetry breaking to generate masses for the force-carrying particles was what Higgs had foreseen in his work in 1964: an additional particle, the Higgs boson, would also be required to exist by this approach.

The fact that gauge symmetry had worked so well to describe both QED and the weak interaction in one theory suggested that the same might be true for the strong nuclear interaction. Indeed, this turned out to be correct, and a new set of theoretical physics pioneers set out to test that idea and successfully developed QCD, the quantum field theory of the strong interaction. The fact that two forces could be unified in this framework also provided hope that a full, unified field theory of the strong, weak, and electromagnetic interaction might be possible. This would be called the "Grand Unified Theory," or GUT. Maybe, if gravity had some quantum manifestation, it too might one day be unified with these other forces. You can see that the seeds of many future ideas to unite the standard model with gravity were planted in this decade.

The development of the standard model, and the consequences of this work, had a number of other far-reaching effects on theoretical physics, they too numerous to discuss in a single book. The reader is encouraged to do some self-study on concepts such as the violation of charge-parity symmetry (CP violation) and quark-mixing to explore one fascinating aspect of the standard model we cannot cover here. Such ideas have consequences for the balance of matter and anti-matter in the universe . . . another continuing mystery not yet resolved.

We choose to focus here instead on the more salient and approachable subjects that you, the reader, might be interested in: the search for the carriers of the weak nuclear interaction (the Z and W bosons) and the Higgs boson.

The Weak Bosons

The standard model contains many *parameters*—numerical quantities whose values are not predicted by the model. They have to be measured. It happens that the model parameters connect to physical measurements that one can make in the natural world. For instance, one can measure the rate at which certain phenomena occur, and from that rate one can extract the value of one or more of the standard model parameters.

While the parameters are not knowable in advance of being measured, the standard model contains a series of beautiful mathematical relationships among the parameters found in the model. This allows physicists to measure one parameter to infer another. For instance, there is a number in the model called the *vacuum expectation value*, v. This tells us about the structure of the vacuum state of the theory, which is intimately tied to the Higgs boson. Its value is not predicted, but it is directly related in the theory to another number in nature, which is called the *weak coupling constant* (or the *Fermi coupling constant*), G_F. This number was already known from measuring the rate at which muons decay, a process controlled by the weak force. Using that number, the standard model predicts the following relationship between G_F and v: $v = (\sqrt{2}\ G_F)^{-1/2}$. Plugging in the current value for the weak coupling constant yields $v = 246$ GeV. Many such relationships occur in the standard model.

In the period between 1968 and 1976, the standard model was largely put onto the footing we teach today in textbooks. The discovery of the fourth quark in 1974 provided additional information for the theory. The mixing of quarks, one into another, by the weak interaction was put on a more complete footing (we'll come to this in a moment). The relationships between force-carrier masses and other parameters of the model were fleshed out. By 1976, there was mounting pressure to develop colliders and design experiments to produce the W and Z bosons. Of course, to do this, colliders needed to know where to "look"—that is, what energy regime to target with their designs. Too little energy and you'd not produce the W and Z; too much, and you'd over-design and thus overspend on the project.

The masses of the W and Z bosons, which are also parameters in the standard model and thus not specified immediately by the theory, are related to other parameters of the model that are more readily measured. For instance, the W boson mass is related to the vacuum expectation value, v, as well as the charge of the electron, and to a third parameter that was hinted at above—the weak mixing angle, θ_W. The W boson mass is given in terms of these parameters by a relatively simple formula,

$$M_W = (ev)/[2\sin(\theta_W)]$$

while the Z boson mass is given in terms of the W boson mass,

$$M_Z = M_W/\cos(\theta_W).$$

You see that once you begin to pull at the unknowns in the standard model, what seemed an impenetrable fortress of unmeasured parameters begins to collapse like a house of cards, yielding one prediction after the next. One need crack just some of the parameters to know where to look for the W and Z bosons. But what is this "weak mixing angle" parameter? How do we understand it?

Let's take a side-step for a moment and look at an earlier moment in the story of the standard model, when geometric angles were first used to describe the underlying behavior of nature in the subatomic realm. Let us look at the work of a brilliant physicist named Nicola Cabbibo (1935–2010).

In 1963, Cabibbo had formulated the idea of a geometric angle to explain the behavior of subatomic particles. Recall the discovery of "strangeness" we discussed in Chapter 5. Here we will see a bit more of the story. What physicists had observed was that particle interactions involving the transition of an up to a down quark (or vice versa), or the transition between an electron and its partner, electron-neutrino, or transitions between muons and their partners, muon-neutrinos, were all equally likely. However, any interaction that required an up or a down quark to transition into a strange quark (or vice versa) was highly suppressed. It was as if nature—specifically the weak interaction, and *very* specifically, the charged current interactions involving W bosons—was not blind to the participants in the transition. There was some kind of preference in nature. Cabibbo found a way to explain this preference.

Cabibbo adopted the view that there are "generations" of matter, with up and down quarks (and electrons and electron-neutrinos) in the first generation of matter (the lowest in mass), and strange quarks (and muons and muon-neutrinos) in the second generation of matter. Interactions that mix (interchange) quarks but maintain the generation—say, a transition between an up and a down quark—are favored over those that mix but jump generations—say, a transition between an up and a strange quark. The mathematics of this suppression looked very much like the geometry of rotating coordinate systems, one into another. In fact, this is precisely how Cabibbo described this, and the angle of rotation between the "weak interaction" coordinate system and the "mass" coordinate system is found to be non-zero. Cabibbo determined this angle to be about 15 degrees from the data he had available to him in 1963, when he published his foundational paper. The modern accepted value of this parameter, known as the "Cabibbo angle," is based on many and more precise measurements and was found to be 13.04 degrees.

Weinberg, when working on the mathematics of the quantum field theory of the standard model, recognized that there is a similar "mixing" that occurs between the pure electroweak states of the force-carrying particles and the real ones we observe in the laboratory—the photon, the Z boson, and the W boson. This relationship could similarly be parameterized by a mixing angle, which is known as the weak mixing angle or the "Weinberg angle."

The information of the day was then taken to the standard model, to interrogate it as to the masses of the weak bosons—the W and the Z. In a paper in March of 1976, the physicists Carlo Rubbia, Peter McIntyre and David Cline used the then-best-known values of these masses to motivate the details of the construction of a proton-antiproton collider that could most easily produce them. At the time, the predicted masses of the W and Z bosons were about 65 GeV/c^2 and 80 GeV/c^2, respectively. Given how early in the application of the model the timing of this was, these predictions are delightfully close to the true and presently accepted values of these masses, known better since that time both by direct measurement and indirectly from measuring more precisely the other parameters of the model. The currently accepted value of the W boson mass is (80.385 ± 0.015) GeV/c^2 and that of the Z boson is (91.1876 ± 0.0021) GeV/c^2. You might think that Rubbia, McIntyre, and Cline were working with pretty inaccurate information about these masses when they pitched their proposal, given the modern small uncertainty on these numbers, but, in fact, they were going "big picture" in their proposal and had put forth designs for a collider capable of probing weak boson masses from 50—200 GeV/c^2. If they were in the ballpark with those predictions, the proposed machine should be able to produce the expected particles.

The weak neutral current interaction was first observed in 1973 using the Gargamelle bubble chamber experiment at CERN. Neutrinos were scattered off a target, and the resulting images were scoured for evidence of a neutrino entering and scattering from the target in specific ways. Indeed, they found about 160 neutrino interactions, confirming this prediction of the standard model. Not only that, but this measurement provided the first estimate of the Weinberg angle. The authors used this data to estimate the angle to have a value between 33 and 39 degrees at the energy level of their experiment (which will change with the level of energy applied). This was useful information for chipping away at the unknowns in the standard model and to make predictions about other parameters. Further experiments over the next few years continued to refine the measurement of the weak mixing angle, arriving at a value of θ_w ≈ 27 degrees. But, what of the prediction that the currents—both charged and neutral—are carried by a new, very heavy pair of fundamental subatomic particles having masses?

That discovery did not occur until the early 1980s, after the construction of the Super Proton Synchrotron (SppS) and its use of proton-antiproton collisions to attempt to produce W, and then Z, bosons. The two experiments at the SppS were designated UA1 and UA2.

This was quite an exciting time for the field, and the adventures (and misadventures) of particle physicists during this period are chronicled in many papers, personal recollections, and books. The short version of the story is that in 1983, the UA1 and UA2 experiments announced evidence for the discovery of the W boson by its predicted decay to leptons—electrons, muons, and neutrinos. The evidence for the W boson placed it at a mass of 80 GeV/c^2, and, once that was known, the standard model was able to make a much simpler prediction about the mass of the Z boson.

If one plugs in the W boson mass, now known directly from experiment, and the estimate of the weak mixing angle from the neutrino-scattering experiments, $\theta_W \approx 27$ degrees, one finds that the Z boson mass is predicted now to lie at 89 GeV/c^2—in remarkable agreement with where it would soon be found and concurring with the currently accepted value of this number! UA1 and UA2 knew exactly where to look, and the signatures to look for that were most distinctive and "clean"—the decay of the Z boson into pairs of charged leptons (such as $Z \rightarrow e^+e^-$—an electron-positron pair). They soon obtained evidence for this force-carrying particle. In a fairly rare move, the Nobel Prize Committee quickly awarded the 1984 physics prize to Carlo Rubbia and Simon van der Meer for developing the methods of detection and production of the weak bosons, and for the subsequent discovery of these crucial aspects of nature.

The 1970s and early 1980s began a period during which one confirmation after another was obtained for the standard model. Aspects of both the electroweak interaction and QCD, the theory of the strong interaction, also present in the standard model, were tested during the years and decades following in a long series of experiments at famous colliders like LEP at CERN and the Tevatron at Fermilab. However, despite having determined a large number of the parameters of the standard model, the aspects of the model related to the Higgs boson continued to be shrouded by experimental shadows.

The Higgs Boson

We've left an important character by the way-side for a bit—the Higgs boson. The last time we discussed it, it was a side-effect of the spontaneous symmetry breaking of the vacuum state of the universe, a necessary act to admit the possibility of massive force-carrying particles. It first was mentioned in Peter Higgs's 1964 paper, though at that moment it was vague and ill-formed. The Guralnik-Hagen-Kibble paper, late in 1964, fleshed out the structure of the "Higgs field" in more detail, but it wasn't until a 1966 paper by Peter Higgs that the crown would be placed on this idea of a new particle in nature.

Higgs calculated, for the first time, the decay properties of the Higgs boson. If it is truly a heavy particle, it must decay to other sub-atomic particles. In his 1966 paper, he provided the essential methods that show how a Higgs boson would decay to a pair of heavy vector bosons —like a pair of Zs or a pair of Ws. In fact, decaying to a pair of Zs was one of the ways in which the Higgs boson was first discovered, forty-six years after Higgs published his paper.

The Higgs boson interacting with other particles is, in the standard model, the physical representation of how particles acquire mass. It is the strength of the interaction between a Higgs boson and, say, a Z boson that yields the mass of the Z boson. That strength is what we perceive in nature as "mass." In effect, we trade the question, "what is the origin of mass?" for the question, "what is the origin of the interaction strength between the Higgs and other fundamental particles?"

This may not seem very satisfying at first, but once it is realized that fundamental mass is a consequence of something like "electric charge"—having to do with the properties of the Higgs boson—then mass no longer becomes a separate part of the theory and a more inclusive theory of nature might then very easily explain the Higgs interactions. Two seemingly independent problems—what it is that sets interaction strengths in nature? and what it is that sets masses in nature?— have been traded into just one problem: what it is that sets interaction strengths in nature?

Let's look a little more at this idea that interaction causes mass to appear in nature. We can do this with a helpful analogy. Let's imagine that we are attending a cocktail reception with a large group of young physics enthusiasts who are evenly distributed throughout a large room. These enthusiasts, who had earlier attended a lecture by prominent physicist, Stephen Hawking, are awaiting his arrival at the reception. The guests have been there a while, so they've gotten some snacks and dispersed around the room. They are analogous to the Higgs field—they are everywhere in space, at all points, evenly distributed. The atmosphere in the room is charged with excitement, analogous to the way that the Higgs field carries a net weak hypercharge everywhere with it.

Let us first consider a case where a particle now enters the presence of the Higgs field. Originally without mass, without inertial resistance to changes in motion, we will see how, in this analogy, the particle *acquires* mass by the strength of its interaction with the Higgs field.

Soon, Dr. Hawking enters the room. His entry immediately creates a level of heightened excitement that rapidly propagates through the group. As he moves into the crowd, people jostle to cluster around him to offer their congratulations, slowing him down and making it difficult for him to navigate the room. From our perspective, it might seem as though Dr. Hawking has just acquired a very large mass, making it impossible for him to move with any haste through the room!

The enthusiasts who are very close to him are eager to greet him and to engage him in conversation, while those not as near continue socializing as the dignitary pushes through the clump that is forming around him. As he moves slowly through the party, this clumping phenomenon continues to happen as new groups, in succession, encounter and engage him, while the previous clump of people disperses somewhat evenly back into the room and goes back to its earlier socializing.

This process encumbers Dr. Hawking as he negotiates the room. This is quite like the action that occurs in the Higgs field among the particles of the universe. The clustering of guests around Dr. Hawking increases his inertia as he crosses the room, slowing him down (similar to acquiring mass) and is much like what happens to a particle in the Higgs field as it gains mass. The top quark, for instance, is the heaviest fundamental subatomic particle yet known, with a mass almost 170 times that of the proton and almost 40% heavier than the Higgs boson itself. The top quark is *very* popular in the Higgs field, and this immense popularity (interaction strength) gives it a correspondingly immense mass (inertia, the resistance to changes in motion).

Let us also now use this analogy to see how the Higgs boson manifests as a result of the field. This will illustrate a point we've glossed over before, regarding the relationship between quantum fields and their corresponding field quanta—the particles that are "excitations" of the fields themselves, when the field is present.

As Dr. Hawking enters the room, a rumor of his arrival begins to spread from that entrance throughout the physics enthusiasts in the room. As the rumor spreads, people clump to share the rumor and then clump again to pass it on, transmitting the rumor through the room by the clumping and chattering and de-clumping of people. This effect is similar to the way in which Higgs particles interact to generate mass for fundamental particles in the standard model. The presence of Dr. Hawking in the room causes clumping, but the rumor of Dr. Hawking, even without his presence, causes some amount of clumping that travels through the room as if it, too, were a particle of its own.

This analogy provides a picture of how this operates for the Higgs field, and how the overall Brout-Englert-Higgs-Guralnik-Hagen-Kibble mechanism can now be entirely explained in terms of the Higgs field.

The Higgs field is special. It is represented by a spin-less particle with no electric charge. All other force-carrying particles, the photon, gluon, W, and Z, are spin-1, and are collectively known as "vector bosons." Unlike the fields associated with vector bosons, the Higgs field exists at all points in space-time. Vector fields, on the other hand, come into existence when one of their interactions occurs and go out of existence at the close of the interaction. The vacuum state of the universe is described in the standard model by the Higgs field. Other fields and matter particles are required to interact with the Higgs. It is these unique features of the Higgs field that give it its role in generating fundamental mass in nature.

The strength of interaction of the Higgs boson with vector bosons and with matter particles was the essential ingredient in discovering the Higgs boson. Once the mass of the Higgs boson is set, its interactions with all other particles are also set in the standard model. Much the same as the relationships among W and Z boson masses, the Fermi coupling constant, and the weak mixing angle are fixed by the theory that when the mass of the Higgs boson is known, then by including the masses of the leptons and quarks, this is also true about its interactions with other particles in the standard model.

There is a problem: the standard model makes no specific prediction about the mass of the Higgs boson. This is, in part, what made it so difficult to hunt it down—there is very little guidance, if any, from the standard model about the mass of the Higgs, leaving it to have any value among a wide range of values.

Once the W and Z bosons were found, that denoted the moment when the hunt was really on for the Higgs boson. Knowing the W boson mass precisely, coupled with a measurement of the top quark mass, allowed physicists to place constraints on the expectation of where the Higgs mass should be. But the top quark was not discovered until 1995 when both the DZero and Collider Detector at Fermilab (CDF) experiments announced that they had detected it in proton-antiproton collisions within the Fermilab Tevatron. Precision measurement of the W mass would have to wait for the LEP experiments at CERN and a later, second run of the Tevatron at Fermilab in the 2000s.

In the 1980s, after the success of the SppS at CERN (and with their later focus on the LEP program in the 1980s and 1990s), CERN began pursuing construction of a larger proton collider that would be designed to cover the huge range of possible masses for the Higgs boson. This was the Large Hadron Collider, which first created physics-quality data in 2010 and required almost thirty years to conceive of, plan, and build. It was designed to provide 7 TeV of energy to each of its twin proton beams.

Also in the 1980s, the first reviews of the United States Superconducting Supercollider, planned at 20 TeV of energy for each of its twin proton beams were begun. That project began construction in the 1990s, but was canceled by the U.S. Congress on October 31, 1993. At that point, it became clear that unless the Higgs boson could be found at the LEP2 collider in the 1990s, or by the Tevatron during its second run in the late 1990s/early 2000s, the world would have to wait for the LHC to come on line in the 2000s.

Indeed, the physicists at the ALEPH, DELPHI, L3, and OPAL experiments at LEP2, and the CDF and DZero experiments at the Tevatron's Run 2 phase, made incredible progress toward discovering the Higgs boson. The theoretical physics community was not idle, either, conceiving of new ways that the Higgs boson might be produced at each collider or conceiving of new techniques for separating even a faint signal of the boson at any experiment. Not knowing what its mass would be, the community devised ever-more complex means of searching for it at higher and higher masses as their colliders raised energies or operated at higher intensities (or both). This was an incredible period during which new experimental and theoretical techniques were devised and the use of advanced statistical and computational methods in the particle physics community boomed.

It would not be until the LHC began operations in 2010 that the kind of data would begin to be available to definitively enable the discovery of the Higgs boson. Let's look at how this was done at the LHC, both in producing the Higgs boson and then observing its decay using the particle detectors, ATLAS and CMS.

Higgsdependence Day

What follows is the personal, first-hand recollections of one of the authors (SS), who was present in the room on the day that the discovery of a temptingly Higgs-like particle was announced at CERN in 2012:

I had joined the ATLAS Experiment a bit earlier, officially becoming a member of the collaboration in 2011. Very large experiments like ATLAS and CMS, engaging hundreds to thousands of physicists, are run very much like corporations. There are too many people and too many physics topics to allow chaos, so the collaboration is organized into large groups. Those groups are formed of ever-smaller and more focused groups until you drill down to specific physics interests centered around specific signatures sought in the particle detector.

I was a member of the Higgs group on ATLAS, but having joined so late, I was not playing a meaningful role at that moment in the search for the standard model Higgs boson. Instead, I was working on searching for cousins of the Higgs boson that might signal an even more encompassing theory of nature beyond the standard model, but that is a subject for another time. That said, the significance of what was ready to happen at ATLAS (and, presumably, also at the competitor experiment, CMS) was not lost on this author. His university had faculty, post-docs, and students, all who had long been members of the search teams for the Higgs, looking for Higgs decays to two-photon pairs, two-Z pairs, and two-W pairs. It was clear that, as more data mounted and as the techniques for analyzing that data rapidly matured between 2010 and 2012, that the discovery of a Higgs boson might be imminent.

Within each collaboration, a member of the ATLAS or CMS experiment has unfettered access to internal information about ongoing physics analyses. Physics collaborations are internally open, but for the purposes of maintaining scientific integrity, there are firewalls between experiments to guard against the omnipresent danger of "bias," the tendency to want to see what your competition is seeing, that can force you to make poor choices about your own data. So, while it was clear within the author's own collaboration, that evidence was mounting for the existence of a Higgs boson at a specific mass, it wasn't so clear across the boundaries of the two experiments (to the day-to-day physics grunts in the trenches), what the other side was seeing.

In December of 2011, a pair of seminars was hosted at CERN, one from ATLAS and one from CMS, during which they "showed their cards." Using the data then available in 2011, there were tantalizing hints that maybe . . . *maybe* . . . both experiments were "seeing something" in the data, but the pattern was not clear. And, there was the danger of experimental bias coming in if one continued to repeatedly spin one's wheels on the same data. So, all hopes were pinned on new data that would come rolling in during the 2012 data-taking period. The energy of the LHC was upped from 3.5 TeV per proton beam (7 TeV total center-of-mass energy) to 4.0 TeV per proton beam (8 TeV total center-of-mass energy). This seems like a modest increase in energy— just a 14% increase—but the rate at which a Higgs boson is produced does *not* scale linearly (e.g., as a simple ratio). This modest increase, coupled with a more significant increase in proton collision intensity in the performance of the LHC, made the probability of seeing the Higgs much larger, and much sooner, than was possible in 2011. It's a testament to the brilliance and commitment of the accelerator physicists who operate, maintain, and upgrade the LHC that they achieved so much, so quickly, in the face of having first been unable to run The LHC at its design energy rate of 7 TeV per beam (14 TeV total energy).

As 2012 progressed, the evidence mounted in the author's experiment, ATLAS, that something interesting was going on at a mass around 120–130 GeV/c^2, but it was feared that this was just the roiling of the oceans of data, causing structure to appear where none might actually be.

Thus we come to July, 2012. Soon was to be the International Conference on High-Energy Physics (ICHEP), a major biennial international conference being hosted in Australia that month. It was assumed that the ATLAS and CMS experiments would engage in their annual talks there. However, with very short notice just before the conference, the venue of the talks was changed to CERN. (You can imagine the changes this caused in travel plans, but such is the lifestyle of the peripatetic physicist.) This caused a great deal of hand-wringing in the particle physics community. Why would the talks be held at CERN and not at the conference? Was an announcement of discovery imminent? All interest turned to CERN.

I was at CERN at the time. Summer is that time when teaching responsibilities fall away for a research faculty member, and is best spent at an experimental site, working directly with colleagues instead of by email or video conference; having coffee in the cafeteria in a kind of social/work ritual that proves more useful than volumes of email; and taking shifts on the experiment to help with operations and data-taking. I recall the day before the seminars, on July 3, in the CERN cafeteria when I noticed a friendly looking, older man carrying his tray to the return area. Struck by familiarity and recognition, I stared—perhaps a bit too much—at this man. When he turned and I got a better look, I confirmed that it was Peter Higgs! What would Peter Higgs be doing at CERN on July 3, 2012???

The presence of Peter Higgs at CERN, and the news from ATLAS that there was evidence for something going on in the data, made it clear that it *was important to be in the room* on the day of the twin seminars. Truly historic events usually come only rarely in a lifetime, and if you want to witness history you have to be in the right place at the right time.

At that time, I was working with a young post-doctoral researcher whom we'll call A. A and I decided that, on July 3, it was best to go home early, sleep, and return to CERN at about ten or eleven that night. A's mother had even come to CERN for this occasion. She wasn't a physicist, but she knew how important this moment could be! We decided to bring food and drinks, and pillows, and things to do, and camp on the floor outside the CERN Main Auditorium, the location of the twin seminars on the morning of July 4, 2012. I recall fondly that A and his mother pulled up in a car outside my apartment at around ten on July 3. As I opened the car door to get in, A said simply, "Let's *do* this!"

When we arrived at CERN, there were perhaps a dozen people in line—nearly 12 hours before the seminars! That night was wonderful. As more arrived, people gathered in small circles on the floor and my group talked all night about the physics of the Higgs boson. Others watched movies, or caught up on email, did physics analyses, or slept. The line grew and grew and grew, until by six or seven that morning, it stretched out the doors of the auditorium, along a wall, bent along another wall, doubled back, and snaked down the stairs toward the CERN cafeteria.

Just then, around seven, a fire alarm went off in the building. *No one moved!* Fire personnel soon came through and cleared the alarm as false, but it left a lasting impression that every one of us physicists would rather die in a fire than give up our place in line.

The doors opened and CERN personnel shuffled people into the auditorium. There was still at least an hour before the seminars were to begin. A large section of the auditorium was roped off for special guests and dignitaries—CERN Council members, key members of the ATLAS and CMS Collaborations, and Francois Englert and Peter Higgs. If that pair of guests wasn't descriptive enough about what was to come, only a sledge hammer would have helped.

The seminars began around nine and were simulcast to the ICHEP conference in Australia, having the ability to interact and take questions from the audience there. The first speaker was Joseph Incandela, spokesperson of the CMS Collaboration, followed by Fabiola Gianotti, spokesperson of the ATLAS Collaboration. The details of the seminars are not important, but I can tell you that the audience sat absolutely mesmerized, waiting for the key data graphs and interpretive plots to be shown. The data were all mounting, at a level far beyond fluke or accident, for describing a new particle with a mass of about 126 GeV/c^2, one that could decay to photons and Z bosons in ways quite similar to those predicted for the Higgs boson by the standard model.

The room exploded in applause, cheers and even tears after each presentation. Such emotion and anticipation broke the dam that day that I will admit to my own tears for such a monumental moment in humanity's understanding of the universe.

It is both right and proper that I express my admiration for my colleagues in both the ATLAS and CMS Collaborations, who sacrificed many things (not the least of them sleep) to analyze the data so quickly and with such rigor and scrutiny. While every person in ATLAS and CMS who worked on these measurements has, no doubt, their own amusing and emotional stories to tell, from the perspective of a physicist who was then still a bit of an outsider to the inner workings of the standard model Higgs physics groups, it was clear to me that this was an achievement that capped almost five decades of theoretical and experimental blood, sweat, and tears conducted by thousands of hard-working physicists. What a wonderful event to have witnessed!

Wrap Up

In 2012 and 2013, the ATLAS and CMS collaborations went on to study in great detail the behavior of this newly discovered particle. I am so proud that, in partnership with the outstanding graduate student mentioned earlier, I was able to contribute to the measurement of the spin-parity quantum numbers of the Higgs, helping to confirm that it is a spin-zero particle, just as predicted for the Higgs boson by the standard model.

Hundreds of other physicists contributed to this and other measurements, finding patterns in the behavior of the particle that continued to confirm that this was exactly what Peter Higgs had proposed in the days from 1964–1966, and all the physicists who built their work upon his, toiled to predict in the decades that followed.

By the end of 2013, the patterns were clear: this was as standard model-like of a Higgs boson that you could have thought to find at the LHC. It was likely for this reason that the Nobel Prize committee quickly, just over a year after it was discovered, awarded the Nobel Prize in Physics to Francois Englert and Peter Higgs, noting the work of the ATLAS and CMS Collaborations in their announcement of the prize. It further illustrates one of the frustrating aspects of the Nobel Prize: it can be awarded to no more than three people. So many more people than Englert and Higgs were part of that incredible period in the

1960s during which symmetry, and the breaking of symmetry to achieve mass, rose to its prominent place in fundamental physics.

Not all of those people lived long enough to see the discovery of the Higgs boson—Robert Brout, for instance, passed away in 2011, just over a year before the July, 2012 announcements. Tom Kibble passed away in 2016, his work and that of Guralnik and Hagen unrecognized by the Nobel Prize committee, but recognized by the earlier award of the American Physical Society's J. J. Sakurai Prize for Theoretical Physics in 2010, awarded to all six of the authors whose papers made such an important stride forward in 1964. This demonstrates the value and the limitation of prizes for scientific accomplishment— they are useful for highlighting important things, which is very good for the general public to note, but they fail to capture the messiness and complexity of the scientific process, with its many people batting about similar ideas simultaneously.

So we have a Higgs boson. We understand now a little better what it means to be a Higgs boson. But what are the implications for the universe? And what causes the electroweak gauge symmetry to be spontaneously in the first place? Why is the universe this way? Why are the interaction strengths of the Higgs set at their particular values? The scientist is rewarded with more questions, and, by shining light on some of the shadows of reality we are rewarded with even deeper questions about the nature of reality.

Such pursuits are the continuing quest of a scientist.

A Sky of Shadows

Do you live in the country? Do you live in the city? Either way, if it's a clear night step outside and look up. Even with a great deal of distracting light from a busy city, you should still be treated to a few gems in the night sky. If you're in the northern hemisphere you may see the bright objects that make up the constellation of Orion. If you're in the southern hemisphere, you may see the bright objects that make up the constellation called "The Southern Cross." So, what are those objects and the universe made from? You might be tempted to think those bright points adrift in a sea of darkness are all that there is to the cosmos. Our sun is a star. Most of those points you see in the night sky are also stars, like our sun. Some are bigger than our sun. Some are smaller, but most of them are bright spheres made mostly of hydrogen, ignited in nuclear fire under the pressure of their own gravity. Perhaps the universe is only made from the stuff of stars.

Astronomers, astrophysicists, and physicists thought this for a very long time. They were wrong. Let's look more closely at the night sky. In doing so— in trying to shine more light onto the shadows of reality we will find that there are more shadows lurking in the sky than we could have dreamed.

Eyes That See What Our Eyes Cannot

Earlier, we had a peek at the structure of the cosmos. We learned that Galileo Galilei, and those scientists inspired by his approach, used the telescope to collect visible light from the cosmos, allowing them to see finer detail of things that are very far from the earth. In doing so, they learned that the earth is only one of many planets and other astronomical bodies that orbit the sun. Later generations of astronomers learned that the sun is but one of many stars. Still later they learned that the stars are gravitationally collected together into whirling galaxies; those galaxies bound together into clusters, super-clusters, and so on. This is the visible universe. Well, it is the universe that is visible to us in those few wavelengths of light to which our eyes are sensitive.

It's not our fault that we're so limited in our ability to see. The most prominent wavelengths of light to penetrate the earth's atmosphere, originating from the sun, are those we call "visible light." If a world is bathed in such wavelengths, then it is natural that being able to see these wavelengths would confer a strong survival advantage to any species able to collect and interpret them. The hand of natural selection is strong here, and it's no accident that organisms with an enhanced ability to interpret these wavelengths would survive to pass along their genes to following generations. We are the product of our environment. Our eyes are one end of a long term of natural selection; not perfect, but pretty good at what they do.

We humans are clever. We are not bound only by the gifts nature has given us. We possess creativity. We have invention. And, over the centuries during which we have embraced more and more technology as a means to enhance our lives, we have learned to see what our eyes cannot.

There are many more wavelengths out there than our eyes can see. For instance, if we were to consider electromagnetic waves with wavelengths shorter than what our eyes can see, we would be in the realm of ultraviolet light (discovered in 1801 by physicist Johann Ritter), and then of x-rays (systematically studied and understood by physicist Wilhelm Roentgen in 1895), and, finally (at the very shortest wavelengths), we would be down to gamma rays (discovered by chemist and physicist Paul Villard in 1900).

Short wavelengths do not easily penetrate the earth's atmosphere. To pursue astronomy with them, one must put telescopes or other such instruments high in the atmosphere, or even above the atmosphere, in an orbit around the earth. If you *could* see these light waves with your eyes, you would see a sky wherein stars glow not only in visible light, but in the ultraviolet light that is caused by their immense temperature. You would see the afterglow of dead stars and the radiation of their death-explosion. If you are lucky you will see the bright x-ray glow emitted by gas that is whipped into a frenzy by the forces of neutron stars or black holes.

All of these things are invisible to the naked eye, but properly equipped, you would see the most cataclysmic explosions known in the universe, gamma ray bursts, whose origins are only now coming into our knowledge. They flare and fade, visible across the inconceivable distances between our galaxy and the other galaxies that litter the night sky.

Now let us consider wavelengths longer than what our eyes can see. We enter first the realm of the infrared wave (discovered by astronomer William Herschel in 1800), then we find microwaves (although *micro* they still are longer in wavelength than the shorter waves we just described) and radio waves (predicted by James Clerk Maxwell in 1867 when he unified the theories of electricity and magnetism, and later actually found by Heinrich Hertz in 1887). If your eyes could be re-tuned to see these long infrared waves, you would see a sky in which the coolest stars (so cool that their red light is very faint) glow bright in infrared; you would see the quick pulsing of spinning neutron stars as they chirp out regular signals in the radio range of wavelengths; you would see the glow left over from the very beginnings of the universe itself—the heat of the "big bang,"—coming at you from all directions as a nearly featureless omnipresent bath of microwaves.

Humans have developed a wide variety of technologies to capture these unseeable longer and shorter wavelengths and transform them into electrical signals. These signals can be fed into electronics and computers, as software, to be reconfigured into numbers, graphs, and images that our eyes can see. We are very good at revealing the unseen and converting it into forms we can see, so that we can understand that which otherwise would elude our senses. This is technological evolution, extending the five familiar senses of the human body into new realms where the paths of natural selection themselves did not take us.

As an example, there is the Fermi Gamma Ray Space Telescope. Launched on June 11, 2008, Fermi has been operating since then in orbit around the earth. It possesses many instruments, one of which is a sandwich of detector technologies that might feel right at home in a particle collider detector like ATLAS or CMS at the LHC. This sandwich is designed to capture gamma rays as they pass through the telescope, allowing the energy in the gamma rays to convert to matter as a pair of particles, an electron and its anti-matter counterpart, the positron. By precisely measuring the energy deposits left by the electrically charged pair, the instrument can determine the original trajectory of the parent gamma ray. This allows astronomers and astrophysicists who use the Fermi data to point very precisely back in the sky to where the gamma ray originated. Fermi, like all other instruments designed to transform the invisible into the visible, is an eye that sees what our eyes cannot. Fermi has captured a huge number of gamma rays this way, and has given us a picture of the Milky Way—and beyond—in color—we were not aware of even a few years ago.

A Dance of Light and Shadow

It is very tempting to assume that all we can see is all that there is. Sometimes, though, when we apply that assumption to the universe we earn a surprise. The universe answers with something unexpected. Such are the beginnings of discovery.

Let us look at a particular discovery. This is the story of gravity and light, and all the shadows these twin messengers have revealed. The story begins with the assumption that atoms, the atoms of everyday life, whose fingerprints are seen in the light emitted by all cataloged bodies in the visible universe, make up everything. The story ends with the revelation that for every one gram of atoms that can be found by using light, there appears to be five other grams of matter that neither emits, nor absorbs, any light.

This story begins with the astronomer Fritz Zwicky (1898–1974). Born in Bulgaria, Zwicky was sent to live in Switzerland at the age of six for the purpose of having him study commerce. Zwicky was to be made into a businessman, but found himself more interested in mathematics and physics. At the prestigious Swiss institution known today as ETH Zurich (*Eidgenössische Technische Hochschule*—the Federal Institute of Technology, Zurich) he earned a degree in experimental physics and emigrated to the United States in 1925 to work with the eminent American physicist, Robert Millikan, famous for performing a painstaking experiment with small oil droplets and electric fields in order to determine the smallest unit of electric charge (a fascinating story in itself!). This put Zwicky at the California Institute of Technology, CalTech, surrounded by some of the most famous names in science. He became known for the discovery of neutron stars, and his work earned him a full professorship at CalTech in 1942.

It is for another piece of work, however, that Zwicky matters to this part of the story. He wanted to determine the mass present in galaxy clusters. Galaxy clusters are vast collections of galaxies, often hundreds or thousands of such bodies collectively interacting via their mutual gravitational attractions. We cannot travel to each galaxy in a cluster; that is not possible. Instead, we must use the known laws of physics and observational methods to infer the quantities we wish to know. Zwicky wanted the mass.

Being clever, he used two independent approaches that should have confirmed each other to make the measurement. Knowing that atoms respond to and emit light, Zwicky measured the amount of light emitted by galaxies in a cluster, converting that number into a number to represent the mass of the cluster. All the stars and hot gas that make up galaxies can be measured using the light emitted by them; the relationship between light and mass is then used to convert one number into the other. The second approach used motion. Because the galaxies in galaxy clusters whirl about each other in a collective gravitational dance, one can measure the motion of the galaxies in the cluster and, using Newton's law of gravity, infer the gravitational mass that is present. Two independent techniques in search of one number—an experimentalist's dream!

But, Zwicky hit a wall. He applied his ideas to the Coma Cluster of galaxies. When he made the two measurements and transformed them into masses, he found that the gravitational mass represented by the motion of the Coma Cluster was about 400 times larger than the luminous mass represented by the atoms emitting light! Since then, other astronomers have repeated these measurements using increasingly more accurate techniques. They do not find a number quite as dramatic as 400 times, but, with our present greater understanding of galaxies and clusters, the mismatch between the gravitational and the luminous mass is still not explained by experimental error or by any other error we can determine that can be described by our present knowledge of light, atoms, and gravity.

Zwicky concluded that the Coma Cluster must be home to an unseen form of matter, one that does not emit nor absorb light readily. He merely called it *dunkle materie* ("dark matter")—a literal description in German of an unknown class (or classes) of matter that are invisible to light.

The invisibility of this dark matter is not limited to the visible wavelengths of light that were more commonly used by astronomers in the 1930s, when Zwicky did his work on the Coma Cluster. Indeed, peering at galaxies using gamma rays, or ultraviolet, or infrared, or radio, these haven't filled in the gap between gravitational and luminous mass. From the perspective of galaxy clusters, we are left with a mystery: what is this dark matter?

The problem isn't just with galaxy clusters; the problem applies also to individual galaxies, which are much smaller and less massive than a cluster of hundreds or thousands of galaxies. Nonetheless, in the behavior of galaxies there is, too, an unseen hand that plays a role in their structure and behavior. To understand this, we need to meet the next important scientist in our story.

Vera Rubin (1928–2016) was born Vera Cooper in Philadelphia and later moved to Washington D.C. There, she developed an interest in astronomy, pursuing it at Vassar College, earning her Bachelor of Arts in astronomy. Intending to continue her work in graduate school, she applied to places like Princeton University. However, at the time she applied, Princeton did not admit women to their graduate astronomy program (they would not do so until 1975!). She enrolled for a Masters at Cornell University, which she earned in 1951. She went on to earn her Ph.D. at Georgetown University, working with George Gamow (famous for his work in understanding radioactive decay and for advancing the big bang hypothesis at a time when it was not considered a serious theory of the formation of the cosmos). She earned her Ph.D. in 1954 and continued to work at Georgetown for eleven years until she joined the faculty at the Department of Terrestrial Magnetism at the Carnegie Institution for Science in Washington. There she met and forged a research partnership with Kent Ford, which eventually led to the study of the rotation of galaxies.

Ford developed an advanced spectrometer that, married with a telescope, allowed for very precise measurements of the atomic spectra of stars at different points in nearby galaxies. Galaxies are huge. The Andromeda galaxy, the one nearest to our own Milky Way, contains about 1.2 trillion stars and takes up an apparent visual area in the sky similar to that taken up by the full moon! Why don't we see this as easily as we see the moon in the night sky? It's simple: although Andromeda contains about 1.2 trillion stars, it's so far away from the Milky Way that the light from those stars, is so very, very, very faint compared to the light from the stars in our own interstellar neighborhood or those in our own galaxy. If you were to aim a hundred super-bright flashlights at a friend ten feet away, that friend would be blinded. Walk those flashlights sixty miles away, and it would be incredibly difficult to see them compared to the bright street lights, window lights, or car headlights that would now dominate the local landscape.

Rubin and Ford used their partnership, she with expertise in the astronomy of galaxies and he with expertise in instrument design, to study the motion of stars in the Andromeda galaxy. With the right telescope connected to the right spectrometer, Rubin could easily see the starlight from Andromeda and fingerprint its atoms using the advanced spectrometer. But it gets better than that. Using the same principle that causes an ambulance siren rushing toward you to be higher in pitch, but moving away from you to be lower in pitch (Doppler shift), Rubin noted shifts in "pitch" (frequency) of the light in the spectra of stars at different distances from the center of Andromeda. This allowed her to create a "rotation curve"—a chart of the speed of rotation as a function of distance from the center of rotation. Since gravity is what binds stars together in a galaxy, those stars and their motion must obey the law of gravity and the laws of conservation of energy and momentum. Together, those laws command that *thou shalt not orbit too fast as you move farther and farther from the center of rotation.*

Consider our own solar system as a model. Mercury, the planet closest to the sun, makes one orbit in about 90 days at a speed of 107,000 mph. Venus does the same thing in about 225 days at a speed of 78,000 mph. Earth makes one orbit every 365 days—what we call "one year"—at a speed of 66,600 mph. Mars does it in about 690 days at a speed of 53,600 mph. Jupiter takes about 4,333 days at a speed of 29,200 mph. See a pattern? The farther out from the sun, the slower a planet orbits the sun. Johann Kepler codified this in his Laws of Planetary Motion, which are really a re-expression of Newton's Laws of Motion that we discussed earlier. (Kepler's work preceded Newton's, so we learned only later the exact *why* of planetary motion—it's gravity!)

And so it should be with galaxies. Stars that are closer to the galactic center should generally orbit more quickly than those farther away. This is the commandment of gravity and energy and momentum. So what did Rubin observe?

She observed that stars at the outskirts of galaxies like Andromeda are moving *nearly as fast* as stars that are closer to the center of the galaxy. This made no sense. According to an accounting of the mass of a galaxy such as Andromeda, using the light emitted by that galaxy, there is too little mass within the orbit of such far-flung stars to permit them to orbit so quickly. They should fly off into intergalactic space. How can it be, then, that galaxies can become so large yet remain stable if their outermost stars are moving so fast?

One explanation is what Zwicky found in his earlier studies of the Coma Cluster of galaxies: there could be an unseen form of matter whose influence is felt through gravity, but goes unseen using light. Non-luminous matter would solve this problem of galactic rotation. Dark matter could be a real thing, and play a more important role, even on the scale of galaxies, than anyone would have guessed.

Other explanations have been offered, of course, for the apparent deviation from what is expected of the gravitational influence of luminous matter. For instance, what if the laws of gravity—general relativity—are not correct when applied to the largest scales (the size of things like galaxies or galaxy clusters)? What if gravity needs to be *modified* in some way to explain what Zwicky, Rubin, and others have observed?

The idea of modified gravity survived side by side with the idea of dark matter for a while, until observations made in the early 2000s. To close out this discussion of dark matter, let's look at two final players in this story: the WMAP Collaboration and the Bullet Cluster.

We begin by considering the early universe. If dark matter exists, influencing galaxies and clusters of galaxies, shouldn't it also have existed in the early universe? This is an excellent question. How might we answer it? Perhaps, if we look at a relic from the very early universe, we might detect the fingerprints of dark matter there. If we do, it will be strong evidence that dark matter is a valid explanation for filling the gap we observe between the motion of stars and galaxies and the motion expected from the gravity of luminous matter.

The cosmic microwave background (CMB)—the light left over from the big bang—is precisely such a relic. In the 1990s, it was established that this light matched exactly a big bang prediction that such light, emitted after the universe cooled enough for hydrogen to form, has been streaming through the universe, unaffected except by the expansion of space-time, for billions and billions of years. But what can we learn by studying this light?

Let's consider some of its history. It was first detected by Arno Penzias and Robert Wilson. They are both American radio astronomers, using radio telescopes to understand the universe. Penzias and Wilson, working at Bell Laboratories in 1965, were using a very sensitive 6-meter radio antenna with the goal of detecting weak signals that had been bounced off of metal balloon satellites. As they pursued their goal, they continually encountered a curious problem—there was a constant noise in the signal, one that could not be removed by any means they attempted—including, famously, scraping the pigeon droppings off their antenna.

At first Penzias and Wilson were unable to interpret their observation. Fortunately, the physicist Robert Dicke (1916–1997) was nearby at Princeton. Dicke, working with the equations of general relativity, had already re-discovered an earlier prediction: that if the big bang was the correct description of the birth of the universe, there should be light left over from it that would, today, have been shifted in frequency by the expansion of space-time into the radio band. As a result of this work, he was one of only a handful of people who could have solved the puzzle and be able to supply the theoretical interpretation that could tie Penzias' and Wilson's observation back to the big bang.

The big bang theory made very specific predictions about the way in which energy is apportioned to each range of frequencies present in the light left over from the beginning of time. This is a famous shape, known as the "blackbody spectrum"—the kind of radiation that is emitted by a body that absorbs *all* frequencies of light impacting it, re-radiating that energy. Was the "noise" that Penzias and Wilson had detected coming from all directions in the sky truly relic radiation from the big bang? To be more confident, one must always test a claim further and further.

To do this, many experiments were devised. One of the most famous and successful was COBE, the Cosmic Background Explorer, a satellite host to a suite of instruments that would reveal the character of the radio waves Penzias and Wilson had detected. Launched in 1989, the NASA COBE mission published its first most definitive results on this radiation in 1992, revealing that the spectrum of this radiation was exactly that of a blackbody spectrum, as predicted by the big bang theory. Not only did this help to cement the big bang theory as a correct theory of the early universe, it initiated a new era of cosmic microwave background science that continues to yield discoveries to this day.

Is the spectrum of this radiation a perfect, smooth, curve? i.e., is it a featureless, continuous, unbroken curve, or is there some fine detail lurking in the light from the big bang? If there is structure, what caused it? What could have perturbed the light from the big bang before it streamed freely through the cosmos, reaching us here on earth billions of years later?

Light and matter interact. Nowhere is this more beautifully summarized than in the standard model, a theory of the interactions of matter and forces. When matter interacts with light it leaves an imprint on the light. Maybe some frequencies of light go missing as a result. Maybe the intensity of the light is affected. When light encounters a region of clumped matter, it may miss the clump but is bent by passing through its gravitational field. This, too leaves lasting effects on the light. In fact, if anything warps space-time and light passes through that warping, this will have lasting effects on the light that can be detected later.

Can we see such effects in the light left over from the big bang? To answer this question, the astrophysics and cosmology communities pushed hard and constructed more precise and ambitious instruments. One of the most famous of those is the WMAP experiment. WMAP stands for the Wilkinson Microwave Anisotropy Probe. A collaboration of scientists that operated and analyzed data from the WMAP instrument observed a fine pattern of temperature distortions in the CMB. These distortions are the very fingerprints we sought. By studying the pattern of these distortions and comparing them to the behavior of space-time, matter, and energy described in the general theory of relativity, one can build a profile of the universe when it was very young—just 380,000 years after its birth—when that CMB light was freed to stream across the universe.

What we learned from the WMAP experiment and its successor experiment, the PLANCK satellite, still ripples through the physics community. Normal matter—the atoms—that are so well described by the standard model, account for just 4.86% ($\pm$ 0.10%) of the energy density of the universe. From all we know, from all that we have studied for thousands of years—all the stars we see, all the galaxies that contain them, all the hydrogen and helium and all the other elements and stuff we know about in the universe—these total to just about 5% of the energy density of the cosmos!

So what is the rest? Well, the good news for the discoveries of scientists like Zwicky and Rubin is that a great deal of it is a form of matter unlike that which is described by the standard model. It is slow-moving. It is cold. There is a lot of it . . . a *lot* more of it than there is of normal matter. It matches well with the observation of a non-luminous matter component of galaxies and galaxy clusters. *It's dark matter.* How much of it is there? The data from WMAP and PLANCK indicates that dark matter presently makes up 25.89% (± 0.57%) of the energy density of the universe! This means that for every one gram of normal matter, there are five other grams in the universe of a form of matter whose properties lie beyond the standard model. Fully 85% of the matter in our universe is a shadow whose immense contribution to the universe, through its gravitational effects, have shaped the very cosmos in which we live daily.

But is it really matter? The evidence from the CMB is one piece of a much larger pattern of evidence that says it is. Let's look at one, last, spectacular piece of evidence. This piece combines all the "greatest hits" of the universe that we've learned about so far: light, atoms, gravity, and galaxies.

This evidence comes from one of a now-large group of collisions called the "Bullet Cluster." The bullet cluster is an astronomical object that is made from two large clusters of galaxies that, long ago, collided with each other and have since continued on their way. The collision concluded about 150 million years ago, when a smaller cluster of galaxies (the actual bullet cluster) passed through a much larger one. While the galaxies in the clusters missed each other, the hot intergalactic gas that follows along with the galaxies collided. This gas also makes up most of the luminous matter present in the cluster. X-ray imaging reveals that the hot, colliding gas lags the actual clusters, which have moved on beyond the gas by a huge distance. Light reveals where the galaxies presently are, and light reveals where the hot gas is presently located. So we can now ask, where is the bulk of the mass from this pair of once-colliding clusters? Is it following the galaxies, or following the gas?

To answer this, we turn to gravity. Regions of clusters and gas that contain a lot of mass will also warp space-time commensurate with its mass. Light passing through such regions will be bent, as if by a glass lens. By looking at the lensing of light that came from behind clusters and has since passed through those clusters and gas, we can figure out where most of the mass is actually concentrated. This has been done, and it reveals that, while most of the luminous matter trapped in the gas, is stuck between the clusters, the most dense concentrations of mass are focused around the galaxies and not the gas.

What does this mean? If a modified gravity theory were the correct explanation, the lensing would mostly be expected to follow the gas because, if there is only normal matter, that is where all the mass (and thus all the gravitational effects) should be found. But that is not what is observed. Instead, the bulk of the gravitational effects follow along with the galaxies, even though the galaxies account for less of the total cluster mass than does the gas. *Curious!* This implies that, as with galactic rotation curves, the motion of galaxies in galaxy clusters, and the fingerprints in the CMB, an unseen and new form of matter is present in galaxies. It interacts primarily via gravity, whereas normal matter experiences all of the forces of the standard model more strongly than it does gravity. Because gravity is the force that seems to be most expressed by this non-luminous matter, it, like the galaxies, just doesn't suffer meaningful collisions as the clusters pass through each other.

The Bullet Cluster is one of at least seventy such systems that have been cataloged. A broad analysis of these colliding galaxy clusters has revealed that the gravitational lensing, and thus the bulk of the mass, follows along with the galaxies in the collision, even when the intergalactic and interstellar gas heated during collisions lags behind the galaxies after the collision. The evidence for a new, non-luminous form of matter—dark matter—has grown stronger as more evidence has been accumulated. The composition of dark matter is an active area of investigation. To date, there is no definitive direct evidence of having detected the constituents of dark matter by any means other than their bulk gravitational behavior.

The astute reader will have noted that if one adds together the observed normal and dark matter energy densities determined from the CMB, one does not arrive at the total energy density of the universe. 25.89% plus 4.86% is not 100% . . . it's just 30.75%. So, what is the rest? If all the matter in the universe accounts for just under a third of the energy density of the entire cosmos . . . what is missing? What makes up most of the universe?

To begin to understand this question (and physicists and astronomers are only *just* beginning to understand this), we have to consider the deaths of stars and the vast distances between things in the universe.

When Shadows Stretched the Cave

How did we know how far away the moon was before humans began visiting it in the 1960s? The answer is a wonderful trick of light and geometry, called parallax. Consider that you are sitting in the passenger seat of a car. Look out the window. Have you ever noticed how objects close to you—like a mailbox on the side of the road—appear to whip by you much faster and rapidly cover a more dramatic distance than do objects that are distant from you?—such as buildings far off that are close to the horizon—or the moon? They appear to stand still. The large motion of close things compared to the much slower motion of distant things when you change your position, is called parallax. It has an exact mathematical description. Using parallax, one can measure the distance to things that are too far away for a ruler or a tape measure to be used.

Parallax is how astronomers measure the distance to nearby things. By "nearby," we mean things like the moon, the planets, and even neighboring stars. But, when you want to measure things that are millions, or hundreds of millions, or billions of years away at the speed of light, you can no longer use parallax. The motion of the thing whose distance you're trying to measure is just *not* appreciable against the background of even more distant objects. Astronomers have developed a toolkit for making such difficult measurements, called the "cosmic distance ladder" that is applied in various ways as the distances we want to measure become larger, and larger, and larger still.

For very distant objects, like galaxies that are hundreds of millions or billions of light-years away, almost nothing works as well as a dying star. Well, a certain type of dying star.

Stars die when they exhaust the hydrogen nuclear fuel in their core and, under the collapse caused by gravity as their outward radiation pressure lessens, begin to burn the heavier elements that are the by-products of their long hydrogen-burning phase. At first the star burns helium, flaring in brightness and size to become much larger than its original size as it feeds on this heavier nuclear fuel. Once the helium is exhausted, it collapses again and ignites a new phase of burning of even heavier elements, going up the chain of nuclei from helium, to carbon, to oxygen, and beyond.

A star like our sun will reach the phase of burning through its helium, producing lots of carbon and oxygen, but will stop there. Why would our sun stop there? Our sun is a middling main sequence star, having a mass that puts it in the middle of the typical stars you find in the universe. At that mass, a star can burn helium but, when it reaches carbon or oxygen, its collapse under its own mass is not sufficient to ignite the next phase of nuclear burning. A star in this stage gently blows off its atmosphere, forming around it a structure called a "planetary nebula," and leaves behind a beautiful white-colored hot core of carbon and sometimes carbon and oxygen. This beautiful stellar remnant is a "white dwarf." In fact, this is how our sun is expected to die. It will bloat when it burns helium, then collapse again, and then hit a wall as to its ability to continue burning nuclear fuel. Its atmosphere will blow off from the heat of its core, and the exposed core will burn bright for trillions of years, slowly cooling from white dwarf to black dwarf over the remaining history of the universe.

It's a beautiful picture. Our sun will end its days like a great nuclear diamond, shining in the night, surrounded by those planets that are left in the solar system after it bloats to become a red giant as it begins to die. This is a quiet retirement for a mid-sized star like ours.

Not all stars are this lucky. Not all stars have a quiet retirement. And, as can be said of some human lives, the trouble begins with an unreliable companion.

It is estimated that about one third of the stars in the Milky Way have at least one companion star, another star that dances around the first under the influence of gravity There is no reason to believe that the proportion of such companions stars is any different in other galaxies. Sometimes, one star will die gently and become a white dwarf. If the companion star is a red giant—either a red giant star from birth, or a red supergiant resulting from the start of the death of the companion star—retirement gets interesting for the white dwarf. Red giants are so large that their outer atmospheres are far from the core, and are more loosely bound by gravity. The neighboring white dwarf, exerting its own gravitational attraction, may begin to slurp the atmosphere off the red giant companion. Over time, the slow march of atmospheric hydrogen from the companion onto the white dwarf causes not only a disk of gas to form around the white dwarf, but for that stolen gas to in-fall down onto the core, accumulating new mass onto the white dwarf.

There is a limit to how long this can continue. If the mass of the white dwarf reaches a total mass that is 1.3 times the mass of our own sun, it comes out of retirement in a spectacular way. The slow burning of the white dwarf can no longer resist the growing gravitational inward pull of its accumulated mass, and the star begins to collapse in a runaway and catastrophic process. This causes a huge explosion, blowing the white dwarf to pieces in the process (and likely destroying the poor companion star as well). This is known as a type-Ia supernova. The light from such an event can be seen with a telescope across the universe. The light that results from this process has a behavior in time and an intensity that is predictable, because the process happens the same way every time: a white dwarf forms, with a mass less than 1.3 times that of our sun; it feeds on the atmosphere of a companion star until it reaches the threshold mass; a runaway collapse begins that, in just seconds, destroys the star and creates the light that we see from our vantage point in another galaxy. Because this process is so regular, you can look at it at any distance and, knowing the measured faintness of the light, correct the faintness back to the original brightness. In doing so, you take into account the vast distance that would make the light so faint, and *bingo*, you have the distance from us to the type-Ia supernova.

Suddenly, the end of the peaceful retirement of a star becomes a meter stick for determining the distances to things in the cosmos. This is an incredibly useful tool!

In 1998, two teams of astronomers used Type-Ia supernova surveys to map distances in the universe as a function of time. They wanted to know how the expansion of the universe, which began during the big bang, has been changing over the course of its history. By looking at Type-Ia supernovas closer to earth, and those very far away from earth, it is possible to see how distance scales have changed in the universe over time. The farther away a supernova, the older it is, and the earlier in the universe it detonated; it took all this time for its light to reach us on earth.

But here, again, general relativity helps us to predict what we should see. General relativity takes into account the geometry of space-time and the matter and energy content of the universe, making definitive predictions about how bright a given supernova should be if its light has traveled over a certain distance in a certain amount of time. If the universe is dominated by matter, for instance, then distant supernovas would be just a little bit brighter to us here on Earth than if the universe were made of both matter and, for instance, a "cosmological constant"—a universal energy density of empty space that can act to push space-time apart faster and faster.

So what did these two teams see? The teams, one led by astrophysicist Saul Perlmutter (1959—) called the Supernova Cosmology Project (SCP) and the other co-led by Adam Reiss (1969—) and Brian Schmidt (1967—) and called the High-Z Supernova Search Team, both cataloged dozens of Type-Ia supernovas, some going back in time as far as about 8 billion years ago. What they found was that the most distant of these supernova were fainter than would have been expected if the universe was dominated by matter, as was the common belief in the 1990s. Instead, a better explanation for what they saw was that the universe was not only expanding, but more recently, that expansion had accelerated, making objects that are far away appear *fainter* than they should be at this particular time in the life of the universe. This was a stunning revelation. Yes, the universe was expanding, but the expansion was not slowing or even remaining constant; rather, the expansion was *speeding up!*

A few years later, when the results from the WMAP analysis of the CMB was completed and, later still, when the PLANCK analysis of the same light was completed with more precision, the picture became a lot clearer. There was another player in the life of the cosmos during the beginning of time and more prominent now. That player behaved, not like matter, but like some kind of energy density that is present in space itself. Einstein would have called this a "cosmological constant," although his purpose in inventing it was to try to hold the universe constant. What he described as his "greatest blunder" now finds a new home in our understanding of the cosmos, not for the purpose of holding its size constant, but, instead, to act like a negative pressure, pushing space-time apart faster and faster as the universe ages, accelerating its expansion. So the missing 69.25% of the energy density that wasn't explained by matter is explained by something far stranger—the seeming energy of empty space itself. Rather than being a subdominant player in the universe, it now appears to be the most prominent player in the cosmos, gradually making the universe bigger and bigger at a faster and faster rate.

We have two mysteries. In trying to shine light on the universe, we have revealed shadows that account for more of its mass than can be explained by the familiar shapes of matter and forces described by the standard model. We have also revealed that matter itself is really just a bit player in the universe, relegated to less than a third of the energy density of the universe. The rest is a kind of dark energy that acts to accelerate the expansion of the universe. Some of these new shadows increase the mass of the universe, guiding its clumping by exerting more gravity on normal matter than normal matter alone could achieve; and some of the new shadows press on the walls of the cave, stretching those walls to make the cave larger and larger, doing so faster and faster.

Epilogue: The Pale Shadow of the Standard Model

If dark energy is but the energy of empty space, perhaps quantum mechanics can come to the rescue! After all, in the quantum realm, particles and antiparticles are free to pop into existence and disappear again. This makes the vacuum alive with virtual particles, whose fleeting existence has real consequences for the universe—including making empty space not empty, but filling it with a kind of quantum energy.

Perhaps this is dark energy! But, alas, the calculations have been done. The results are surprising. You might have expected that we would tell you that the standard model doesn't offer enough virtual particles to come into and go out of existence. Rather, *the opposite is true*. The standard model would predict *too much* dark energy, more than is observed to exist in the cosmos. This calculation doesn't overshoot by a little; it over-predicts *spectacularly* badly. The calculations overshoot the observed amount of dark energy in the universe not by a factor of two, or even ten, but by more than 100 factors of 10 multiplied times each other! Such a spectacular miss cries for an explanation.

As we leave the puzzles of dark matter and dark energy for other topics, we must keep these twin modern mysteries in mind. The standard model may hold no particles that can explain dark matter, but a more all-encompassing theory of nature must. This is a stiff demand on an expanded theory of nature. In addition, that theory must explain why dark energy exists, and why it must be so much smaller than would be predicted by the vacuum energy of the standard model alone.

Echoing the beginning words of this chapter, do you live in the country? Do you live in the city? It doesn't matter. If it's a clear night, go outside and look up. Don't look at the stars. The temptation of their light is a distraction. The universe is ruled not by those bright gems, but by the voids and the shadows that lie between them.

## Chapter 8

What Do Physicists Really Know Right Now?

In the early parts of this book, we endeavored to tell some of the story of how we know things, what things we know, and when we learned about them. Some of the discoveries noted so far may already feel to you to defy belief: subatomic particles that cannot be isolated from each other (quarks); invisible matter that keeps galaxies from flying apart (dark matter); a kind of negative pressure pressing on space-time that causes the universe to expand faster and faster (dark energy).

These are not matters needing your opinion as to their believability, rather, we expect that, having learned something about the scientific method, you have come to understand that scientists have produced evidence for these phenomena and that this evidence is not about esoteric things that don't affect your day-to-day life. The development of superconducting magnets during early particle physics experiments spun off an entire industry manufacturing these magnets which are now at the heart of every Magnetic Resonance Imaging (MRI) scanner in the world. Anti-matter is used to detect tumors in hard-to-reach parts of the human body, thereby reducing the need to do exploratory surgery, helping to extend and perhaps even save lives (PET scans). Particle beams, once used only to explore physics, are now used to treat cancers (beam therapy). Asking fundamental questions about the nature of reality fosters results that extend to everyday life.

It is useful to pause at this moment and reflect more deeply on what physicists know, or think they know, about the present universe. This chapter will bridge from the more established and familiar aspects of the universe into more speculative territory. We are about to enter into the frontiers of human knowledge and experience, where ideas bubble up and compete to be the correct description of reality.

Before stepping off into these frontiers, let us pause at one last way station of strong familiarity.

Let us begin with our world, the world of human beings. Matter appears to humans to come in many different shapes, sizes, and textures. It manifests as liquids, solids, and gases. These different forms have a common thread: matter is made up of molecules and molecules are made up of atoms. Atoms impart the characteristics to the material of which they are a part. These characteristics are a reflection of the subatomic particles within the atom that are bound together by the electroweak force. An important structure within the atom is the nucleus which controls the attributes of the atom. The nucleus contains protons and neutrons bound together by the strong nuclear force.

While gravity has so far played no detectable role in the realm of the atom (and smaller realms), the gravitational force might be conveyed through the action of a particle called the "graviton," a particle that has not yet been detected.

Figure 8.1 illustrates particles and subatomic particles showing the manner in which they are thought to relate to each other. Six levels of hierarchy are shown in descending order. The top level shows the material level that looks quite like what the material is; cloth, metal plastic, etc. The second level shows molecules, making up the structure-like fabric of the material. To see what the molecules are made from, we descend to the third level—the world of the atom.

The atom includes the nucleus (protons and neutrons) and electrons. Other particles play a role here: the neutrino is emitted from the nucleus when it undergoes a weak interaction; the photon is emitted or absorbed by electrons, changing their energy levels in the atom. If we descend again, we enter the fourth level, that of the subatomic world—quarks and all the other leptons. Quarks make up the protons and neutrons that are stuck together by a dense mass of gluons.

Electrons, quite active in their movement, are located at distances from the nucleus that are hundreds of thousands to millions of times the size of that nucleus and move around the nucleus in a cloud at great speed (about a hundredth of the speed of light). The presence of quarks is revealed by particle ac-

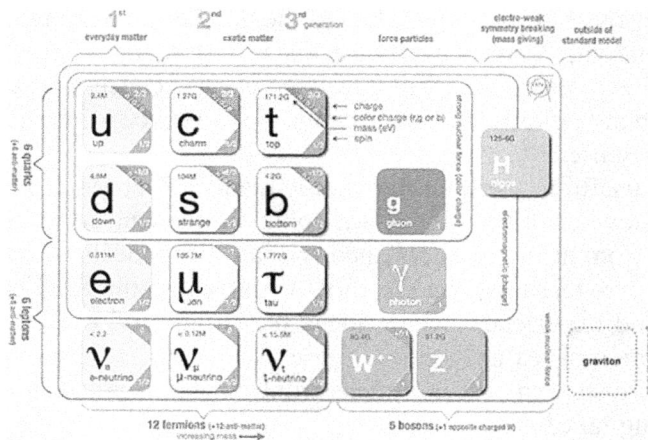

Figure 8.1 The Standard Model of Particles

celerator experiments. We presently perceive no structure inside the quarks or the leptons, but this may be simply because particle accelerators cannot "see" this far as yet. At this moment, quarks and leptons behave as if they are not made of anything else. But, if they are made of something else, that would help us to understand a lot about the oddities of the standard model and, perhaps, even gravity.

For decades, physicists have been increasing the energy of particle accelerator experiments as they approach deeper levels of reality. At the moment, we might say that physics is stuck between floors and we are not sure when the next one will be reached. But there is hope that there are more levels beneath the subatomic realm, levels at which we would see even more fundamental structures of natures—perhaps tiny, spaghetti-like vibrating objects called "strings," whose silent, tiny "music" is the true source of the shadows on the cave wall.

It is important to keep in mind, as we cross the line into the substructure of quarks and leptons, that we leave the scope of what physicists have seen in the laboratory to now enter the domain of things that physicists only *speculate* are there. Mathematics is their tool in this realm. We should also keep in mind that when a physicist uses the word "seen," it generally means that the phenomenon has been detected by instruments or other nonvisual means (see Chapter 5).

Peering into the Frontier

We will return to a deeper discussion of strings and the large theoretical framework that describes them later in this book. For now, note that, if strings could be seen, they would appear as open-ended or closed filaments that vibrate in a manner that results in conveying to us the properties of the known particles (e.g. their masses and quantum numbers). By assuming different shapes, modes, frequencies of vibration, and other characteristics, they create the forces and the matter that can be detected as various particles—a collection of spaghetti-like energies, each vibrating differently to produce particles. Strings are not the only strange idea we will encounter as we move forward into the frontiers of physics. Incorporated in the idea of strings are extra dimensions of space, and whole objects that occupy those extra dimensions ("branes"). Strings and branes are closely related to each other, and, by their existence, would resolve many of the mysteries of the standard model

To describe how these speculative components act in the real world, physicists have created many mathematical models (that differ from each other) about how the four forces of nature, together with matter, exist in a mathematically unified description. These descriptions can then be used to predict the outcome of past, current, or future experiments. These concepts then need to come together in a self-supporting theory that meshes with the insights and successful discoveries that were achieved during the twentieth century. What was accomplished in that period, all of it demonstrated by actual measurement, well-describes the universe. The scientific method requires that any new theory must demonstrate itself to be consistent with previously established results.

Particle theory research has been successful because the theories it has developed were testable by observation. String theory, as described presently, and, seemingly into the foreseeable future, is not testable by direct observation. Future work in colliders that involves increasing their operating levels to tremendous energies beyond what they can presently attain will be required.

Let us seek to understand the scale of the energies (and the corresponding sizes of things) that would have to be achieved to directly observe the activities of fundamental strings. We earlier discussed the physicist Max Planck. He attained many notable achievements, and a notable number of them are named after him such as Planck's Constant, h, which is related to the smallest unit of spin angular momentum that a subatomic particle can possess. Planck developed a system of units that relates many of the most fundamental constants to each other. In doing so, it seems also to define the smallest unit of anything that physicists think can exist in the universe. Just as there is a limit to how little angular momentum (spin) a particle can possess, there are limits to the size of space and the length of time that can minimally be probed or pass, respectively. The Planck length, the smallest physically meaningful unit of length, is defined as:

$$l_p = (hG/(2\pi c^3))^{1/2} \approx 1.6 \times 10^{-35} \text{ m} \approx 10^{-33} \text{ cm}$$

where G is Newton's gravitational constant, h is Planck's constant, and c is the speed of light in empty space (the fastest that anything can travel). While there is no well-established meaning to this length (no instrument can yet measure a distance this small—a distance that is one one-hundredth of a billionth of a billionth of the size of a single proton!), in many mathematical theories that unite quantum field theory and gravity, this (the Planck length) is the distance at which the quantum behavior of gravity is most easily revealed—there are no distances shorter than this that have physical significance. In order to probe a distance this small, one must reach a correspondingly high energy (representing a short de Broglie wavelength) for an accelerated particle. That energy, the Planck energy, is given by

$$E_p = (hc^5/(2\pi G))^{1/2} \approx 1.2 \times 10^{28} \text{ electron-Volts}$$

In comparison, the Large Hadron Collider (LHC), the most powerful particle accelerator that humans have ever constructed, reaches just 1.4×10^{13} electron-Volts at its design energy. The most powerful *natural* accelerators known in the universe that beam accelerated particles into the cosmos (some of which interact here on Earth), achieve particle energies in the range of 10^{18} electron-Volts.

Such energies are achieved, or so the evidence suggests, when the supermassive black holes at the centers of galaxies go through a violent "feeding" period, gobbling up gas and stars that in-fall to the galactic center while spewing out some of that energy in jets of particles that can be seen across the universe. This is paltry by comparison to Planck energy, the energy level that one would have to achieve to probe the Planck length.

The difficulty in demonstrating string theories is that the energies required for direct observational tests must approximate Planck energy—energy on the order of 10^{28} electron-volts. Such energies would have been present shortly after the creation of the universe, but are not feasible using current or even planned technology. Thus, how can one confirm that the mathematical theory of strings works and is correct? This may seem like an impossible situation doomed to take us into a non-scientific philosophical debate. However, long before these energies would need to be encountered, string theories should pose implications that reveal fingerprints of their existence.

This is not the first time that science has faced such a circumstance. When atoms were first proposed, it was inconceivable they would ever be directly observed. The case was similar when quarks were first suggested. In fact, the best understanding of the laws of physics as they pertain to quarks suggests they will never be directly observable. Nonetheless, their presence is inferred by observations made at high-energy physics laboratories around the world.

We'll come back to strings in a bit. Let's take a look at some of the ideas that are connected to string theory, but could also exist on their own without a need for string theory to be true.

The structure of the space and time dimensions of the cosmos may also explain the puzzles that have been left to us after centuries of observation and explanation. Perhaps we have misunderstood the limits of space and time! For instance, there is the possibility for a family of models called "braneworlds" that can provide a basis for collider experiments at the Tevatron at Fermilab and the LHC at CERN to probe the idea of hyperspace—dimensions of space beyond the three we are familiar with. Braneworld models may provide the possibility for experimental verification of extra-dimensional ideas, but they would take place at energy levels that may be experimentally accessible now, in contrast with those needed for direct testing of string theories.

There are also special limits of the mathematics of string theory called supersymmetry and supergravity that create opportunities for observing some of the inferences of string theory through experimentation by revealing the existence of "fraternal twins" of the familiar standard model particles. These are called "super-partners," a consequence of a new symmetry in nature—supersymmetry—which, while not part of the standard model, is nonetheless a viable symmetry of nature that connects matter particles (those having half-integer spin) to force-carrying particles (those having integer spin). This would, at minimum, double the number of particles in the cosmos, creating a set of twins of the known particles and forces.

Currently there is no evidence of such super-partners. However, even without this evidence, the mathematics of supersymmetry has pointed toward an understanding of why the mass of the Higgs particle is as small as it is. Recall that the standard model makes no strong preference known for the mass of the Higgs boson. It is a largely free parameter of the model, constrained by other parameters that are not sufficiently well measured to know whether the standard model is entirely self-consistent. Nothing prevented the mass of the Higgs boson from being heavier than 125 GeV/c^2 at which it was observed. Or is there? Supersymmetry offers an explanatory framework for the Higgs boson's mass to be relatively small as compared to the possibilities left open to it in the standard model.

The mathematical constructs of string theory are part of what physicists began searching for after the initial successes of the standard model. If the standard model is a unified field theory, and if uniting the strong and electroweak forces is "grand unification," then string theory and its mathematics represents something grander still,—a "theory of everything,"—that brings gravity and all of the other known forces under one roof.

But, what do we do about the additional six or seven dimensions many physicists believe string theory demands? What became of them after the big bang? Why don't we see them today? Did these dimensions become so small that they cannot be seen even using the carriers of the forces of nature? Are they really dimensions? Or are they entirely different attributes of reality? Will these constructs permit humans to understand them and do they offer realistic quantitative descriptions of what goes on in the universe? If these dimensions are curled up, can they be larger than 10^{-33}cm? If so, will their sizes be sufficiently large as to be detected in experiments? (Sizes as large as 0.1 millimeter would still be described as small, but they would be detectable in a collider.)

While string theory opens a new pathway to explore what makes the universe tick, it also presents another, more human question: the issue of whether it is physics or philosophy. Can the theory be tested? This has been a requirement of modern scientific research, articulated by all of the practitioners of this pursuit, and fervently so, by figures such as Galileo and Newton. Can string theories, and their associated ideas and mathematics, offer tests that can be done using existing technology? How long would humanity have to wait to assess these ideas?

Convinced that quantum field theory, which had been developed using the point particle concept, but augmented by quantum wave function notions, could take them no farther, a small group of physicists seceded from the main body of the community during the 1980s to embrace string theory in the belief that this mathematical approach had great promise and should be developed. Decades later, stretching well into the modern era and after many bumps, bruises, and triumphs, these researchers, substantial in number, continue to develop a knowledge that is largely independent of the community of physicists still working to see how far the quantum field theory idea can really go, and whether it's yet reached the limit of its applicability in the standard model; or of the physicists still stressing the limits of general relativity, with its view of the cosmos separate from the standard model, focused on the overall cosmic structure of space-time and all the constituents trapped therein.

The collision, if you will, of the more traditional physics community and the more string-theory physics community is truly a frontier—one replete with the all-out bar room battles seen in depictions of the American wild west. Among many physicists, string theorists are criticized as being quasi-physicists or quasi-mathematicians. That said, it should be noted that other branches of science and engineering often begin their research by formulating their problems mathematically in order to more efficiently and definitively guide their experiments toward posing questions to which they must find answers. Particle physics, for example, is largely guided by mathematical theory.

Early on, physicists would simply turn on an accelerator having a detector, measure what happened, and wait for nature's beneficence to provide them with a new curiosity. It was like going on safari. All one had to do was trek to an exotic land and it was guaranteed that new curiosities would be seen. That era in particle physics experimentation is largely over (for now), with the pace of new particle discoveries having significantly slowed even as particle accelerators have continued to rise in energy. The community has transitioned from a period in the 1930s–1970s when haystacks were littered with needles to the period after the 1980s when haystacks had only occasional needles. This has altered the strategy for physics research at particle accelerators, lest they miss a crucial needle hiding inside a stack of established physics processes.

Searching for the Higgs particle is an example of such a cyclic probe by the experiments at the LEP Collider at CERN, the Tevatron at Fermilab, and later, the LHC at CERN, as researchers doggedly probed to narrow down the mass regime(s) where this particle might be located. As we have seen, that search ended with the discovery of the Higgs, but in doing so has raised new questions about the "why" of this player in the cosmos. The Higgs boson is a new playground for ideas about physics that is beyond the standard model, and even for other models of physics including exotic phenomena like extra dimensions of space and supersymmetry. We'll revisit those ideas later in a synthesis chapter, in part two of our pursuit of the nature of the Higgs boson.

Black holes and neutron stars may be the next significant playground whereon theory and experiment could meet again on these frontier subjects. The detection of colliding black hole pairs via detection of gravitational waves in 2015 and 2016 opens a new door on the study of space-time. In addition, 2016 also witnessed the first observation of a quantum effect near a neutron star. The observation of the polarization of the vacuum, where strong magnetic fields from a neutron star caused empty space to gain optical properties that then affect light, may give us a handle on studying standard model physics in the presence of very strong gravitational fields. Whenever physicists have probed into regimes of extremes—extremely small, extremely large, extremely fast, extremely slow, extremely hot, extremely cold—they have learned something new about the cosmos. What can be learned from looking at extreme quantum phenomena in the presence of extreme gravity?

While the braneworld model has not led to new insight into the physics of black holes, string theory holds the record for being first to produce new results for describing the mathematics of black hole behavior. The black hole is a great testing ground for theories that require understanding of the unification of forces. String theory shows the promise for achieving unification of the forces because, in this class of constructions, behavioral descriptions that work for black holes may work equally well in describing the universal behavior of matter and energy. Black hole physics combines the quantum behaviors of all matter contained in a black hole with the macro-physical behavior it reveals in its environment. If its state is to be correctly described, the thermodynamics of the black hole must take into account its behavior in the presence of gravity, the nuclear forces, heat, and other conditions.

In today's practice of particle physics, theorists are forced to play the role of guide to experimentalists about where in general to look, and what sort of signals should be sought, by developing a mathematical description of the region to be experimentally probed prior to the commencement of experimental observations. With the aid of powerful computer programs, theorists can even predict the form that signals from undetected particles will present in the detectors. This is a major change from the earlier traditional path that first conducted experiments to obtain the reality of the problem before formulating it mathematically. We'll discuss more about this concept in a later chapter, showing how modeling of physical phenomena in real experiments can be achieved by a marriage of mathematics and computation.

The Shadows Take Shape

We've learned so much about the universe already. Consider the electromagnetic force. It is essential to everyday life. The bonding of atoms—all of chemistry—is enabled by this force. In enabling chemistry, electromagnetism makes possible all of biology—and thus, all of life itself. The photon, the quantum of the electromagnetic field, is the primogenitor of all of this. Without the photon, day-to-day realities—the meanderings of aerobic bipeds like ourselves—would not be possible.

In equal sense, we can point fingers into all the corners of the standard model and say something similar. There would be no stability—no certainty—in the universe, without the strong nuclear force, its bond a guarantee that atomic matter will continue on for ever and ever. There would be no warm home called Earth, without the weak nuclear force, which permits the processes that power the core of the sun and all other stars. The fundamental mass of all matter and force particles is guaranteed by the Higgs particle—the primogenitor of mass.

What about gravity? It is tempting to take the mold of the standard model and imagine a quantum gravity, just as there is a quantum of electricity and magnetism. This quantum would be called the graviton—the carrier of the gravitational force. If it exists, it has long eluded experimental detection, thereby stymieing the unification of the laws of the universe in the sense that electricity, magnetism, and the nuclear forces are unified. The authors of this book expectantly await the detection of a graviton, as the detection of such a particle would support a crucial moment in the understanding of the totality of the cosmos.

The period 1960 to 1970 saw a breakthrough in unification with the emergence of the Glashow-Weinberg-Salam model explaining the weak nuclear force. Spurred on by these developments, unification was a target of intense study leading to the notion of a "grand unification" which would provide a consistent description of the strong, weak, and electromagnetic forces. During this period we also saw the emergence of the standard model. It embodies the ideas of unification right down to its core, and it is tempting to want that kind of success to continue.

Meanwhile, collider technology has continuously improved, resulting in the discovery of new quarks and leptons in the late 1970s and early 1990s, the W and Z bosons in the early 1980s, and the Higgs boson in 2012. Every time the standard model has offered a prediction, it has turned out to be correct. The standard model is a highly accurate description of reality that has generated new knowledge about the universe.

Then there is that other big and very successful description of nature, Einstein's theory of general relativity. This theory is all about the non-quantum mathematical description of space, time, energy, and matter. It has no quantization of these things built into it. This makes general relativity inconsistent with the standard model. But, general relativity is not inconsistent with reality, and that is what is so tricky about it. Like the standard model, general relatively has generated new knowledge. From it we learned of the bending of starlight by the warping of space-time, which was confirmed soon after the theory was finalized. It taught us that extremely dense mass can create regions from which light cannot escape—black holes—that were confirmed beginning in the 1970s. It taught us that, when bodies orbit each other in a certain way, energy is carried away from those bodies in space-time ripples—gravitational waves. This was confirmed indirectly and, in 2016, verified directly.

Let's pause for a moment and reflect on just how impressive this last observation, gravitational waves, has been. Waves of gravity were predicted as a consequence of Einstein's work in 1916, but even Einstein was skeptical that they would ever be observed. Why? Consider two black holes orbiting each other in a kind of death waltz. These objects are the most extreme forms of matter ever predicted by any theory of nature. They can possess masses of just more than one sun to millions of suns, depending on how they were formed. Such extreme creatures of the universe, whirling about each other in gravity's embrace, causes space-time to ripple in response. These ripples, gravitational waves, are about as big as one might imagine making them. And yet, with all of this mass and all of that whirling, space-time is expected to create a ripple no larger than a millionth the size of a proton. It is no wonder that, for a very long time, the possibility of seeing them was considered impossible.

Nonetheless, what made it possible to see these waves directly was the very same technology that set Einstein on his quest to understand space and time: the interferometer, a device that produces a pattern when two sources of light are merged. It was such a device, constructed and operated by Albert Michelson (1852–1931) and Edward Morley (1838–1923), that revealed the invariance of the speed of light even when the source of the light is in motion. This is what prompted Einstein to throw out the constancy of time in favor of the constancy of light's speed.

A modern version of this instrument, called LIGO (the Laser Interferometer Gravitational-wave Observatory), is a revolutionary leap in this kind of technology. Michelson and Morley's most advanced device, operated in 1887, fits on a large tabletop. In contrast, LIGO consists of two related instruments (called Fabry-Perot Interferometers), each made from a pair of 4 km-long arms (almost 2 1/2 miles). To identify and filter out local disturbances, one interferometer is located in Hanford, Washington and the other in Livingston, Louisiana. By sending light back and forth along the long arms that are at angles to each other, it is possible to detect ripples in space-time. By noting identical disturbances (ripples) that are detected in the two devices that are thousands of miles apart and ignoring non-identical disturbances (that are caused by sources local at each interferometer), gravitational waves can be detected.

The first detection of gravitational waves was announced in February of 2016, with a follow-up announcement of a second detection several months later. LIGO, and its partner instruments Virgo and GEO600 in Europe, are part of a larger scientific collaboration working to upgrade their instruments to make more and more detections of these phenomena. A hundred years after the birth of general relativity, one of its most challenging predictions, waves of space-time, was proven to be absolutely right!

The standard model is not wrong. General relativity is not wrong. How do we then bridge these successful cornerstones of reality?

A new formulation is required. Something like superstring theory, or some other as-yet-unknown construction will provide this. A quantum theory of gravity predicts a graviton and a similar formulation should make it possible to unify general relativity with the standard model. This is the dream that began with the unification achieved in the standard model—to fully define gravity's interactions with matter particles in a quantum mechanical manner.

There are presently many open issues about general relativity that make it impossible to achieve unification. Let's consider one of these. Recall that Einstein's equations take advantage of Riemann's metric tensor. Although a natural for describing gravity, this mathematical construct—used to create a description of the gravitational force based on the curvature of space-time and the matter density of the universe—is drastically different from the mathematics that provides the foundation of the standard model. Blindly rushing ahead to create a quantum version of general relativity is the equivalent of taking a square mathematics and cramming it into the round holes of a totally different mathematical framework. It is going to make a mess!

The Twentieth Century Transition

The standard model, with its particles shown in Figure 8.1 (see page 148) is, perhaps, the greatest paradox in science. On one hand, it has been tested many times and found to pass every examination. In some areas, its predictions match those of observation to better than one part in one billion. There are no other such statements that can be made in any other area of science in which experiment and theory have been so rigorously tested and found to agree. The large number of particles and force carriers within the standard model, and the forces it describes, are quite accurately consistent with quantum physics—at least with respect to three of the four known forces of nature.

By the end of the twentieth century, the standard model stood as an accomplishment par excellence. It gave (and continues to give) a quantitatively successful model of the behavior of the physical universe of subatomic particles and forces. It is solidly accurate and has had more testing through experimentation than has any other piece of science . . . *ever*. It describes quantum mechanics with great precision, inferring how the early universe behaved from a time just shortly after the big bang right up to this very time, a span of 13.8 billion years.

That said, the standard model is known to be incomplete as it does not incorporate gravity, nor does it explain the composition of dark matter, and it grossly overshoots on the prediction of dark energy (among its many other flaws). To put that last issue in perspective, all of the beautiful machinery of the standard model—gauge symmetry, spontaneous symmetry breaking, and field theory—comes together to predict the existence of the Higgs boson, but then, when applied to the energy of empty space, it disagrees with the observed effect by more than one hundred and twenty orders of magnitude! This has been called the worst prediction of physics in all history. How do we reconcile this paradox?

Where the standard model and general relativity might begin to intersect, concerns the idea of mass. In the standard model, fundamental mass is intimately tied to the Higgs boson. Without the Higgs, all quarks, leptons, and vector bosons would be massless. Mass does come from places other than interaction with the Higgs boson, something that is also explained in the standard model. However, the reason for disparities in masses—the masses of different quarks, for instance, or the masses of states made from quarks—is not well understood. For instance, the mass of the proton and neutron is not due to the Higgs boson, because the quarks that make up these particles are very light compared to the mass of the resulting bound state. Protons and neutrons are mostly made of gluons, and gluons are massless force-carrying particles. So how is it that protons and neutrons can have masses that are hundreds of times larger than their heaviest constituents?

This is one of the beautiful tricks that nature plays, summarized in Einstein's famous equation, $E = mc^2$. The binding energy of the gluon manifests as the mass—the inertia, the resistance to the change in the state of motion—of the proton and neutron (and any other such bound states of quarks). So, when we speak of the mass of, say, an atom, we have to recall that more than 99.9% of the mass of an atom is derived from its nucleus, and nearly all of the mass of the nucleus arises from the binding of gluons and quarks, not the mass of the quarks. In the scheme of the kind of mass that general relativity cares about, the masses of planets, stars, galaxies, galaxy clusters, and such, better than 99.99% of that mass has nothing to do with the Higgs boson. Einstein taught us that mass is an important attribute of gravity. Yet there is no apparent relation between the Higgs particle and gravitation! Yet, without the Higgs boson, the universe simply ceases to be. There can be no stars or planets, and likely, no clumped material structure at all.

So there is a place where the standard model, with its quantum realm, and the theory of general relativity, with its space-time fabric interacting with matter and energy, may intersect—extreme mass. If one could create a quantum state of extreme mass—masses capable of observably warping space-time—while preserving the quantum aspects of wave behavior and field theories, we might have a playground within which to learn how gravity and the other forces of nature intersect with each other. Perhaps this means firing enough energy into a subatomic particle to pierce the veil of our three dimensions, where gravity is weak, and to expose the other dimensions of space, where gravity might be very strong. Perhaps this could be at the edge of a black hole, where gravity is so strong that not even light can disobey its command. Perhaps, in these places, we might glimpse the graviton.

The gravitational force field is extremely weak compared to the other three forces. Why is this so? This is part of the hierarchy problem, which can be stated most simply as: why is the weak force a million million million million times (1 with 24 zeros after it) stronger than gravity? This is another roadblock, making it difficult to combine the forces in a single mathematical formalism. One has to bridge this gap in strength, and not just a gap in the mathematical structures that underlays each successful theory of nature.

Experimental work and celestial observations continue to seek to tie the standard model to the cosmological behavior of galaxies, stars, black holes, and other physical objects. While the standard model contributes significantly to our understanding of the formation of matter-based structure after the big bang (after the first 10^{-11} seconds, which is roughly the period of time that can be recreated with present accelerator technology), it has failed to reveal what happened just immediately after the big bang before that moment (10^{-11} seconds) in time.

Historically, each time collider energy levels are improved, new particles are found. Since we do not know the more fundamental theory of the quantum realm beyond the standard model, we have no good guideposts telling us whether we can expect that trend to stop or to continue. Paralleling the quantum quandaries, there are many other questions in cosmology to be resolved. Dark energy and dark matter requires greater understanding.

The Roads to the Frontier

We are about to leave our comfortable inn on the road of the past and head into the frontier of human knowledge. There are many roads to choose from, many shadows that cloud our ability to choose the right way to fully describe nature and present a clear view to the horizon where answers await. It is wise to consider as many of the possibilities as one can, so we will endeavor to sample among many of these ideas in the remainder of the book, providing examples as we go, of places where ideas and methods intersect. Experimental tests may lie at those intersections.

In the late 1960s, members of the physics community, having reached a stumbling block in their search for a comprehensive understanding of the universe, sought other approaches to unifying the laws of the universe. Initially, a small number of researchers, continuing to search, returned again in the 1980s to something we'll soon explore in more detail, the "Kaluza-Klein theory," a concept that had lain dormant for almost thirty years. This return was primarily motivated by the curious discovery of a mathematical theory called "11-dimensional supergravity," a means to marry concepts in the standard model with those in gravity at the cost of adding more dimensions of space. It was felt that it held the mathematical potential for inspiring a new attempt to continue the conventional process of analysis, experimentation, and corroboration that is the hallmark of the physics research process.

Another group of physicists during the 1960s and 1970s worked on quantum physics, cosmology, black holes, worm holes (tunnels through space-time), the fundamental nature of space and time, as well as the Kaluza-Klein theory, and an even more bizarre extension called "superspace," as they sought the means to find a new thrust in physics research. It was this idea of superspace that led to the discovery of 11-dimensional supergravity.

These activities eventually morphed into something found in the depths of the equations that physicists were using—a new pathway called string theory. String theory is a significant departure from the now-conventional physics of the standard model. During the past forty years physicists have developed a multifaceted analytical framework, rooted in the original string theory ideas, to accommodate and build upon the results of twentieth century experimental work. The challenge to this work is that it continues to suffer from the inability to corroborate its results through experimental means.

Problems and questions about string theory will become apparent as we discuss this further. For example, the number of dimensions required to describe string theory is often described to be a number greater than the four dimensions of our universe (length, width, depth, and time).

While all of this was going on in the 1970s and 1980s, there were also efforts to independently develop a theory of nature based on a concept also tied to string theory: supersymmetry. This allowed for mysteries of the standard model to be explained using just the usual four dimensions, but at the cost of adding more particles to the universe. That being said, supersymmetry opened another door to another investigative pathway, a four-dimensional string theory that we will discuss later, that is being developed in fits and starts by a small band of dissident physicists (among them one of the authors, SJG).

Currently, physicists are contending with the surprises and conflicts presented by the data from colliders, satellites, and telescopes (across a broad spectrum of electromagnetic radiation), information that was obtained over the past few decades, that will multiply in size during the coming decades as experiments take in more data at a faster and faster rate given better instrumentation. This has caused many more questions to be asked than answers have been produced. The pace of technology poses an interesting challenge to the theoretical efforts that want to make sense of the cosmos. As in theory, a unified approach, bringing experimental and theoretical physicists closer together, may present the best chance of meeting this challenge of chasing the shadows away with the liberal use of a bright light.

Evidence suggests that quantum physics and cosmology must be unified to describe a comprehensive theory that will agree with what we observe. In a way, Isaac Newton set us on this scientific path by unifying what happens in the heavens with what happens on earth. The moon keeps in orbit around the earth for the same reason that the apple falls from the tree to the ground. That unification continued in the 1800s, with the brilliant work of the theoretical physicist James Clerk Maxwell, who united electricity and magnetism into a singular electromagnetic force. Einstein tried, in the last part of his life, to unify electromagnetism and gravity, but he did so in an intellectual shelter that ignored the discoveries of the quantum revolution he himself had helped to set in motion, the findings of nuclear forces and the zoo of new particles produced in particle accelerator experiments. The physicists of the 1950s and 1960s recognized that the electromagnetic and nuclear forces had things in common, and that the things that they held in common were more important than what divided those forces, so much so that it was possible to unify the electromagnetic and weak forces into a singular electroweak force. A grand unification of the electroweak and strong forces feels inevitable, although the path to it is not clear. A theory of everything, finally harmonizing the standard model and gravity, feels even more necessary, but has proven even more elusive.

By the conclusion of the twentieth century and continuing into the early twenty-first century, the physics community had analyzed and tested Einstein's general relativity, the modern description of gravity, and the standard model of quantum theory. These accomplishments separately achieve a comprehensive and accurate description of the world of both the large and the small. Together, they are well-developed descriptions of cosmology and quantum physics that yield a battle-tested (although perhaps incomplete) table of particles and interactions.

Large numbers of unanswered issues create a demand for new approaches to new directions. Gravity is so weak compared to the forces of the standard model. The known particles and forces have a pattern of mass, connected to the pattern of Higgs boson interaction strengths, whose ultimate cause is unknown. Dark Matter is totally absent from the standard model. Dark energy, if it's the energy of the vacuum of space, is grossly over-predicted by the standard model. The mathematics of general relativity is a square peg addressing the seeming round hole of the mathematics of the standard model.

How do we begin to cause any of these problems to unravel? We will now set off for the frontiers of human knowledge and, using the twin lights of theory and experiment, see what sense we can make of this universe. Get ready to leave the familiar (if strange) world that we know to enter realms of thought that can have stunning consequences for the nature of reality.

Chapter 9

The Walls are Multiplied

When you spend as much time as our species has in trying to make sense of the shadows of reality, it is easy to forget the basics—those shadows play on the cave wall. One could spend a lifetime thinking up ways to shine light on them and then test each idea. But, what if the shadows can only be understood by rethinking the cave itself? What if we have made a fatally flawed assumption? What if the cave wall, as we perceive it in our day-to-day lives, is not the only surface that takes part in this shadow play?

The physicists and mathematicians of this and the previous two centuries are not the only ones who have ever considered that there might be more to space than just three dimensions, nor that there might be more to time than just one dimension. Inspired by the possibility that many spatial dimensions might bring solution to the problem of unifying all known forces and thereby explain more of reality, it is the physicists and mathematicians during the last 150 years who have delved most deeply into this problem, far more than any who came before.

For a long time, the goal has been to achieve the simplest possible framework for understanding as many natural phenomena as possible. The twentieth century was, perhaps, the most prolific period in history for the development of the physical sciences. The study of particle behavior and the unification of the laws of physics was pursued by many, some of whom we have already met. Let's meet a new one.

The Finnish physicist, Gunnar Nordström (1881–1923) accomplished some remarkable things during his short life. He is known for being one of the few people who adopted non-Euclidean geometry as a means to better understand space and time (another of those people was Albert Einstein). Working at the same time as Einstein, and publishing his ideas slightly earlier than the final form of Einstein's general theory of relativity, Nordstrom made two unsuccessful attempts at explaining gravity. The second one is especially notable because it represents the first time that gravity is described as arising from the curvature of space-time. Simultaneously, Einstein was building his theory of general relativity on this very notion. Nordstrom is also unique because (in 1914!) one of his theories of gravity included an extra dimension of space! He discovered that this would seemingly allow him to bring electromagnetism and gravity into a singular framework.

162

Nordstrom's work was published in the Swedish language. As a likely result, it did not circulate as widely as it could have. He admired Einstein and is even reported to have nominated him twice for the Nobel Prize. Nordstrom's zeal illustrates just how much fervor there was in that time for geometry, especially exotic geometry, and how it and physics play a stronger role together.

Another such explorer was Theodor Kaluza (1885–1954) who sought Albert Einstein's endorsement in 1919 for the publication of a modified theory of general relativity. As with Nordstrom's work, the goal was to unify gravitation with the electromagnetic force. Kaluza was not the first person to consider a unified description of the force of gravity together with the force of electromagnetism. Maxwell, who first gave the complete mathematical description of electromagnetism, had made earlier unsuccessful attempts.

Kaluza, born in what is presently Poland, studied as a mathematician but became deeply interested in general relativity. He sent his own work to Einstein in 1919, seeking approval from "the master." When Einstein *did* take public note of Kaluza's work and provide the endorsement, it is likely why Kaluza is remembered today for these ideas while Nordstrom is not. We will see that Einstein waffled a bit before accepting Kaluza's ideas.

Kaluza achieved a mathematically successful unification of Einstein's theory of gravitation with Maxwell's theory of light by imagining that our universe possesses a spatial dimension in addition to the usual three. Kaluza realized that the introduction of a mathematical fifth dimension enables unification. Thus, by 1919, the "hyperspace" concept became part of the mathematical toolkit that physicists use to increase understanding of the universe. This chapter discusses using multiple extra dimensions as a means for finding better understanding. (We will learn in a later chapter that supersymmetry offers yet another pathway to achieving that end and that the two can be combined to achieve even deeper understanding).

Hidden Extra Dimensions

Einstein, who reviewed and recommended prospective papers for publication in key scientific journals, was initially skeptical of what Kaluza had sent him. At first, he rendered a negative judgment because Kaluza's paper was so controversial in suggesting hyperspace. Einstein let the paper languish for two years as he considered its implications. He was finally convinced that Kaluza's equations were correct, that the theory had merit, and recommended in 1921 that Kaluza's novel idea be published. Adding more dimensions into the discussion of ongoing developments in physics theory made the consideration of a fifth dimension very controversial. After all, where is there evidence to support such a concept? The reader is reminded that added dimensions are a construct created by mathematics. Proof of their existence remains to be found.) A few years after Kaluza's 1921 paper, another physicist suggested a resolution to this dilemma.

Oskar Klein (1894–1977) was born near Stockholm. In 1926, he suggested a way to improve on Kaluza's hyperspace. Klein suggested that the new dimension could be made minuscule ("compactified") to so tiny a radius that the fifth dimension would be very small and of Planck length (10^{-33}cm). Because it was so small, it would make physically detecting a fifth dimension extremely difficult by experimental observation and would help explain why no one had yet noticed this extra way of moving in space. It was too small to have meaningful daily consequence.

Why does size matter so much? To answer this, it is useful to construct a "thought experiment," a method for which Einstein was famous.

Let us imagine a perfectly flat floor on which a ball is free to roll. Now ask the question, "How many independent dimensions are there in which the ball can roll?" We begin the answer using our intuition about day-to-day motion: first, the ball can roll *leftward* or *rightward*. We can define *left* and *right* by having a person stand with their arms outstretched such that they take on the shape of a cross. We then continue our answer: second, the ball can roll either in a frontward direction or a backward direction. We can define front as the direction in which the person sees when looking straight ahead; this is then conveniently perpendicular to the directions in which the arms are pointed. Backward is then the reverse of this new direction. Any other motion of the ball along the floor is a combination of the two. By this means we have a way to measure the dimensions of the space defined by the floor; it is a *two-dimensional surface*.

Now let's take a tiny needle and punch incredibly large numbers of regularly spaced holes into the floor. After this, we will now perform some experiments to measure what is the observable number of dimensions of the surface of the floor by rolling the ball on it again. Once more the floor will be found to be two dimensional, because the holes are so tiny that the ball does not "feel" them – it doesn't noticeably bump up and down as it rolls over the tiny holes, because the holes are smaller than the point-of-contact of the ball. So, what needs to be done in order for the ball to detect the presence of the holes?

Let's imagine making each pinhole increasingly larger. At first this does not make any difference to the path of the ball along the floor. However, when the holes become a good-sized fraction of the size of the ball, it will be noticed that the motion of the ball along the surface of the floor becomes "bumpy." In addition to its left-right and front-back motion, it will bob up and down as it moves along the floor, thereby exhibiting motion in a dimension that we could not perceive before!

Measuring the dimensions of space works in exactly the same manner as this thought experiment. If there were directions in space that are smaller than any atom, or even any subatomic particle, then these directions would not be apparent by observing the motion of subatomic particles, atoms or the motion of things that are larger. But, were one to use something smaller than atoms or subatomic particles, there *would be a way* to detect new directions in space caused by dimensions previously unseen by us. What are we familiar with that is smaller than the size of an atom? The wavelength of light (or, to keep this as broad-minded as possible, the wavelength of any quantum of either matter or force)!

The wavelength of light (and matter) can, as far as mathematics is concerned, be made so small as to be without limit. This is a theoretical statement implying that there is no fundamental limit to achieving this. There is, however, a practical limit. One would have to construct a device capable of producing light with incredibly tiny wavelengths. The world's record (set in 2011 and still valid today) for the shortest wavelength of light (produced by lasers) is about 6×10^{-11}m, just a fraction of the size of an atom. This was achieved with an x-ray laser, the SPring-8 Angstrom Compact free electron LAser (SACLA), at the RIKEN advanced science institute and the Japan Synchrotron Radiation Research Institute (JASRI). This represents the best that humans have ever been able to achieve at making light "see" smaller and smaller things.

There are, however, particles in nature having smaller wavelengths. Matter possesses a wavelength that can probe deeper than x-rays. The electron, whose physical size has been determined by experiment to be smaller than 10^{-18}m (far tinier than the best x-ray laser) and whose associated matter-wave behavior possesses a wavelength that can be made smaller and smaller by particle acceleration is a great probe to use. By the careful measurement of electrons, the best measure is that extra dimensions would have to be less than one-ten billionth of the size of a yard stick (approximately a meter). At that level of size there is no sign of extra dimensions of space, thus making Klein's idea workable.

Following Klein's idea that, being so small, the effects of a new direction would not be "visible" in the known physical behavior of the universe, his theory did not win favor in the physics community because it was considered to be a mathematical "vehicle of convenience" inasmuch as its concepts appear to fly in the face of observation. After all, if a thing cannot be detected— if there are no consequences of its existence—what would the use of its existence be? What would a fifth dimension mean to the physical world?

The world, for a time, would move away from the exploration of hidden extra dimensions in the universe. The quantum theory took hold in the 1920s and developed as the new pathway that physics would take. Its concepts were tied to research that could be verified with laboratory observation of behavior in the atom, and eventually in the subatomic realm. While Einstein, Nordstrom, and Kaluza began their work in the era after World War I (an era when there was just gravity and electromagnetism), by the early 1930s it was clear that there was more at work in the universe than just those forces. The nuclear forces would be discovered and characterized, and there would be an explosion of new particle discoveries. The Kaluza-Klein theory was not then pursued as a path to unification because there were more forces at play than the familiar electromagnetism and gravity.

All research related to this idea was pretty much set aside. Indeed, for the next few decades, the majority of physicists let the idea of electromagnetic and gravitational unification wither as an unexplored idea. As physicists saw it, the classical physics of Newton's three laws along with Einstein's relativity was deemed sufficient to understand nature at scales greater than the level of the atom. Quantum theory and relativity took care of understanding things at the scale of the atom and smaller, so it was further built upon and seems to work just fine in four dimensions, demanding no more than that framework of space and time. Everything is under control, so who needs more dimensions?

Einstein described his vision of the universe—his geometrical description—as marble; smooth, and elegant. He sought to achieve a smooth solution to the theory of everything. He thought God would want an elegant solution that was simple and easy to understand, but this was not to be achieved in his lifetime.

Einstein likened quantum physics to wood—a crude, rough, incoherent mass of matter and forces that would forever remain a jumbled mess. This view was based on his inability to accept an approach that rested on a concept that uses a highly probabilistic approach to understand nature. He believed that we ought to be able to do better than that. But, because physicists generally followed Galileo's teaching that science must be based on observation, Einstein's wish for physics to move beyond the "wooden era" was thwarted.

Einstein wanted to define a theory of everything that was built on the foundation he first developed in 1916 in his theory of general relativity. No probabilistic nonsense would rear its head in *that* marble-like work. But this was not to happen, even after thirty years of pursuing this Holy Grail. It would be the probabilistic approach of the quantum, something he himself helped to start, that would be the path the field chose as the pennant to point toward progress. However, the Kaluza-Klein theory, "blessed" as it was by Einstein, remained lurking in the background in his world of marble that someone else would have to find. The Kaluza-Klein idea would later make a remarkable comeback.

The quantum path forward, chosen by many, completely embraced probability, leaving Einstein as a bystander to his beloved physics. Mathematical quantities, established by the ideas of Newton, represented particles that were replaced by new and different mathematical quantities that displayed the attributes of particles and the attributes of waves, depending on the circumstances of any given experimental observation. Newton's concept was based on the notion of particles that were envisioned as idealized billiard balls. Quantum theory would replace this notion with the idea of "the wave function." The pinnacle of this line of reasoning, reached in the middle 1970s, was the creation of the standard model of the elementary particles.

The Allure of Extra Dimensions

Are we really talking about dimensions? What are dimensions? Are they simply other directions in space? The theories of most interest in the present string theory community require up to eleven dimensions. The equations that have been developed with this method neatly unify the particles and forces in a way that the standard model cannot. The extra dimensions (or they might more appropriately be called extra directions) leave room (provide latitude) to incorporate the equations that are required to describe the forces and particles of nature, including gravity.

This framework, and that is the best way to describe what it is, while it can stand on its own (and, as you will see, provide its own unique and attractive features) can also be part of an even more complex theory of nature. We will come to string theory in more detail later, but it is sufficient to say now that the extra-dimensional framework will allow physicists to learn how strings work under various constructs. The ultimate goal of this is to make predictions about the natural world. As is common within the great cycle of science, if this framework can be pushed enough to make testable predictions about the familiar 3+1 dimensional world, then it may be possible to learn which is the correct model of nature and its dimensional content. Measurement can refine theory, and we can continue to probe deeper into reality.

Present measurements cannot demonstrate the credibility of an 11-dimensional construct, because, as noted by physicist Michio Kaku in his book *Hyperspace*, "theory projects that the unification of all forces occurs at the Planck energy, or about 10^{19} billion electron volts, which is about one quadrillion times larger than the energies currently available [in colliders today]." Further, the string, the proposed most fundamental quantity of physics, is so small that there is no instrument presently in existence that can observe it. However, other research suggests that extra dimensions may be large enough to be seen in future collider experiments at energy levels not much larger than presently available. We will discuss this later in this chapter.

The foundations of physics are based on observation, mathematical formulation, and experimentation. At present, the observation of and experimentation using string phenomena are not possible, but maybe we can lean a little closer to the infinitesimal world by finding ways to detect other dimensions of space.

Making Sense of Dimensions

From the observation of nature, there are three spatial dimensions (length, width, and height) and one time dimension. These have sufficed for the millennia during which humans have made measurements and observed behavior in the universe. It does not appear necessary from direct observational experiments that there needs to be more dimensions to be able to define the state of existence of a system or a process.

Do additional dimensions exist, and, if so, why are they not in evidence? This question has been pondered by many for years. Edwin A. Abbott (1838–1926), a teacher, author, and scholar, sought to make sense of dimensionality in a book

Figure 9.1 The Flatlander Duck

entitled *Flatland*, written in 1884. As far back as several centuries ago, mathematicians, physicists, and philosophers have manipulated dimensions in their investigations, dealing first with the three spatial dimensions to then add a fourth dimension, time. By the time physicists began to play with the number of dimensions in the early twentieth century, they were just the latest group to do so.

Let us consider concepts from Abbott's book to think about dimensions, and what it might mean to perceive of a *higher* dimension. Consider two-dimensional beings—Flatlanders—who live in a flat, two-dimensional world. The Flatlander duck (Figure 9.1) appears as it does in the illustration in this book. It lives on a flat plane. The duck can only recognize two-dimensional objects. When a three-dimensional object such as a beach ball enters its flat-surface world, the duck can see only a two-dimensional slice of it.

For example, see Figure 9.2. As the beach ball descends, the duck will observe a flat slice of the ball's shape growing uniformly wider and subsequently shrinking uniformly as the ball passes through the space between sky and ground. The illustration depicts the sequential size changes viewed as the beach ball descends onto Flatland. Each of its slices is seen as a circle, the diameter of which increases until it reaches the largest dimension at the equator of the beach ball. From there it diminishes in size as it describes its shape while descending to Flatland.

Figure 9.2 shows these sequential slices physically separated. However, the Flatlander would see a single disk growing from a small size to its largest size

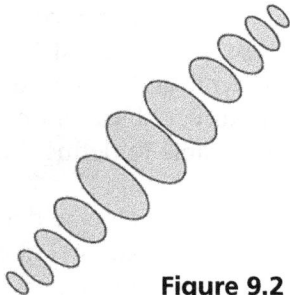

Figure 9.2 A Beach Ball Entering Flatland

as the ball reaches its equator. From there, the disk would shrink until it disappeared from view. This would be quite a stunning event to our Flatlander duck! Going about its day in its universe, it is witness to the sudden appearance of a disk that grows in size up to some maximum diameter, then shrinks again and vanishes. From the duck's perspective, something has come into existence out of nowhere, increased in size, then decreased in size, and disappeared into nowhere again!

Were a human, a three dimensional individual, to parachute to Flatland, the duck or any Flatland inhabitant would not see that third dimension - the person's height. The Flatlander would see only cross sections – tiny slices - of that person as he or she floats down to Flatland. An upright person (seen from beneath) would appear as cross-sections of the human body as if it was a series of still photographs from the bottoms of one's feet and ending at the top of one's head, the image changing shape as the body moves through the 2-dimensional space. First two oblong shapes followed by two circular shapes would appear as the person's feet and legs descended into Flatland, followed by various outlines of waist, shoulders, and head (ignoring the parachute).

Hold onto this image, and let's try to apply it (it will feel strange, we assure you) to our own 3-dimensional universe. Think now about what we would see viewing a five-dimensional object such as a *hypersphere*—a sphere in four, not three, dimensions—entering our four-dimensional universe. To us, trapped in our three spatial dimensions, it would seem as if a small ball appeared out of nowhere in front of us, growing in size to some maximum, then shrinking again until it disappeared. What a curious event that would be! A rash person would want to throw out the law of the conservation of energy during that moment, assuming it to be nonsense! Of course, no conservation laws will have been violated as there are now four spatial dimensions in operation. The hypersphere merely passes through our 3-dimensional slice of a 4-dimensional universe. You would not see the five dimensions of the hypersphere and would not be able to discern its true shape.

Thus, a 3-dimensional object is not perceived by the duck, or any other Flatlander, as anything but slivers of the whole. Portions of objects from a 3-dimensional universe might be sensed, but they would not be observable in their entirety. To us in our 3-dimensional universe, perhaps wrongly thinking that this is all there is to reality, we would experience a similar eerie and confused sensation were a higher-dimensional object to pass through our slice of the bigger space. This is how we can think about higher dimensions, and how we might experience them even if we cannot access them.

The Einstein Hypotenuse

We are not accustomed to dealing with dimensions beyond the usual three, so it appears that we cannot grasp the existence of more than three dimensions.

Mathematics, however—that beautiful language of the universe—is a tool that allows us to gain further understanding of multiple dimensions. Let us use a conceptual device that was created for non-physicists by one of the authors, SJG, that he calls "Einstein's hypotenuse." It was introduced by him in *Superstring Theory: The DNA of Reality*, a DVD lecture series produced by The Teaching Company.

The analogy illustrates multiple dimensions using simple mathematics. Einstein's hypotenuse is a mathematically visual way to describe how multi-directional behaviors can be related to problems in the dimensions encountered in everyday life. We will begin its description with a geometrical figure familiar to most high school students, the right triangle. Using the triangle we will develop the concept of multi-directions. This will help us to better understand how theorists and mathematicians grasp the possibility of extra dimensions of space-time and multiple directions beyond those that we observe in nature.

Let us refer to Figure 9.3 and recall a lesson from high school. Consider the right triangle shown in that figure. What makes it a right triangle is that two of the three sides form an angle of 90°. The side of the triangle opposite that angle is called the hypotenuse (always the longest side, it is denoted in the figure by a heavy, solid line), and for the sake of brevity let us denote its length by the letter H. The other two sides have lengths as well—let us denote them by L_1, the triangle's base length, and L_2, its vertical side's length. How do we relate the lengths of the sides of the triangle? For this, we learned the Pythagorean theorem: in a right triangle the square of the length of the hypotenuse is equal to the sum of the squares of the other two sides. In other words:

$$H^2 = L_1^2 + L_2^2.$$

To aid in the visualization, the figure is drawn inside a coordinate system of the three axes (the axes are three directions at ninety degrees to each other) labeled from left to right (moving in a counter-clockwise direction) as: X, Y, and Z. In this system, the triangle lies in just one plane of the coordinate sys-

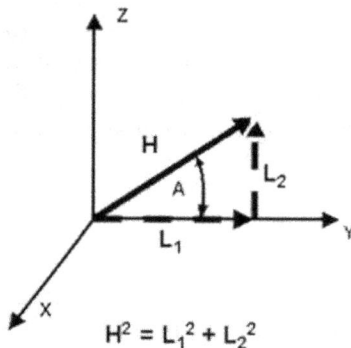

$$H^2 = L_1^2 + L_2^2$$

Figure 9.3 The Right Triangle

tem—that described by the Y-Z axes together. These coordinate axes are a convenient way to label the directions we see in our universe. In this case we use X as the forward-backward direction coming out of the page; Y is the left-right direction; and Z is the up-down direction. Only one quadrant of this universe is shown in the figure. The right triangle is called a *two-dimensional figure*, since its entirety occupies just two of our three spatial dimensions (Y and Z, in our case).

The dimensions of the sides of the triangle are expressed in one's measurement of choice; for example, inches, feet, miles, etc. The directions of the three sides can be described by their relationship to the X, Y, and Z axes.

In general, the sides, L_1 and L_2, can take on lengths varying from 0 to H. When the hypotenuse length changes, it causes the lengths of L_1 and L_2 to adjust accordingly; the angle between the hypotenuse and the base of the triangle, denoted by A, can vary from 0 degrees to 90 degrees.

Each side of the triangle can be used to construct a square. It is possible to construct one square that is H x H in area. It is possible to construct a second square that is L_1 x L_1 in area. Finally it is possible to construct a third square that is L_2 x L_2 in area. These areas (call them A_H, A_{L1}, and A_{L2}) are each equal to the square of the lengths of the sides of the triangle. Thus, each side of the triangle can be associated with an area that is equal to the square of each side's length (H^2, L_1^2 and L_2^2), respectively.

This picture helps us to understand something deep about this triangle. If you increase the length of one of the sides—vertical or base—but maintain the hypotenuse at a constant value, you are required to always also (and correspondingly) change the length of the third side (and the angles between the hypotenuse and the base or vertical side). If one holds H constant—and thus, holds its associated area, A_H, constant—you can scale the other sides around but only in a way that maintains the rule about the relationship of the lengths of the sides. Otherwise, you break the triangle and it ceases to be a triangle. From this picture, we can also now see that the Pythagorean theorem is really a statement about the areas of the squares associated with each side of the right triangle:

$$H^2 = L_1^2 + L_2^2 \rightarrow A_H = A_{L1} + A_{L2}$$

Pictorially, this relationship is shown in part (b) of Figure 9.4.

As these lengths change, their associated areas change with them. Let us change the color scheme of the figure so that the areas of the hypotenuse are differently colored, keeping the hypotenuse area solid black. Let us also adopt the generic notation for the areas associated with the base or vertical side: A_r will signify an area associated with each of the sides, with the subscript, r, taking on a value coincident with the particular side length.

Figure 9.4(a) shows only two areas: The black area is associated with the hypotenuse of the triangle. Its angle, relative to the Y-axis, is zero, in this case, so

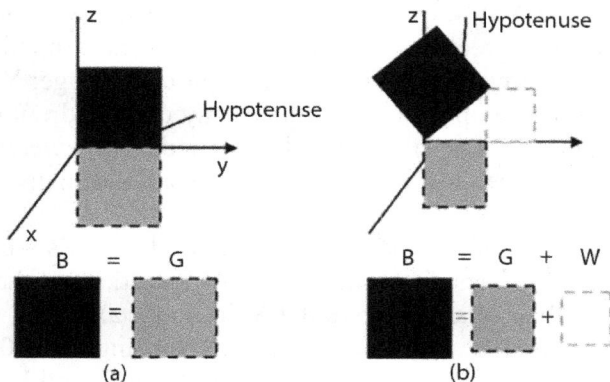

Figure 9.4 Pythagorean Hypotenuse with Areas

that the vertical line of the triangle is zero height. The black area is shown lying on top of the gray base of the triangle in (a) (L_1 in figure 9.3) for the case where the angle A between the two areas is zero. The vertical side of the triangle (L_2) has zero length in this case. These two areas are equal to each other as shown in 7.4(a). This result is denoted by the graphic above its triangle showing that the black area is equal to the gray area (G = B). Since the right end of the triangle is at the same height as the left end, there is no side with which to associate an area and, in this special case, the related area is zero. The hypotenuse area (black) is the principal area in this discussion. The black area will, in the continuing discussion, always be referred to as the hypotenuse.

When the angle of the hypotenuse is greater than zero, 7.4(b), three areas are shown. We see clearly here that the areas associated with the sides depends on the angle, A, between the base and the hypotenuse. As the angle on the left side of the triangle is allowed to change, the gray and white areas of the triangle change, correspondingly.

The area associated with the vertical side of the triangle is white and the area associated with its base side is gray. Note also, that as the angle of the hypotenuse changes, the areas of white and gray change with it, but the area of the black hypotenuse always remains the same. This is the key point, so we will emphasize it clearly:

The size of the black area will remain constant, while the other two sides of the triangle change.

That the black area does not change is the simplest example of what physicists mean by the word "invariant." There are many invariants in physics—these are numbers whose values in nature are robust against the conditions under which they are measured. The speed of light in empty space is an excellent example. Einstein's acceptance of experiments by such physicists as Albert Michelson and Edward Morley, as well as the consequences of Maxwell's equations, led him to treat light's speed in empty space as an invariant of nature. From this, he arrived at the correct description of space and time. Finding an invariant is a powerful tool in understanding the universe.

Let us return to our exercise. The hypotenuse, and its associated length (and thus the area associated with it) are taken as an invariant. Consequently, the equation below the triangle in Figure9.4 (b) shows that the black area (B) is always equal to the sum of the gray area (G) plus the white area (W). As the hypotenuse angle increases or decreases, these two areas change correspondingly so that their sum always adds up to the size of the black area.

The descriptions of the hypotenuse of the triangle and its sides have been used to describe the triangle's characteristics and may now be arranged to mimic other physical situations. If you can grasp the essentials of this exercise you are well on your way to understanding why some theories of nature, like string theory, seem to demand the existence of an extended hyperspace in which our 3-dimensional world is embedded.

Let's extend the discussion as if this were a game. Imagine that we take three squares; a blue one, a brown one, and a purple one. We can add up their areas and create a green square whose area is the sum of the areas of the other three. This situation is shown in Figure 9.5. It is important to keep in mind that, in the Pythagorean theorem that is taught in school, the area of the triangle itself is not part of the discussion. For this discussion, *what happens to the squares is what is important*.

Once the area of the green square is fixed we want it to stay invariant (unchanged). It is possible to do this by making the area of the blue square a little smaller while simultaneously increasing the area of the brown square and holding the area of the purple unchanged. Or we could increase the area of the

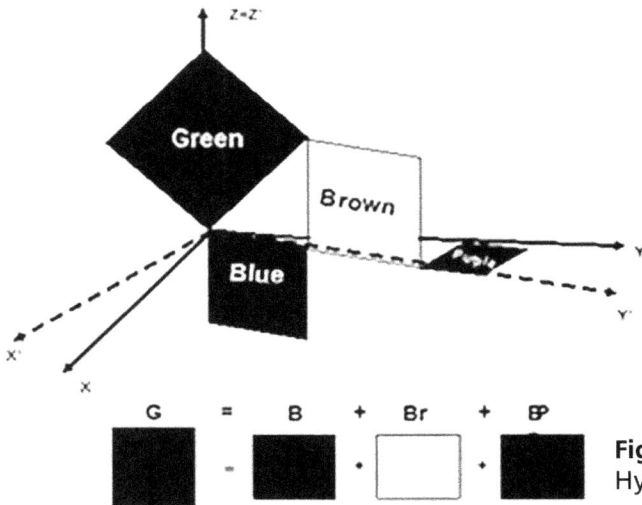

Figure 9.5 The Three-Dimensional Hypotenuse

purple square while simultaneously shrinking the areas of both the blue and brown squares enough to keep the green area fixed. It is probably obvious that there are a myriad ways to play this game, all the while keeping the area of the green square fixed.

It may not be obvious, but the "game" we just described with three squares whose areas can vary is actually *three dimensional trigonometry*, just as the previous game with two squares whose areas were allowed to vary is the usual *two dimensional geometry taught in high school!* This game of invariants is a powerful tool for stepping gently into a universe of extra dimensions.

This technique of adding the areas of 1 to N squares to form the area of associated with a hypotenuse can be extended to where N represents any number one chooses. We can write the general equation that applies to any situation describing the hypotenuse of the triangle for the static case shown in these figures. For more directions we simply add more terms that make the equation behave like the multi-directional version of the area associated with the hypotenuse and its triangle, by denoting them as areas (Ar_N) of squares (always excluding the area of the triangle in these applications), where N is a number defining each direction to which it pertains.

Let us take stock of the lesson by using mathematics and equations to express what we have learned. The Pythagorean theorem taught in high school, which is implicitly (or explicitly) applied on in 2 dimensions (Fig. 7.4b), can be represented using our area-color scheme as:

$$\text{BLACK} = \text{WHITE} + \text{GRAY}$$

and using our notation for the areas as:

$$H^2 = L_1{}^2 + L_2{}^2.$$

If we now consider our extended game, moving from the original 2 dimensions to the 3 dimensions of Fig. 9.5, we arrive at something quite similar:

$$\text{GREEN} = \text{BLUE} + \text{BROWN} + \text{PURPLE}$$

or, again, using mathematical notation:

$$H^2 = L_1{}^2 + L_2{}^2 + L_3{}^2.$$

These areas are designated as new area functions based on lengths; such as (L_N). For example, by squaring the lengths of each area with lengths L_1, L_2, L_3, etc. of the triangle, the above equation repeats the Pythagorean area equation. We can then generalize this to any number of directions, N:

$$H^2 = L_1{}^2 + L_2{}^2 + \ldots + L_N{}^2$$

or

$$Ar_H = Ar_1 + Ar_2 + Ar_3 + \ldots + Ar_N$$

This describes an N-directional configuration as the sum of N areas and extends this concept to include any directional set greater than three directions. Thus, for something like string theory, where the fewest number of total spatial dimensions is 10, the above would apply.

These equations describe all areas associated with length directions assigned to make up the hypotenuse for any static (non-motion) physics problem. We have not worried at all about what those extra directions *mean*—but we can quantify their impact on a mathematics problem without that knowledge. This is merely a way of generalizing the mathematical description of the problem.

Einstein intuitively understood that, in the problem of seeking to understand how light travels at the same speed in a vacuum for all observers, he needed to use the game we just played using four squares whose areas were allowed to vary. Now, this is not the way he thought about the problem (after all, he was a genius) but this exact form of the mathematics is buried in his equations. He made one critical change that we will get to later, and because of this change we call it the "Einstein hypotenuse" instead of the "Pythagorean hypotenuse."

Motion, the very thing that Einstein first concerned himself with when studying space and time, changes things from our static picture above. Specifically, the key word is "momentum." Everyone uses the word "momentum"—for example, you might say something like, "the Rockets have finally taking the *momentum* of the game." This is a different use of the word. Momentum, as used in this book, is a scientific term used when a mass or a body is in motion. When a body is in motion we know that it is in an Einstein realm; that Einstein's special relativity applies. Because the body is in the space-time frame of reference, and time is just another dimension, it must be folded into the mathematics. This, too, can be summarized by a kind of hypotenuse in a 4-dimensional space—3 space dimensions and 1 time dimension.

Einstein wrestled with the problem of motion, and how to relate space and time observations to each other. He discovered that motion has a way of tangling up the dimension of time with the dimension of space, but that no matter what, the speed of light could be held constant in the process. In motion, the speed of light is the "hypotenuse" of the problem—the invariant. Space and time—or, more to the point of motion, momentum and total energy of a moving body—are the sides of the many-dimensional triangle. If the speed of light is the invariant as space and time are tangled up by motion, then, when it comes to energy and momentum, what is the corresponding invariant? What is the "hypotenuse" of movement?

You might be tempted to rush ahead and write down something like this:

$$H^2 = (E/c)^2 + p_x^2 + p_y^2 + p_z^2,$$

where E is the total energy of a body in motion and p_x, p_y, and p_x are the amount of momentum of the moving body lying along the X, Y, and Z directions of space. But it is his hard work and the genius of Einstein that sets him apart from our rashness. Einstein recognized that the hypotenuse of motion, whether we are talking about space, time, and the speed of light, or momentum, energy, and something else, required a key change from the Pythagorean approach as we sketched it above:

His genius about nature and the constancy of the speed of light suggested that sometimes the areas of the squares are subtracted rather than added to obtain the invariant—an analog of the hypotenuse of the problem.

The equation we wrote becomes modified—the momentum components are *deducted* from the energy component as we arrive at the form that Einstein actually worked out:

$$H^2 = (E/c)^2 - p_x^2 - p_y^2 - p_z^2.$$

And what is this invariant? For energy and momentum, the corresponding invariant in nature is the mass of a body viewed in the frame in which it is not moving— the so-called "rest mass." Thus we arrive at the beautiful relationship that Einstein worked out between mass, energy, and momentum:

$$m^2c^2 = (E/c)^2 - p_x^2 - p_y^2 - p_z^2.$$

If one sets the momentum to zero—that is, makes the body lie at rest—we obtain the famous Einstein equation relating mass and energy of a body:

$$E = mc^2.$$

The Einstein hypotenuse was provided here as a tool to help non-physicists understand how theorists conceive of and calculate the properties of geometries when there are more than four dimensions in play. The theorists do not themselves use these analogies because the mathematics are clear to them. The use of mathematics enables solutions that are molded to fit into the standard model of particles and interactions, a model regarded by physicists as sacrosanct because of its unparalleled success in describing the quantum realm. Mathematics allows the analysis and solution of problems without requiring one to demonstrate the reality of extra-dimensionality while finding solutions that follow the rules.

Dimensions: Great or Small?

In our example of how to think about extra dimensions—as extra *directions*— we were glib about caring about the character of those dimensions. It was useful to empower you, the reader, with a basic tool for extending calculations into additional dimensions, without regard to their shape or size or even what it means to move about in such dimensions. Now, however, we will turn our attention to the nature of those dimensions. Are they always only small, so small that we haven't noticed them? Or, perhaps, is it possible that they are large but there is some good reason why we haven't detected them yet? Do either of these scenarios have consequences? As we shall see, this subject not only provides some definite clues to the implications of size, but allows modern physicists to predict new phenomena in nature that, if detected, might signal an entire hidden universe in which we are embedded.

The Einstein hypotenuse exercise was a fun mathematical exercise, but we never made it clear, beyond the story of Einstein's focus on motion, whether it was useful for some other purpose. A great challenge for theoretical physicists is to not get lost in elegant mathematics. By this it is meant that however esthetically pleasing mathematical formalism may be, the real challenge is to use the mathematics to make predictions that are, in principle, falsifiable (are able to be proven false by some plausible test). Many different approaches to this challenge have been taken. One which became the focus of intense study by a large segment of the particle theory community early in the new millennium was created by physicists Lisa Randall and Raman Sundrum. Their ideas marked a bold step into the twenty-first century.

Randall and Sundrum, like Nordstrom and Kaluza and Klein, were interested in extra dimensions and the physical problems that such dimensions might solve. Our earlier analogy of extra dimensions imagined a floor on which a ball could roll, seemingly locked in two dimensions. But in that analogy we created a tiny, undetectable third dimension (the pits in the floor). The work of Randall and Sundrum imagines not tiny dimensions, but large dimensions. In fact, their ideas imply the existence of spaces, like our 3-dimensional space and 1-dimensional time, occupying its own 3-dimensional surface embedded in a much higher-dimensional space—possibly containing other lower-dimensional surfaces like our own perceived universe. This is the universe of the "braneworld."

In the decade that followed the year 2000, Randall and Sundrum's paper on this work was the most highly cited paper in that arena. It made testable predictions for present and near-future experiments. Those experiments have been searching ever since. However, just as the predictions of supersymmetry have not been observed in experiments such as the Large Hadron Collider, predictions of braneworlds have also, so far, not been observed in nature. Although there has been a marked decrease in braneworld study among theoretical physicists, we still see it to be useful as a demonstration of the manner in which the field operates to make progress.

In her popular-science book, *Warped Passages,* Randall presented the braneworld model (known formally in the physics community as the Randall-Sundrum model) for non-physicists. This construct comes closest to describing the real-world behavior of our universe (as compared with competing ideas, such as the Horava-Witten model, a similar construct we will explore later).

The key concept underlying the "braneworld" view of extra dimensions is that of the arbitrary-dimensional membrane, or, simply, "brane," This concept also plays a central role in string theory, and, in fact, the concepts of Randall and Sundrum's original paper connects directly to those concepts in string theory because it was in 1999 that the paper was published. As we will soon see, "branes" come in all shapes and sizes; the braneworld takes a particular view of those shapes, sizes, and orientations in order to reproduce some of the features of the known universe, with testable consequences for experiments that serve up predictions having the potential to be falsifiable—or *revolutionary*, should they prove to be correct.

Brane games

Randall describes branes as "fully loaded" with particles and forces. "Brane" is verbal shorthand for "many-dimensional membrane," a geometrical object that occupies a space of fewer, or potentially many more, dimensions than the three space dimensions with which we are familiar. The activities of particles and forces, like those described in the standard model, can be shown as confined to a host brane. From the perspective of brane-bound particles, if it weren't for the existence of gravity or bulk particles with which they *can* interact, the world might as well consist only of the dimensions of this one brane. From the perspective of particles trapped on a specific brane, the weakness of forces like gravity might be due to the fact that gravity spends more of its time in the bulk, (the space within which branes exist) where it is strong, and leaks only weakly into our brane. As a result, not only might one come to understand the relative weakness of gravity, but perhaps that weakness is a sign of extra dimensions." An extension of string theory, the concept of the brane was put forth during the 1980s. At first it had no particular distinction. It was thought of as yet another behavior of a string. The concept of "p-branes" (where p denotes some arbitrary dimensionality) was developed independently by Paul Townsend, who studied them at Cambridge University in the 1980s. Because branes can contain forces and particles, they are a kind of sub-universe on which humans (and all other observed aspects of the cosmos) can reside.

The concept of a "p-brane" is generic. There is a special sub-class of these objects, called "D-branes," that were mathematically discovered and articulated in 1989 by Jin Dai, Rob Leigh, and Joe Polchinski at the University of Texas, and, independently, by Czech physicist, Petr Hořava. Unknown to any of them, even earlier, Warren Siegel (currently a professor of physics at the Yang Institute at the State University of New York), had proposed their existence. These geometric, many-dimensional structures had a role in solving "the problem of too many dimensions" that seemed to plague early efforts at a coherent string theory of nature. The "D" in "D-branes" is a reference to the special mathematical condition associated with these structures, known as a "Dirichlet boundary condition."

Peter Dirichlet (1805–1859) was a mathematician who articulated the first solutions to an equation known as "Laplace's equation" (named after the mathematician who discovered it). This class of solution applies widely to many problems. The solutions assume certain behaviors of their mathematical forms at the boundary of the problem, and thus take the name "Dirichlet boundary conditions." D-branes come loaded with this assumed condition, specifically that the strings that connect with such branes obey those conditions. This has extremely useful properties, as we will explore later in the string theory discussion.

D-branes, like the more general class of p-branes, are designated by the number of spatial dimensions they contain, from 0 to 9. That designation is assigned by naming them as "Dp-branes," where D is the type of brane (Dirichlet), and p designates the number of spatial dimensions on that brane. A D0-brane defines a point particle, a D1-brane is a string, a D2 brane is a 2-dimensional membrane, a D3 brane is a 3-dimensional brane volume, and so on. Branes are a more generic class than strings, which themselves can be thought of as a kind of brane.

According to Randall, particles (and even strings) that are bound to a brane are like a train bound to railroad tracks. A train can roll along the tracks but it cannot leave them. Similarly, particles in a brane can move around in all of the dimensions of their brane, but they cannot leave it. Branes are host to forces, such as electromagnetism, and particles, such as electrons. The constituents of matter and forces can have properties such as charge and mass (in part, a consequence of the Dirichlet nature of the brane). When there is more than one brane there can be more forces and particles. We humans, if branes are really an accurate description of the cosmos, appear to live our lives trapped on a D3-brane, experiencing the dimension of time as we live and age, enjoying all the benefits of the standard model and gravity without having yet become aware that there could be additional dimensions of space—and perhaps even additional branes with their own particles and forces distinct from our own.

Our brane traps the particles and forces of the standard model. If string theory is the correct description of nature, then our brane must also be host to a class of strings whose vibrations result in the observed forces and matter particles. We tease this idea now, to whet your appetite for later chapters of the book.

Figure 9.6 illustrates the idea of a brane, with its own particles and forces, separated by "bulk dimensions" (those extra dimensions beyond the three on our own brane). We live on a brane in this universe. Open strings can connect from one brane to another. Interestingly, gravity might be the only force we presently know about that also travels through the bulk space between branes.

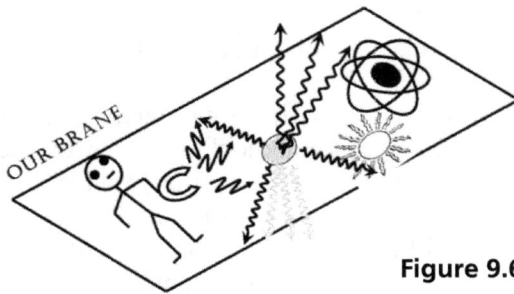

Figure 9.6

Perhaps this is what makes gravity so apparently weak on our brane in comparison to the forces described by the standard model.

The graviton, the quantum gravity carrier, is a gauge boson that can only reside on a closed string, as will be made clearer later. Open strings can start and end on a brane, but closed strings can only reside in the bulk space. Therefore, the hierarchy problem consigns the graviton to reside in the bulk, located, generally, close to a brane, imparting its force-carrier characteristics to the brane and to the bulk space by "leaking" into all the dimensions in both places.

Branes are not passive players in this view of the cosmos. Like the idea of strings that preceded braneworld models, branes, too, have tension. For strings, tension determines the types of particles it produces. The higher the string tension the heavier are its particles. The lower the tension a string has, the more flexible the string becomes and the more sensitive they are to the forces they possess. They can be taut or they can have flexibility, just as would the surface of a trampoline. When taut, branes tend to stand still, but when they are flexible, they can be pushed and will respond in kind to their charges and forces while producing interactions. This is in much the same way that one experiences the reactions on a trampoline. Branes should also enjoy the rich possibility of phenomena influenced by their inherent tension.

The bulk space between the familiar brane we live on, and other branes, is not empty. The bulk space, too, can be described as a brane. The bulk brane can have more than three dimensions so long as the bulk contains at least one more dimension than the brane we live on. In the braneworld model we are developing here, these bulk dimensions are not constrained to be very small or very large; it can be assumed that the higher dimensions are curled up in small volumes at each point in our space, or they can be very large (relative to particle sizes). In either case, these extra dimensions could have remained hidden from discovery by earlier experiments conducted on our own brane.

Communication between branes is not impossible, but if it were happening all the time and out in the open we should have noticed that by now. Instead, let's think about the communication between branes in a braneworld using the following analogy.

Consider a basketball tournament where four teams are playing in simultaneous semi-final games. The winner of one pair of teams will go on to play the winner of the other; thus, each pair of teams has a deep interest in how the other game in the tournament is going. If the two games were being played side-by-side on adjacent courts, then it would be possible for each pair of teams to see, in real time, the progress of the game on the other court. This would be like direct communication by strong forces (those in the standard model) between two branes; the denizens of each brane would be directly aware of each other all the time.

But instead, it's more typical that the two games are being played in entirely different geographic locations (e.g. different high school gyms or different arenas). The players have no direct information about what is going on in the other arena. But players on the bench in one game might send text message updates to their player-friends also resting on the bench in the other tournament. This is a weak form of communication, a leaking of information between the courts that are separated by a vast distance that prevents strong, face-to-face communication. A few characters of text message will suffice to send status updates about each game between the players. Most people, fans, coaches, and other team members alike in the arena, won't even be aware that this communication is occurring, unless they are closely scrutinizing their favorite players on the bench and notice that they are texting.

This kind of weak and hard-to-notice communication is analogous to the kind of universe we think braneworlds describe—one like our own universe. Branes exist (the basketball courts) that are separated by bulk space (different geographic locations in the higher dimensions of the universe). While direct communication is not possible, perhaps there are ways to transmit information through the bulk; e.g., via gravitational waves, taking advantage of gravity's possible ability to move in the bulk, and pass messages between branes. In this manner, branes can get information from each other about interactions; that information manifesting itself as behaviors on their own branes.

We see that this braneworld idea provides a hypothesis. A universe is more than just the 3 space and 1 time dimensions we know for certain are real. The existence of a higher-dimensional "braneworld" helps explain why the forces of the standard model are so strong in comparison to gravity. This model has many consequences and, expressed in the detailed mathematics developed over more than a decade, has potentially measurable outcomes. Observing these outcomes would offer proof in favor of this hypothesis. Failing to observe them means that this model is wrong, or is oversimplified, or is not yet accessible, given the present tools. Let us look at the predicted consequences of such models in a running experiment, the Large Hadron Collider.

A Fingerprint of Extra Dimensions

If we start with development of the S-matrix during the 1940s, the mathematical behavior of strings and the related constellation of geometric phenomena has been under investigation for more than seventy years,. During this period, string theorists have consistently demonstrated that using more than three dimensions makes it easier to describe strings and even solves some of the early, pernicious mathematical difficulties in the theory. But no tests yet have confirmed this extra-dimensional behavior because the extremely small dimensional characteristics that strings exhibit cannot be measured with present equipment. Such tests would require extremely large energies to detect them, likely very close to the Planck energy. While some string behaviors may be revealed in collider experiments in the next five to ten years, the next decade or so may be sufficient to establish whether extra dimensions *do* exist. This would be a very important revelation lending credence to the viability of string theory as an important mathematical construct of the universe. If there are extra dimensions, they are sure to leave an imprint during an experiment—something akin to the appearing-disappearing disks of Flatland we discussed earlier, the fingerprints of higher dimensions left on a low-dimension space.

String theories appear to demand additional dimensions beyond the ones we already know. Physics experiments have not ruled out the existence of such extra dimensions. We suggested earlier that the Large Hadron Collider is a prime tool for hunting these extra dimensions. How can this be? So far, discussion of extra dimensions has assumed that they cannot be seen directly. They are unobservable because they are stated to be compactified into extremely small surfaces (like Calabi-Yau manifolds) everywhere in space, or contain our D3-brane where the standard model lies trapped. They could be so small that they are not visible with even the strongest of instruments, or large, but inaccessible to the particles of our D3-brane. But even were these extra dimensions to be *that* invisible there might be some residual effect that would present itself in an experiment that would indicate the existence of those higher dimensions. Indeed, the Randall-Sundrum model reveals the fingerprints of such a universe.

The fingerprints would be a new class of yet-undiscovered particles called Kaluza-Klein (KK) particles, in honor of that earlier and pioneering work by Theodor Kaluza and Oskar Klein. The energies of proton collisions at the Large Hadron Collider provide a new opportunity to reveal the existence of KK particles. KK Particles are the equivalent of the appearing-disappearing disks in Flatland. The physicists at the LHC should be able to spot them in the debris of proton-proton collisions. Let us understand what these fingerprints look like and why they are likely to occur in a higher-dimensional world. Extra-dimensional clues will help to piece together the messages physicists get from collisions, decoding them, seeking to determine whether higher dimensions are really there.

Many models of nature predict a new "zoo" of subatomic particles that could manifest at the LHC. Other theories of nature, like supersymmetry, also predict a zoo of new particles that could be found by the LHC. What distinguishes a zoo of KK particles from that of supersymmetric particles? Finding a bunch of new particles is not, by itself, proof of extra dimensions. Let us look more closely at the predictions of KK particle properties, to see what special pattern they should exhibit."

In suggesting how branes and particles might communicate their characteristics to each other when separated, it was suggested that two separated branes might influence each other through messenger particles traveling in the bulk; a brane is able to be influenced by another brane, although very weakly. Recall the analogy of the two basketball semi-final games in which no direct observation between the two games is possible, but where weak messages (telephoned text messages) between a few players in each semi-final game allows for information to be passed between them. You have to search hard and know what the weak communication will look like.

But how weak would such communication be? If communication between branes is possible, then it will happen. This is very much the spirit of the quantum world, where even unlikely things are possible and a kind of "anarchic principle" reigns. If it can happen, it will happen. How could brane-to-brane communication be permitted without it being so obvious that we should have already noted it?

While it is possible to consider Randall-Sundrum models with supersymmetry, Randall and Sundrum also considered researching the breaking of the universe's supersymmetry and what that might do for understanding our D3-brane. This they achieved by using sequestering as the means of preventing unwanted influences of the supersymmetric fraternal twin particles on the particles in our D3-brane. After all, no experiments to date have detected the long-sought fraternal twin particles, the supersymmetric shadows of the standard model. Perhaps that is a feature of a braneworld, rather than a problem.

Inasmuch as we have not seen the supersymmetric twin particles, if supersymmetry is real in nature, we can say that it is "broken"—that it is a symmetry that exists but is not demonstrated in nature now. Randall and Sundrum's notion of sequestering some particles to one brane and others to a different brane might explain why supersymmetry appears broken on our D3-brane. The standard model particles are confined to our brane; their superpartners are confined to a different brane. Communication between the branes is weak, at best. But there must have been some communication in order for that symmetry breaking—that sequestration—to have occurred. It could have happened at a much earlier period of the universe, perhaps just a billionth of a billionth of a second after the big bang, at a time when the universe was much smaller and far hotter than it is today.

This kind of effect caused Randall and Sundrum to consider models in which, during a supersymmetry-breaking process, they sequestered particles from our Standard Model D3-brane. Super-partner particles are stuck on a nearby brane, one that is hidden or sequestered. These two branes are separated by a fifth space-time dimension (four spatial dimensions plus one time dimension). Any particles that could influence supersymmetry-breaking would reside on the sequestered brane. This would help explain why we have been so far unable to detect superpartners. This is one of many examples of problems that sequestering can help resolve, and motivates the existence of at least two branes separated by a bulk higher dimensional space.

But the solution to the missing superpartners has consequences. If there are extra dimensions, this might influence the behavior of known particles, those familiar to us from previous experiments. The question then becomes, what are the expressed behaviors that we can expect?

We said earlier that if a brane has (3+1) dimensions, the bulk space must be of a higher dimension, meaning that it must have at least four spatial dimensions. If there are higher dimensions, how can one determine that they are there, and to what degree can they be evaluated? Just as one can mathematically reproduce any sound characteristic that a violin string can make by superimposing its many resonant modes, one can produce higher-dimensional particle-behavior by replacing them with appropriate Kaluza-Klein (KK) particles.

KK particles can fully characterize higher-dimensional particles and the higher-dimensional geometry in which they travel. In order to mimic the behavior of higher-dimensional particles, KK particles need to carry extra-dimensional momentum. Each bulk particle that travels through a higher-dimensional space is replaced in our four-dimensional description by a KK particle having the correct momentum and interactions that mimic the higher-dimensional object. A higher-dimensional universe hosts both four-dimensional particles and their higher-dimensioned KK relatives.

The relationship between mass and momentum imposed by special relativity tells physicists that extra-dimensional momentum would be exhibited in our world as additional mass. KK particles, then, are particles like the ones we know in the standard model, but have heavier mass that reflects their extra-dimensional momenta. Particles such as electrons and quarks, therefore, would show up as copies of themselves in a collision, but have heavier masses than those found in lower energy experiments.

Where does such extra mass come from? To understand the origin and mass of KK particles, you must move away from the intuited picture of tiny, ineffectual, curled-up dimensions as have previously been described. It is the extra dimensions themselves that lead to this extra mass, and more specifically how the original particle and the extra dimensions interact with each other. Particles that can travel in higher dimensions have additional ways of moving, and that means additional momentum—momentum that, in the 3+1 dimensions where the standard model lives, would manifest as extra mass. This was hinted at in the example of the "Einstein hypotenuse"—extending momentum into additional dimensions has consequences for the mass-energy relationship of special relativity.

Other familiar properties of particles in the standard model, such as charge, would not be changed by this picture. We would only note that new particles appear with masses that are greater than those in the standard model, but be otherwise similar to the standard model particles we already know. The key observable detail that is revealed by this model of the universe is that for each particle we know about in our four-dimensional world, there should be a KK particle with the same charge, although each would carry a greater mass. This greater mass would indicate that they exist simultaneously in an extra-dimensional environment.

The reader might pause here and ask, "Wait, why can't the KK states of the electron, for instance, have any mass that they want to have?" Remember that we are dealing with the quantum realm and, as in the energies possessed by electrons orbiting the nucleus of an atom, not all conceivable energies are allowed. For KK particles, the mass is determined by how many wavelengths of the particle's wave function can be fitted into the extra dimension(s). This, in turn, is linked intimately to the geometry of those dimensions. It's just as in violin string vibrations: only the resonant modes of vibration can manifest, and the fundamental note will be determined by the length of the string. Quantized extra-dimensional momenta of KK particles depend on extra-dimensional size and shape; only the quantized states—integer numbers of wave crests—will result in the extra overall momentum and thus the apparent extra mass from the perspective of our D3-brane.

You might already have found some additional value in this analogy. If we can know the spectrum of notes—the fundamental and the harmonics—that are possible to emanate from an unknown-length violin string, it is possible to determine the properties of that violin string without directly measuring the string! Likewise, as has been pointed out by Randall and Sundrum, if we observe a series of new particles at the LHC, having a regular and predictable pattern of masses starting out from some fundamental mass (e.g. the electron mass), then, if the cause of this pattern is extra dimensions, physicists should be able to work out the properties (the number and shape) of those extra dimensions.

Such an approach would be similar to the one used by Niels Bohr (1885–1962) to determine quantized electron orbits in the atom. Bohr, like every other physicist of his day (and decades to come), would never lay eyes on an electron in orbit around an atom, but they could see what the electrons were doing around that atom by measuring a set of properties of the atom. These were the atomic spectra—the patterns of light and dark emitted by atoms when they are stimulated by an external energy source. These patterns are the fingerprints of the quantized orbits of electrons zipping around a nucleus. The mathematics of particles moving in extra dimensions leads to a similar distinct pattern, not in light emitted by something, but in the system of masses that result as the wave function of a particle vibrates—once, twice, three, or any number of integer times in an extra dimension. Even if there was just one circular extra dimension, such behavior could occur. There are no half-measures in the quantum world. A wave function vibrates once or twice, but never one-and-a-half times.

Such models make a further prediction: that vibration in an extra dimension requires energy. If you want the wave function of, say, an electron to vibrate even once in an extra dimension, you need to give it a minimum amount of energy that makes such behavior possible. If we want to encourage the wave function to move around in extra dimensions, we have to put a lot of energy into particles to get them to go there in the first place. A tool like a particle accelerator might be a perfect means to do this, and why there is such excitement about searching for new particles at the LHC. The LHC develops the most energy that humans have ever been able to put into particles, and even if causing a particle to wrap around an extra dimension is a rare event, the LHC has the intensity to reveal such rarities in the universe.

If the universe does contain additional dimensions, and particle accelerators achieve sufficiently high energies, they will produce these heavier particles. Heavier particles that carry nonzero extra-dimensional momenta will be the first real evidence of extra dimensions. In our example of even just a fifth dimension, those heavier particles are associated with waves that have structure along the circular additional dimension.

The lightest KK particle is one that vibrates exactly once in the circumference of the extra dimension. We've made the point a few times in this book that higher energies correspond to shorter distances. This is also true about moving in extra dimensions. The smaller the physical size of the extra dimension, the higher the momentum of the resulting KK state. (Think about wavelengths of light that are shorter, such as ultraviolet. These are more energetic than those having longer wavelengths, such as radio.) The more times a wave function vibrates in an extra dimension, the higher its momentum and thus the more mass it appears to have in our D3-brane. This also means that the smaller the extra dimensions are, the more energy we have to put into the particle to make it move in the extra dimension. That means that if we fail to observe such a ladder of massive particles at the LHC, all we can do is infer that the extra dimensions, if they exist at all, are smaller than we are able to probe with the energies of the LHC. It would not rule out the idea of extra dimensions, but it might make the idea less attractive to solving the known problems of the day.

Wrap Up

Mathematicians and physicists in the 1800s and early 1900s set us on a path to the frontiers of human knowledge, along roads leading into this strange space filled with ideas. They sought a means to describe, in a general mathematical sense, what it means for there to be extra dimensions of space. Physicists recognized implications for the unification of known forces as early as the early 1900s, although, when more forces were discovered (the nuclear forces), this effort was abandoned to explore the more fruitful quantum field theory approach with its gauge symmetries and broken symmetry.

However, the success of the standard model leaves many questions unanswered. Why do the matter and force particles in the model have a strange pattern of masses? Why are the electroweak and strong forces so different in character than the gravitational force, and why is there such a vast gulf that seems to separate them? If other, more general models of nature, such as supersymmetry and string theory, are correct descriptions of the universe, why are their consequences not more manifest in our observations of the cosmos?

Could it be that the allegorical cave wall, on which we have been watching the shadows of reality dance, is not only a host to those shadows, but, being multiplied in more ways than we can tell, is casting some shadows of its own? Extra dimensions may help to reconcile some of the big questions left from the successes of the standard model. Postulating that our known universe, with its 3 space and 1 time dimension, are embedded in a much higher-dimensional universe begins to shine strange light on these questions. The idea of a hidden universe, just next door in an extra dimension, adds a depth of wonder and awe to our understanding of the cosmos. But is it the *correct* idea, and how would we even know?

Models of extra dimensions, like the Randall-Sundrum "braneworld" model, offer not only interesting reasons for the observed universe, but testable consequences that can be assessed with existing instrumentation such as the LHC. If LHC physicists happen to observe a zoo of new particles with heavy masses, those masses separated by a regular and predictable pattern, this pattern might be the fingerprint of the standard model's particle wave functions "leaking" in quantum ways into extra dimensions. Such a pattern, much like the atomic light emission spectrum, would help us to chart those dimensions even though we cannot just open a door and walk into them.

There are even more ideas than these out there—ideas that portend equally compelling explanations for why the standard model is the way it is and answer some of those questions left open by the standard model. These other ideas could intersect with the braneworld idea, but provide a construction different from the one envisioned in the Randall-Sundrum model.

There is much to explore. Let's continue our journey.

## Chapter 10

The Shadows
are Multiplied

There is a very specific problem in the standard model, one that interferes with its otherwise beautiful reputation for uniting the electromagnetic, weak, and nuclear forces—they don't actually unite. To be clear, electromagnetism and the weak force absolutely *do* unite. Nothing demonstrates that more beautifully than the photon and the Z boson. In many ways, these two particles are very much alike. They are spin-1 vector bosons. Neither possesses electric charge, but they are dissimilar in that the photon has no mass while the Z boson does. However, this poses no problem! Recalling the discussion of the Higgs boson, this comes about because there is a broken symmetry in the universe, specifically in the vacuum state of the universe. Cranking up the energy restores that symmetry, and so it should be (and is, according to experimental evidence) that, as one raises the temperature of the universe, the photon and the Z boson become more and more indistinguishable; they become but two aspects of a single, spin-1, neutral quantum field. This is the beating heart of the electroweak force.

So what about the gluons, the force-carrying particles of the strong interaction? One would expect that as one cranks the energy up even more, making quarks freer to move and thus weakening the strong interaction, it, too, would become comparable in strength to the electroweak interaction. Then, would not the three forces all meet somewhere? But, in the standard model, they miss. As though it were the mournful couplets of some Shakespearean tragedy, the standard model seems a cruel tale of lovers separated for 13.8 billion years, never to be reunited in symmetric bliss, ever only to come teasingly close to this reunification.

This is one of the most frustrating aspects of the standard model. You can almost forgive it for not being able to explain the pattern of the masses of its particles. You can almost forgive it for not explaining dark matter. But unification was to have been its job. The promise of gauge symmetry and quantum field theory to unite disparate forces under a single umbrella was long the expectation. And yet, it doesn't quite do its job.

You can see why finding a new symmetry of nature, one that would fix this irritating problem, would prove to be a great temptation. Such a symmetry has been proposed, and it comes with the superlative name, "supersymmetry." We've mentioned it before in passing. In this chapter, we will focus on supersymmetry, which we will refer to in shorthand as "SUSY" (the jargon name used by particle physicists for supersymmetry and its constellation of associated ideas). If that symmetry results in multiplying the problem of there being "too many particles" in the universe, you might consider forgiving it for its ability to help with the problem of unification in the standard model.

Doubling the number of particles is considered. . .well. . .an aesthetic sin. Physicists, in their desire for a simpler universe, want there to be one particle (such as a superstring) that explains *everything*. SUSY explains *more* things, but, in doing so at least doubles the number of particles in nature. In doing this, some physicists are willing to trade simplicity in the number of fundamental particles in order to solve some problems in the standard model. It's a difficult trade. Does it help? (Keep in mind that nature does not require that nature be simple. That's an aesthetic bias that may turn out to be wrong when we finally figure out the real "theory of everything.")

We'll explore the origins of and expectations for SUSY. We'll look at how experiments continue to search for evidence of its existence. We'll see why the relatively low mass of the recently discovered Higgs boson made SUSY theorists very excited in 2012 and 2013. And, we'll explore how this idea might ~~be seen to~~ interact with other ideas about the universe such as extra dimensions, string theory, and whatever else might be next for this long-searched-for theory of nature.

The Tragedy of the Standard Model

In addition to the problem of unification, there is a technical problem about the standard model that SUSY elegantly swoops in to save—the Higgs boson tends to want to run out of control.

As we said in our earlier discussion of the Higgs boson, the standard model makes no prediction about the actual mass of the Higgs. However, the standard model can tell us how the apparent inertia of the Higgs boson is affected when it interacts with other particles such as top quarks, bottom quarks, or weak bosons. Recall the analogy about the Higgs field and the Higgs boson. The field is like a room full of physics enthusiasts awaiting the arrival of Stephen Hawking and the Higgs boson is like a rumor spreading through that room. When Dr. Hawking enters, the enthusiasts crowd around him, slowing his progress and making his inertia appear larger, the crowd causing him to have a bigger mass. The Higgs boson is like a traveling rumor. A rumor has no physical form, but in learning that Dr. Hawking is about to enter the room, the people near the door clump, share the rumor, and pass it to people farther from the door. The rumor gains physicality in the process. The act of clumping to share the rumor and unclumping after the rumor is passed makes it appear that a knot of animation is moving through the room. This is like the Higgs boson. The speed at which the rumor travels through the room, *unaffected* by other possible interactions of the crowd represents its inherent inertia—what we would call its "rest mass" in the language of special relativity.

This analogy can also be used to understand how other particles in the standard model can affect the apparent inertia of the Higgs. If the standard model is a reliable theory of nature, we should find that such effects exist but are finite and well-regulated by the theory—that no infinities should result. Consider the analogy again. The rumor of Dr. Hawking's arrival is traveling through the crowd when some among the group of people who are passing the rumor realize that they are standing next to Lisa Randall, another famous physicist. That smaller part of the group is distracted as they greet Dr. Randall excitedly, but, in a flash, off she goes to see if she can find Dr. Hawking. The knot of animation resumes as the rumor of Dr. Hawking's imminent arrival continues to be passed along, but the rumor's spread has been slowed by its interaction with the other famous physicist who was suddenly there and immediately gone. In this analogy, we see how the Higgs boson, described by its inherent rest mass, can be expected to appear to be further slowed in its journey through space. In the language of quantum physics, when virtual particles pop into existence, have an interaction with the Higgs (thus adding to its apparent inertia), only to disappear again, the Higgs will continue on its way but is slowed briefly, making its apparent mass seem larger than its rest mass.

In the language of quantum physics, this is called a mass correction. All known particles obtain such corrections to their mass in the standard model. However, with the exception of the Higgs boson, all other particles have corrections that settle down to stable numbers—ones we can measure in the laboratory. Using the standard model one can calculate what the apparent inertia of a Higgs boson will be if it interacts with another particle. The answer is not useful. When interacting with the virtual particles in the standard model, the inertia of the Higgs shoots off to infinity.

That can't be right. We know from observing the Higgs boson that its inertia, its mass, is *not* infinite. It's a reasonable 125 GeV/c^2, a bit heavier than the W or Z boson and lighter than the top quark. It's not too heavy, and not too light. What is it that maintains this balance? What prevents the Higgs' interaction with other particles described in the standard model from describing trouble for the universe?

Theoretical physicists have wrestled with this for a long time. It turns out that there are mathematics that one can apply to the standard model to address the problem. One can hand tune the mathematical calculations to prevent the Higgs' inertia from climbing to infinity. But this is problematic! When a theory of nature has to be hand tuned by a theorist to make sure it does its job by making sensible predictions, you have evidence that the theory is flawed. This is known as a fine-tuning problem that signals that you have hit the limit in the applicability of a theory. It tells that there must be some more general theory somewhere—one that supports present knowledge while providing a mechanism that resolves the fine-tuning problem.

You can see why physicists sometimes feel strongly that there must be something just around the corner to ride in and save the universe from these unruly actors in the standard model. Many ideas have arisen over the decades. Perhaps the Higgs boson is not a fundamental particle, as it is assumed to be in the standard model. Perhaps it's something else, some new substructure yet to be discovered. That would help solve this inertia problem. Perhaps there are other kinds of as-yet-unknown particles that act to cancel out the runaway interactions we've described. That could also solve the problem.

Let's take a look at one of these ideas, one that has been around for many decades and has been alternately touted as a savior and a sinner: SUSY.

The Twinning of Particles

It should be stated at the outset that the standard model inconsistency discussed above was *not* the motivation for the development of SUSY. Supersymmetry is an independently conceived physical notion, the utility of which for solving aspects of the hierarchy problem cannot be denied.

SUSY is a symmetry that interchanges fermions and bosons. Quantum field theory is built, in part, using operations that exchange things for their opposites or counterparts, so you can expect how SUSY might have emerged from that type of framework. For instance, there are operations in quantum field theory that swap left and right, up and down, and forward and backward. They swap particles and antiparticles and swap forward in time for backward in time. These operations come with symbols to make it easier to talk about them. Reversing coordinate axes so that front becomes back is called parity transformation, and is denoted by the symbol P. Swapping particles and antiparticles is called charge conjugation, denoted C. Swapping time's forward axis and its backward axis is called time reversal, which is denoted T.

What is simplistic about the quantum world and its mathematical representation is that there is nothing that prevents any of these operations from happening. It is absolutely okay, from a quantum perspective, to replace all the particles with antiparticles or reverse the axis of time.

What is interesting, and what field theories can help us to understand, is whether or not there are consequences to these actions. A physical process that is invariant under any of these reversals—C, P, or T—is said to respect a symmetry associated with these operations. For example, if one replaces all the particles in the universe with antiparticles (or vice versa) and nothing noticeable changes about the appearance of the universe, then the universe is said to be invariant under the C operation.

Physicists have been studying such symmetries of nature for a long time. It was assumed early on that all physical reactions in nature were invariant under P—that the universe wouldn't report a difference if you flipped all coordinate directions in the cosmos. However, a brilliant experimental physicist named Chien-Shiung Wu (1912–1997) did an experiment in 1956 using a radioactively decaying cobalt nucleus to show that nature, does in fact, know the difference if you reverse all of its coordinate directions.

Let that sink in for a second. There are physical reactions in nature that, if our universe was reversed left-to-right, front-to-back, and up with down, would look different from what they did the day before! You can tell the difference between a universe with our parity settings and one that is parity-reversed. The violation of P symmetry is a cornerstone of the weak nuclear interaction. Its violation was fundamental in the construction of the standard model in the 1960s.

Let's see if we can make this seem more real. Let us imagine that, before going to bed you look up at a clock on the wall of your bedroom. It has two hands, one to mark the hour and one to mark the minute. They point at numbers on the face of the clock—12 at the top, 3 at the right, 6 at the bottom, and 9 on the left. You observe that, as the minute hand moves, it does so in what we call "clockwise" fashion—from the small numbers to larger numbers. You go to sleep. During the night, an impish deity with great powers to change the underlying structure of the universe reverses up and down, left and right, and front and back. In physics, we call this a "parity transformation." When you wake in the morning, will you notice this change?

Newton's Laws, and, in general, the laws of the macro world, say that you will not. Under those laws, a transformation of parity will leave the observable features of the universe unchanged. When you look at the clock, the minute hand will still travel around the face of the clock from the small numbers to the larger numbers. But, what if the laws of the universe were not so unchanging under a parity transformation? What if they could "feel" the change? This would be akin to waking up, looking at the clock, and realizing that the minute hand is now running backward—from larger numbers to smaller ones. How weird this would be. Is it possible to detect such a change?

The answer turns out to be yes, but one must look to the quantum world to see the change. It was experimental physicist C. S. Wu, and her observation of the decay of cobalt nuclei, who revealed that the weak nuclear force is sensitive to the parity of the universe, and can tell when parity is transformed. Indeed, this preference for orientation in space is a cornerstone of the weak interaction, and affects the way in which weak bosons, like the W boson, interact with particles like electrons. If the electron's spin is oriented too much along the direction of its motion, the W boson's interaction is less likely than if the spin is oriented the other way. That may seem strange, but it's a confirmed feature of the weak interaction. This represents a fundamental violation of parity symmetry in nature. Nature knows a normal universe from a mirrored universe.

Similarly, other symmetries in nature have been observed to be violated. The violation of charge conjugation, C, is observed in the interactions of neutrino particles. While the combination of C and P, called CP, was believed to be respected in nature, the violation of this symmetry, too, was observed in 1964 by the experimental physicists James Cronin (1931–2016) and Val Fitch (1923–2015). CP violation plays a key role in why the universe is predominantly matter, and not equal parts matter and antimatter because the weak interaction favors the production of matter over antimatter, contributing to a "tipping of the balance" in favor of matter, at the birth of the cosmos.

The violation of T symmetry is much more difficult to observe, but it's not impossible!. A first observation of this phenomenon was reported in the behavior of kaons (relatives of pions constituted from a strange quark and either an up or a down quark) during an experiment called CPLEAR at CERN in 1998, although within the particle physics community there has been scientific debate about the interpretation of these observations. An independent observation of T violation was made in 2012 using, not kaons, but a heavy cousin of theirs, B mesons (containing bottom quarks in place of the strange quarks of the kaon). This observation was reported by physicists on the BaBar Experiment at the SLAC National Accelerator Laboratory and had great statistical significance. It is safe to say that nature supports many of these symmetries—and violations of these symmetries.

Quantum field theories have a core tenet that makes them viable—*thou shalt not violate the combination of C and P and T symmetry (CPT)*. Were all three together to be violated, the quantum field framework would cease to make useful predictions and would have to be replaced with something else. To date, no CPT violations have been seen.

We've seen that there are discrete symmetries in nature. Are C and P and T all of them? No. There are other discrete quantities in nature that can be swapped around, one of them related to spin-angular momentum.

Recall that spin is what distinguishes the fermion class of particles from the boson class of particles. In the standard model, fermions are associated with matter (quarks and leptons) while bosons are associated with forces (photon, gluons, W, Z, and Higgs). Matter and forces are separately represented by their own class of spin-carrying particles. Fermions have half-integer spin, while bosons have integer spin. This is a situation akin to reversing directions or swapping matter and antimatter. Perhaps it, too, seeks an associated symmetry.

What is now SUSY was first conceived in the 1970s by a number of theorists working in small groups, independently of each other. Their work was entirely in the context of quantum field theories, which (with the rise of the standard model) was a hot topic during that time. Jean-Loupe Gervais and Bunji Sakita, as well as Yuri Gol'fand and Evgeny Likhtman, along with Vladimir Akulov and Dimitrii Volkov, hit upon what emerged to become SUSY during 1971 and 1972. Each pair was studying a specific aspect of quantum field theory when they found something more fundamental about the symmetry of nature.

Gol'fand and Likhtman were working to introduce P violation into quantum field theory. A paper by Akulov and Volkov sought to determine whether there could be Goldstone fermions in addition to the Goldstone bosons we discussed earlier. In both cases, each pair of physicists hit on a central idea, the symmetry of nature reflected in special relativity might be more encompassing than originally believed.

Among its other statements, Einstein's theory of special relativity is a theory of invariants. The speed of light is an invariant. The rest-mass of a particle is an invariant. Space and time measurements are altered by relative motion, but there is an anchor, an invariant. to which one can reliably turn. This allows observations to be related among different frames of reference, even when space and time get themselves tangled up by motion. There is a symmetry at work in special relativity that protects the invariants. That symmetry is described by a group of transformations known as the Poincaré group.

Henri Poincaré (1854–1912) was born in France, his father a professor of medicine and his family generally wealthy and influential. His mathematics teacher called him the "monster of mathematics" due to his prowess in this subject. Late in his life, around 1904–1905, as Einstein was about to have his "miracle year," Poincaré corresponded with Hendrik Lorentz about a space-time transformation that Lorentz had developed seeking to understand the motion of bodies through space. This transformation, the Lorentz transformation, was to become a cornerstone of special relativity.

Poincaré wrote Lorentz about an error he found in one of Lorentz's papers, a paper that Poincaré felt to be of great importance in understanding space and time using mathematics. Lorentz sought to reconcile the behavior of the laws of electromagnetism and the laws of mechanics under the case of relative motion, and, in doing so, derived the Lorentz transformation. Poincaré pointed out that Lorentz's treatment of the way time was transformed in these equations had a greater meaning. It made the Lorentz transformation part of a mathematical object known as a "group." Groups are special in mathematics, they have a set of well-defined properties and are firmly tied to mathematical symmetry. In 1906, Poincaré noted that combinations of space and time coordinates, akin to the Einstein hypotenuse we introduced earlier, are invariant. The group of transformations, of which the Lorentz transformation is a part, became known as the Poincaré group in honor of Poincaré.

The Poincaré group includes transformations that allow certain properties to remain invariant under transformations of systems in space: translations (sliding the entire system along a coordinate axis); rotations (changing coordinate axes by turning the system around), and boosts (changing the system's velocity). Although any degree of translation, rotation, or boost is possible, each can be described by a series of small transformations. One can approximate any chosen translation, rotation, or boost by very small, finite, step-wise translations, or rotations, or boosts.

Theoretical physicists of the early 1970s were exploring transformations, and whether or not the Poincaré group was complete—whether *all* transformations akin to translations, rotations, and boosts were described within the group. It was realized that there was another transformation, one that, when it was added to the Poincaré group, resulted in a kind of super Poincaré group. The new transformation emerged from the question of whether fermions and bosons can be swapped (by swapping spin-angular momentum quantum numbers). This mathematical exploration has consequences not just for quantum field theory, but also for a parallel theory that emerged in 1971, string theory. We will explore this further in coming chapters.

The connection between the super Poincaré group and a formal field framework was made by two theorists in 1974: Julian Wess (1934–2007) and Bruno Zumino (1923–2014). What emerged from their collaboration is the Wess-Zumino model, a four-dimensional quantum field theory (3 space and 1 time dimension) that contains supersymmetry. The model was fully renormalizable, which means that it behaves itself and allows precise calculations to be made without having to fine-tune each step.

As the study of SUSY has evolved since that first model, let's see if we can understand it better. What is SUSY?

Particles possessing spin-angular momentum are described using a mathematical construction called a spinor. If you think of a direction in physical space as a a vector, a collection of three numbers—one lying along X, one lying along Y, and the last lying along Z—then in a more abstract space, the spinor tells us direction in that space. First discovered in 1913 in a purely mathematical context, physicists found them to be useful in describing particles such as the electron (during the 1920s). What SUSY does is to articulate a transformation among ways of representing spinors. It's another type of transformation of a mathematical object. If the object exists, that transformation should also exist. What is intriguing about this idea is its physical implications when applied to a quantum field theory.

SUSY, in a field theory context, transforms fermions into bosons and vice versa. As a result, a full quantum field theory (ala the Wess-Zumino model) describes both matter-like fermions and bosons, and force-like fermions and bosons. The standard model describes only matter-like fermions and force-like bosons. From supersymmetry's perspective, the standard model is half of a theory.

Introducing SUSY into quantum field theory has the effect of (at least) doubling the number of particles the theory describes. If SUSY is confirmed and found in nature, and, if nature is accurately described at the quantum level by field theory, a complete field theory must then contain supersymmetry. Let us look to determine the implications for nature and, using a "canary in the coal mine," the Higgs boson, see what happens when we pursue the addition of SUSY to the standard model.

The Double-Edged Sword of SUSY

To be tongue-in-cheek: the beautiful thing about SUSY is that it multiplies by two the number of fundamental particles in nature; the horrible thing about SUSY is that it multiplies by two the number of fundamental particles in nature. We'll soon see why SUSY cuts both ways, but let us first reflect on one of the great and attractive features of SUSY: it adds more particles.

Let us go back to the particle physics reception analogy of earlier. Recall that, as the rumor of Dr. Hawking's arrival travels through the physics enthusiasts in the room, a clump of enthusiasts suddenly encounters another famous physicist. They stop to greet her, but she is suddenly off to look for Dr. Hawking and the crowd continues to pass the rumor. However, in the process, the motion of the rumor is slowed, increasing its inertia. In the standard model, this effect is amplified to infinity, grinding the spread of the rumor to zero speed. Why? Because in the standard model picture of this analogy, there are so many famous physicists in the room that, as people try to spread the rumor, they get more and more slowed down trying to interact with one famous physicists after another, after yet another, and the activity becomes runaway, halting the spread of the rumor altogether.

This is where, in supersymmetry, extra particles ride to the rescue. Imagine that, as the rumor spreads, physicist Lisa Randall appears beside the clump of people spreading the rumor. Just as that clump notices her and stop to go over to greet her, in swoops her also-famous colleague, Dr. Raman Sundrum to remind her that they need to talk soon about an important detail in their latest paper. Before the clump of people can get to them, Dr. Randall and Dr. Sundrum have retired to a corner of the reception engaging in conversation detached from the activities of the physics enthusiasts in the room. The rumor's spread is unaffected as it was in the original scenario, traveling with a finite inertia through the room.

SUSY provides extra particles, whose role is to quiet the otherwise unceasing chatter of particle interactions with the Higgs boson. It cancels out those interactions, returning the Higgs' inertia to a finite and calculable value. For instance, for every spin-1/2 electron in the standard model, there is a spin-0 superpartner of the electron, a "selectron," that cancels out any troublesome interactions with the Higgs boson. Similarly, for every spin-1/2 quark there is a spin-0 "squark."

This cancellation process removes the manual fine tuning that was demanded by the standard model, returning simplicity to the mathematics. The

Higgs mass is thereby stabilized in the process. You can see why this was such an attractive feature! Once, where there was a troublesome corner of the standard model, we now have a more well-behaved calculation. Expanding the standard model in this way, with as minimal an extension as possible, is known as the minimal supersymmetric standard model.

The MSSM is attractive, as well, for other reasons. It makes specific predictions about things that are not at all predicted by the standard model. For instance, in the MSSM a relationship develops between the Higgs boson mass, the Z boson mass, and some parameters of MSSM that result from adding SUSY. Physicists in the 1970s and 1980s were quite excited about one of these predictions, that the mass of the Higgs boson should be less than that of the Z boson, less than 90 GeV/c^2.

The astute reader, having 20/20 hindsight, might be more alert right now. The Higgs boson's mass is not less than the Z boson mass. Indeed, it's almost 40% larger than the Z boson mass. So why are we still considering SUSY if that key prediction was not confirmed by the work done by the LEP experiments and the LHC discovery in 2012?

We haven't provided the full story yet, and we purposely did not do so to illustrate the danger of oversimplifying assumptions. The original prediction of the limit on the Higgs mass was done in the simplest possible context, and in so doing ignored contributions to the Higgs mass that arise from an even richer tapestry of virtual particle interactions than those we have described in our simple analogy.. The original prediction is considered a naive prediction, the kind one would make quickly, without taking into account what the consideration of greater complexity might do to the result.

A useful analogy involves guessing someone's weight. One's instinct might be to carry the person to a scale, weigh him or her, and the task will be completed, but that would likely surprise the person (and not in a positive way) and seems socially complicated for this simple task. Instead, you might try to make an educated guess based on facts—be a theorist! For instance, if this is taking place in the United States, you might know that the average weights of male and females is 195 pounds and 165 pounds, respectively. So you could make a guess based on those averages. This is known in physics as a first-order approximation. One uses the most readily known information to see where it leads you.

However, most first-order approximations are wrong. In this circumstance, working with a specific random individual, you are likely to be incorrect because averages are just that, a single number derived from a large sample of numbers based on different individuals, that vary in direction (over or under). If this is a friend and you wish not to offend her (by stating a number higher than the true value), you will want to add complexity to your calculation and achieve greater accuracy than is possible by using averages. You might factor in other things, like whether or not she is physically active and how much snacking she does. You can then adjust up or down from the average accordingly. To flatter them, you might even lower the adjustment a little, as one's weight is a socially complex issue in the U.S. involving cultural as well as factual considerations.

This kind of added complexity is called a higher-order calculation, and takes additional (hopefully small) modifying factors into account. Of course, the only true test is measurement, but, if we are being theoretical about a subject, this strategy has been shown to work quite well in mathematical frameworks where one first considers the greatest effects and then increment the result using small effects to correct the original prediction.

One can do this with the MSSM by calculating the corrections to the Higgs mass that occur from interactions with super partner particles like the stop squark. Using such factors, the MSSM predicts that the Higgs boson mass has to lie somewhere below 140 GeV/c^2.

Now one sees why theorists with an interest in SUSY were excited about the Higgs boson discovery in 2012. Many SUSY models preferred a still lighter Higgs boson, but there were also many models in which the Higgs boson mass went as high as 140 GeV/c^2 and those results were still okay with SUSY. Some theorists have claimed that this was the first successful test of a SUSY prediction, yielding a positive result supporting SUSY's existence. This is certainly scientifically debatable as, at best, it is but indirect evidence of SUSY. The best direct evidence would be to physically discover stop squarks and/or other super partner twins that are predicted by the theory. Let's look more closely at this issue, because SUSY has failed, so far, to supply proof of the existence of these particles.

The particle constituents of the MSSM are shown in Fig. 8.3. As you can see, the number of things described in nature by adding SUSY is quite a multiple. The force-carrying particles in the standard model have partners in the MSSM named with "-ino" added to the word's end (pronounced ee-no). For each W boson, there is a wino (wee-no) fermion; for each gluon (a boson), there is a gluino (glue-ee-no, a fermion). While there are relationships among the expanded set of parameters in the MSSM, part of the job of the SUSY theorist is to propose schemes for intelligently guessing the values of the core unknown parameters—those values that must be known before predictions can be made. It is these guesses that have led to models, some of which have survived the discovery of the Higgs boson, and some that have perished.

Many MSSM models predict a hierarchy of masses in which the lightest standard model particles result in the heaviest SUSY partners, and the heaviest standard model particles result in the lightest SUSY partners. For instance, consider the top quark. It is by far the heaviest quark in the standard model. The stop squark, on the other hand, would be the lightest (and thus most easily produced) squark. The Z and the W and the Higgs are very heavy bosons, but the Zino, Wino, and Higgsino would be very light. (This leads to an attractive aspect of SUSY—its expectation for the existence of a dark matter-like particle.

We'll return to this idea soon.)

We are jumping over a number of stages in the evolution of SUSY from its earliest forms in the 1970s to the present forms that survived the purge caused by the discovery of the Higgs boson. We respect the blood and sweat metaphorically spent by hundreds of theoretical physicists to get the theory to where is today, but our goal is to focus the reader on SUSY's modern efforts to understand the cosmos. For that purpose we need to focus on the features of the theory that are under assessment by current experimentation.

The most striking prediction of SUSY is the new particles that should be discoverable. Clearly, selectrons are not found in our everyday world. If that were so, the idea of a fundamental spin-0 particle being part of nature when the Higgs boson was proposed in the 1960s would not have seemed so alien. There has never been the observation of a selectron, or other scalar partner of quarks or leptons since the time this idea bubbled up in quantum field theory. This suggests that supersymmetry, like other symmetries of nature, must somehow be "broken." Breaking of this symmetry would permit the particles of the standard model to remain relatively low in mass but that their shadow partners, the twins of SUSY, would be shown to have high masses, masses that were beyond the possibility for appearance in experiments done during the 1970s.

Because we do not know the values of the parameters SUSY adds to the MSSM, we cannot know if or when we will discover SUSY. It has been assumed that SUSY is "just around the corner," that the next time the energy of a collider experiment is increased, the energy threshold will be crossed into the SUSY world and particles will rain from the collider much as mesons and baryons rained from colliders in the 1940s, '50s, and '60s. There was much optimism in the early 1980s, when the UA1 and UA2 experiments at CERN began to hunt for the W and the Z bosons, that SUSY particles would be right there, waiting to be discovered. UA1 and UA2 turned up empty-handed on SUSY particles. So did the Tevatron at Fermilab and the LEP collider at CERN. Thus far, the LHC too has seen no definitive evidence of these particles, but SUSY predicts such a varied spectrum of possible outcomes that it continues to be safe to say that the physicists at the LHC have not exhausted all corners of the data where SUSY might be lurking. Poking about in those corners will take years, if not decades.

So we see how SUSY is a double-edged sword. It quells some of the bad behaviors of the standard model, such as the quantum corrections to the Higgs mass. It also has the attractive feature of mostly solving a problem in the standard model, the unification of the electromagnetic, weak, and strong forces into a single force, a grand unified force. SUSY allows this unification to become possible, and instead of the interaction strengths of the three forces meeting each other in various spots, they instead come very close together in one concentrated location. Unification is temptingly within reach in a SUSY-enhanced standard model.

But SUSY proves equally frustrating in that its most striking prediction—that standard model particles have shadow partners, twins with slightly different spin-angular momentum—has failed to be confirmed. This is smoke from the fire of SUSY, and if it is ever to be taken seriously as a reliable theory of nature, this prediction has to be realized.

We commented earlier how these super partners might be hidden on a nearby brane, and how the separation of standard model particles from the super partners on two branes might point to the mechanism by which SUSY was broken as a symmetry. However, for that to be a satisfactory explanation, physicists would also have to detect the presence of extra dimensions, another prediction that has eluded experimental evidence. That said, a clever theoretical physicist can readily contrive a model by which super partners are hidden, but those models have measurable consequences which eventually will have to yield to experimental tests and survive the presence of experimental evidence. To be a scientific idea, those ideas must ultimately be falsifiable. It is not clear yet whether SUSY is that kind of theory. Its power has been in its ability to calculate real effects in experiments that we can run presently and in the future, and is what has helped it to survive as a tool to study new data.

Before we close our discussion of SUSY, let's look at its intersection with the quantum realm using a problem from the cosmic realm, dark matter.

SUSY in the Dark Cosmos

As we have noted, the night sky is ruled, not by the bright stars that dance in our vision, but by all the dark voids in between. Dark matter is observed to make up about 85 percent of the matter content of the universe. What is it? Frustratingly, no one knows. There is no direct evidence of particles that make up dark matter, and the standard model is quiet on the matter, offering only an insufficient number of neutrinos to explain this phenomenon.

The MSSM, however, presents greater promise. It predicts the possibility of there being a low-mass and electrically neutral spectrum of particles. The lightest of these is generically named the neutralino, and it stands for the lightest of either the bino (a combination of the zino and photino), wino, or the Higgsino, or some combination of those, manifesting as a physical particle. Regardless of its exact identity, the MSSM predicts that there is a lightest one and, more interesting, that it is possible for this particle to be stable on the time scale of the universe, 13.8 billion years. It could have been there at the beginning of time, and there still in the universe today. This would line up remarkably well with observations from the cosmic microwave background (CMB), the light left over from the big bang, whose pattern of hot and cold spots tell us that there was dark matter present near the beginning of time. This would line up well, too, with the observations of the present cosmos, where galaxies and galaxy clusters move in ways that suggest an unseen form of matter is at play that maintains and guides their structure.

What is most exciting about this is that the MSSM predicts that there are interactions between the traditional standard model particles and super-partner particles. These interactions would be quite weak and rare, rarer even than the interactions of the weak nuclear force, but they would still be stronger, when they occur, than quantum gravitational interactions would be expected to yield, making these interactions observable with present and near-future experiments. Such interactions would explain why dark matter "barely notices" normal matter, mostly giving us a pass except through the collective gravitational influence of large clumps of cosmic dark matter. Again, this would be consistent with observations from structures like the bullet cluster, which we discussed earlier.

If there were a dark matter detector to see evidence of interactions that cannot be explained by standard model physics, it would be a crucial step toward knowing whether or not SUSY is a correct theory of nature. For instance, if the XENON100 or XENON-1T detector in Italy, or the LUX or LUX-ZEPPELIN (LZ) detector in the U.S., or the PandaX detector in China, were to observe such interactions it would be possible to measure the mass of the particle and provide information about its interactions with the detectors. This would help constrain SUSY theories, and determine whether they are viable in lieu of these results and independent results from the LHC and other experiments.

Wrap Up

When Einstein formulated the theory of special relativity and published it in 1905, it contained a key mathematical idea, the Lorentz transformation, that is the tip of a beautiful and fascinating iceberg. Henri Poincaré recognized that the Lorentz transformation was representative of a group of transformations in space-time that preserved certain quantities as invariants. Through this, we came to understand that special relativity is a theory of symmetry and invariants. If one extends the Poincaré group by incorporating symmetries that transform spinors, representations of physical states with internal spin-angular momentum, into each other, one arrives the the super Poincaré group and the basis of supersymmetry. Many theoretical physicists converged on this idea in the early 1970s, focusing on interesting and fundamental questions about quantum field theory. Wess and Zumino formulated the first complete quantum field theoretical model based on this idea, and since then, SUSY has been on a quantum roll.

Enthusiasm for this idea, because of its power to unite the three forces of the standard model and to show us how the Higgs mass can remain a small and finite number, is tempered by a lack of evidence for its most stunning prediction, that standard model particles have shadow partners with different spin. SUSY seems as broken as the gauge symmetry of the vacuum state of the standard model or as disrespected as the symmetries of parity, charge conjugation, and time reversal. Nature is just not, apparently, big on respecting symmetries. Nonetheless, we seem to owe our existence to such broken symmetries, and maybe we will find evidence for SUSY particles and come to understand how and why this symmetry became broken.

As we will come to see shortly, SUSY didn't find itself a hero of the standard model. Indeed, other parallel. and perhaps more encompassing, theories of nature benefited from the recognition of this symmetry. In following chapters we will explore string theory, its many incarnations and underlying frameworks, and the role that SUSY played in saving it as a theory of nature.

For now, a decades-long quest to find the SUSY twins of the standard model particles continues at experiments such as the LHC. Physicists from around the globe are poring over data from this collider, hunting in all the corners, lit and dark, of this massive store of subatomic information for evidence of SUSY. In addition, dark matter hunters are using large detectors seeking to catch dark matter particles. They are in search of electrically neutral and possibly heavy stable astronomical particles, remnants of the breaking of SUSY that were passed down to us from the big bang. There is presently no compelling evidence for the existence of SUSY in nature, but the allure of the idea, much as in the allure of extra dimensions, is that, from something so simple as a relationship between fermions and bosons there might emerge a rich and complex universe, and, perhaps, a new ray of light to play on the multiplying shadows dancing on the cave wall.

## Chapter 11

The Smallest Shadow
Part I: Early String Theory

How did the universe begin? What were the laws that governed the universe back to its very first moment of existence, the time that is defined as Planck time, about 10^{-44}s (seconds) after the beginning of the universe? How did the universe carry forward in time to become the universe that is described by the standard model, which describes the quantum realm quite accurately after about 10^{-11}s past the big bang? How did the universe then become the grander cosmos that is described by the theory of general relativity? When did general relativity part ways with the standard model in the structure of the cosmos?

These are a few of the questions that physicists have posed since the quantum field theory revolution in the second half of the twentieth century. During that same turbulent period in the 1950s and 1960s, when gauge symmetry, broken symmetry, and field theory was emerging as a framework for describing quantum reality, another parallel idea was being developed in competition to the explanation of the properties of emerging particles such as kaons, pions, and other matter being produced in colliders. This idea, the Veneziano theory (more about him later), would be shown to be the wrong path for describing those discoveries, but it later became the seeds of what physicists know now as "string theory."

In this chapter, we will discuss the early roots of string theory, name some of the physicists who played important roles in its development, and list the basic ingredients of string theory. We'll also show how physicists struggle with mathematics, wrestling with paradoxes that often emerge from an incomplete understanding of a new idea. You will see how supersymmetry emerged from the ashes of the first string theory and, as did the phoenix of ancient myth, rose in a more glorious and powerful framework tempered by those early struggles. We will view, with the same eye we earlier brought to extra dimensions and supersymmetry, the requirements necessary to a framework such as string theory—predictability, with particular relevance to the universe, as we know and understand it, while also predicting discoveries specific to the theory. String theory is a story of beautiful mathematics, hefty skepticism, and a view of the universe that, if it is real, would be far stranger than one could dream it to be.

In this chapter, we will focus on the early ideas and the first struggles. In the following chapter, we will reframe our discussion, aiming it at the more modern versions of string theory, M-theory (and, maybe, F-theory), and the challenges that modern string theorists must meet in order to make progress with this framework. Later still, we will see where it is that string theory and the cosmos might first have come to meet—black holes. For now, let's start at the exiting decades of the second half of the twentieth century, the birth of string theory.

The Birth of the String

String theory has evolved over many decades as theoretical physicists grappled with the mathematics describing these fascinating structures. The first phase was based on describing bosons, as bosons are the fundamental particles of the four forces of nature. Early string theory showed that particle characteristics were governed by bosonic behavior. However, the mathematics of this approach had a fatal flaw. It did not reflect how the universe works. The first bosonic string theory had mathematical inconsistencies. Such inconsistencies are equivalent to proclaiming that $2 + 2 = 5$! Not unusual, this kind of nonsensical mathematics can arise in quantum physics. It is described by the term "anomaly."

As we noted earlier, the particle accelerator and particle detector technology led to an explosion of the discovery of new subatomic particles. Many physicists sought to make sense of the patterns revealed during the appearance of these particles. What was successful along the way to the development of the standard model, was improving the concepts of quantum numbers, gauge symmetry, and quantum field theory. Realize, though, that this main pursuit was not the only attempt during that time to make sense of the particle zoo.

A then-young theoretical physicist named Gabriele Veneziano (1942–) worked in the late 1960s on exactly this problem. What was the ordering principle of the subatomic particle zoo, and what framework would explain that order? Veneziano, on leave from the Weizmann Institute of Science in Israel, working at CERN in Switzerland, hit on a mathematical relationship that appeared to describe the manner in which mesons (what would come to be understood as bound pairs of quarks—although this was not fully established at the time) scattered during collisions. We know now that this is governed by the strong nuclear force, but this force and especially its associated quantum field theory, QCD, was not fully articulated in the 1960s.

In the late 1950s, a theoretical framework for describing the rate of scattering and how it related to the momentum of the scattering particles, was developed (using Newtonian mechanics ideas; i.e., for low velocities) by the physicist Tullio Regge (1931–2014) that came to be known as Regge theory. Veneziano wanted to find the mathematical form for the degree of scattering, the function that could relate momentum to probability for strongly interacting particles. 1968, when Veneziano made his discovery and published his famous paper on the subject, is the year that the physics community marks as the historic moment of the origin of string theory. As others studied his work they found that the formula describes string-like structures inside the mesons. When the string-like structures interact with each other (meet, join, split), the mesons scatter. The string's micro-behavior leads to the meson's macro-behavior. The equation describes string-like structures that are responsible for the underlying work. While he did not realize it at the time, he had planted the seeds for a long mathematical quest that developed an entirely new quantum theory of the gravitational force, string theory.

As QCD rose to the fore in the 1970s as a means to make predictions about experimental observations of particles charged with color, Veneziano's earlier view of the strong interaction fell to the wayside of lesser interest, but it was neither entirely forgotten, nor entirely ignored.

Leonard Susskind, Holger Nielsen (1941–), and Yoichiro Nambu (1921–2015) were inspired by Veneziano's discovery to propose that mesons behaved more like extended objects than they acted like point particles. Called "strings" and described by what would become known as "string theory," each object behaves like a filament, a length of spaghetti that vibrates in the manner of violin strings (Figure 11.1). There are two kinds of strings: open strings (think of a violin string) and closed strings (think of the percussion instrument, the triangle).

The Properties of Strings

Unlike point (0-dimensional) particles assumed by the standard model of particle physics, strings are one-dimensional extended objects that can have open or closed configurations. Strings vibrate much like air vibrates in an open-ended

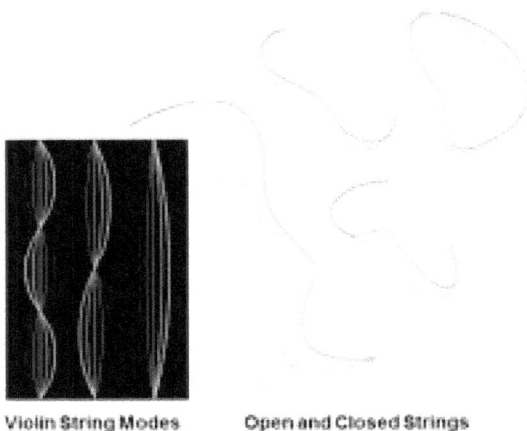

Violin String Modes   Open and Closed Strings

Figure 11.1 General String properties and Characteristics

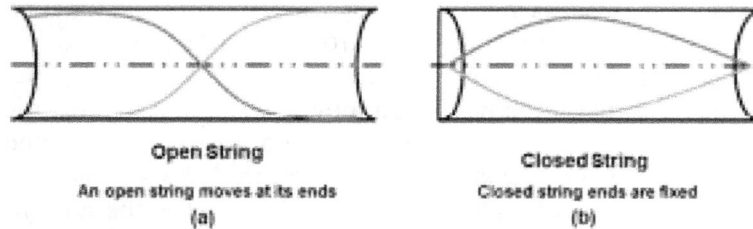

Open String

An open string moves at its ends

(a)

Closed String

Closed string ends are fixed

(b)

Figure 11.2   Open and Closed String Vibrations

or closed-ended organ pipe (Figure 11.2). The dark-shaded thread in 9.2(a) is one-half of the open string's vibration cycle and the lighter-shaded thread is the other half of the cycle.

The different ways that vibrations exist in these two kinds of strings have physical consequences, much as the restriction of certain kinds of vibrations in open or closed organ pipes will alter the character of the notes emanating from the pipe. Taking this musical analogy a bit farther will help us to understand the mathematics of strings (and pipe organ waves).

Certain musical instruments behave like closed strings and others behave like open strings. This results in instruments that produce different tones. A clarinet or a harp produces what musicians describe as *anharmonic overtones* (having only non-integer multiples such as 0, 1/2, 3/2, 5/2) of the fundamental frequency. On the other hand, bassoons, oboes, and saxophones produce *all possible overtones*. Similar to the way string theory works, open strings are like clarinets and harps while closed strings are like bassoons, oboes, and saxophones. (One must have a trained ear to detect these differences, but they are there!)

The closed fundamental string vibrates and interacts in a way that is almost uniquely specified by its mathematical self-consistency. This seems to be the secret by which it is capable of forming a valid quantum theory of gravity.-

Open strings have some advantages over closed strings. The open string of figure 11.2(a) allows for electric charge to be placed at its ends. The vibration frequencies and modes of a string describe particle characteristics such as spin, mass, and other similar properties. Strings are thought to be *ur-structures* of the point-particles of the standard model. ("Ur" simply means that which comes before.) Thus, strings come before; i.e., they are the fundamental, most elementary structures from which particles and energy are made.

One often encounters the erroneous expression that strings are made of energy. This is precisely opposite of the reality described by their mathematics. Energy and everything else would, in fact, be made *from* strings (should it be that they provide an accurate description of nature).

Let us look more closely at how strings would interact, and what that would mean. Figure 11.3 shows two situations: open-ended strings (top group), here vibrating at four different frequencies; and closed-ended strings (bottom

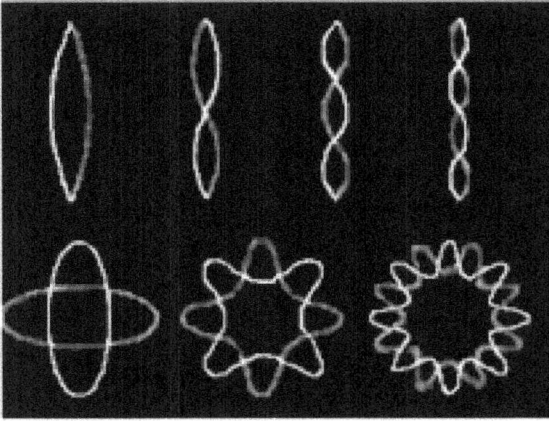

Figure 11.3   Various String Behaviors

group), here vibrating in three different modes ("frequency of vibration" and "mode" are identical concepts, one denoting the other). Each of the seven pictures represents a different particle. It should be noted that there are an infinite number of vibrating modes; for the purposes of this illustration, we discuss only the ones shown.

The tension on a real (a muscial) string affects its frequency of vibration. That is why one either tightens or loosens the strings when tuning a banjo, guitar, piano, violin, or any musical instrument that produces notes using strings.

The mathematical strings with which we propose to describe our reality behave like the filaments of a musical instrument that vibrate and spin in specified manners leading to the identifiable characteristic properties of particles. An example of this behavior is the graviton, a closed-end string, with spin equal to two, represented here by the lower left graphic in the figure. (Later, we will look at the graviton and how it was first visualized in the mathematics of string theory.)

The ends of open strings are not attached to each other and can terminate on branes (surfaces or volumes on which different particles and forces can be defined). Being open, their characteristics are different from those of a violin string affixed to a violin. A violin string's vibration (frequency) changes based on where the musician presses with a finger on the real (non-mathematical) vibrating string. Finger placement on a violin string controls its effective length. The location at which the end of a string terminates alters its effective length; this location can be controlled by the nature of the spatial dimensions available and what is present in those dimensions. This causes its vibrating frequency (and its resonances) to change, producing different tones. Closed strings vibrate too. However, these vibrations create different modes of the closed string that determine a resulting particle's properties—mass, charge, spin, etc. The graviton will be seen to be made from a closed string.

Arbitrary vibrations of violin strings are not necessarily resonances. Natural resonances sustain themselves over time. Resonances can be heard in the shower, when one sings. The confining shower walls cause the sounds to bounce back and forth (echo) at a specific frequency in the space between them. The frequency or tone of the sound is what sustains the sound if it is a resonating sound. If the distances between the waves are just right, the sound amplitude and frequency of vibration are sustained and reinforced, "blooming" into a resounding sound as though one had turned up the volume control.

These many possible prolonged responses are the resonances. Therefore, it would appear that there are many matter particles—maybe an infinite number. As we learned from the chapter on extra dimensions, if such dimensions are present we can expect even more particles (over and above ordinary particles) due to Kaluza-Klein resonances. In general, the higher the string's frequency of vibration, the greater is the mass of the corresponding particle produced.

The string occupies the dimension of the Planck length, 1.6×10^{-33} cm, a size that is 10^{-20} times the size of the nucleus of an atom. This is a very small distance—so small that it cannot be measured with the instrumentation of today, and perhaps, as some argue, cannot with any imaginable instrumentation in the future. Remembering that wavelength and energy are inversely related to each other (short wavelengths or distances correspond to large energies) the energy levels required to precipitate such a small particle in a collider are staggering. To provide a sense of scale, an energy level sufficient to reveal a string of about 0.1 millimeter is presently achievable at the LHC.

Strings are proposed to be the sub-structure of everything in the universe, including space and time. It is not that strings move through space and time; they create it so that particles can move through it. A good analogy is to consider waves on the surface of a body of water. The waves cannot exist unless there is water to support their existence. Strings are like the water; particles are like the waves. One cannot have the waves without the water. If string theory is accurate, one cannot have space, time, matter, or energy without the strings.

The types of strings shown in Figure 11.3 present comparative characteristics of the fundamental strings and their features. Strings can meet and combine to become extended versions, or they can break up into smaller lengths as they travel through space-time.

Point particles travel through space-time forming a one-dimensional world line. Strings can exist as constituents of these particles, too. Open strings move through space-time forming a 2-dimensional world sheet, while closed strings move, forming a 3-dimensional pipe/cylinder as they move. Their meeting and joining together create behaviors that produce everything in nature.

These kinds of "stringy" interaction pictures are akin to the Feynman diagrams we showed earlier. Because it is possible to create interaction pictures for strings (and associate clear mathematical rules with those pictures), string theory is quite similar to the standard model in that it allows one to see whether calculations, having more and more interaction complexity, remain finite and stable. The techniques that make that possible in the standard model apply here, and theoretical physicists long ago found that string theories, as mathematical frameworks, have lots of the "good behavior" that is cherished in the standard model.

Bosonic String Theory

Let us look at the early kind of string theory that was formulated right after Veneziano's work. It will help us to understand how physicists evolved the idea and, having dug deeper into the mathematics that describe it, confronted the first challenges to using string theory as a physical theory of reality. The first generation of string theory, initiated in part by the work of Susskind and Nambu, involved the bosonic string. That string was incapable of describing fermions, it could only work in a universe that had twenty-five spatial directions and, worst of all, it was mathematically inconsistent.

The bosonic string has two types of mathematical anomalies. These inconsistencies are subtle and it took a while to discover that these nonsense statements were buried beneath the complicated mathematics. Interestingly, one of these nonsense mathematical statements disappears when space has twenty-two more spatial directions than are apparent in the three of our universe.

Let us focus on the problem of many dimensions. The very first string theory, called "Type I bosonic string theory," was found to have the ability to make meaningful calculations only if the minimum number of dimensions in the universe was twenty-six: twenty-five in space and one in time. That many dimensions allowed for enough freedom in the theory to achieve the conditions for mathematical usefulness—at the cost of having to postulate more than eight times more spatial dimensions than are presently observed. A lessening of this problem was achieved by two of the most important tools that physicists have at their disposal: coffee breaks and chance meetings.

It was theoretical physicists John Schwarz and Michael Green who, in work conducted in the early 1980s, sought to eliminate the anomalies of the 26-dimension Type I string and many other anomalies that had crept into the mathematics of string theory during its initial stages of development. According to journalist Aida Edameriam in a *Manchester Guardian* article, physicists Michael Green and John Schwarz would not have collaborated and solved some of these problems had Schwarz not wandered into the CERN canteen in 1979 for a coffee break at a chance moment when Green was there. As a result of the serendipitous collaboration between Schwarz and Green, they reduced the number of dimensions required by the bosonic string. This and many advances accompanied the first string revolution in 1984. While their collaboration created a theory that nicely followed Einstein's gravitational theory with the emergence of a massless graviton on a closed string, it did more than that.

Capability, opportunity, time, and chance come together in unpredictable ways, and sometimes that is how theoretical physics makes advances. Brilliant breakthroughs cannot be ordered up on a regular schedule. Understanding the cosmos is not clockwork, predictable labor—it's messy, and if there is any lesson here it might be this: if you can't see your way through a problem, take a break—you might meet someone to help you see the problem in the right way.

Through this chance meeting in the CERN cafeteria, there began an intense dialogue addressing the most significant anomalies (i.e. the mathematical inconsistencies) of string theory—those of combining (unifying) matter particles (fermions) with forces (gauge bosons). They reduced the minimum number of space dimensions in string theory from twenty-five to ten.

The Tachyon Monster

But even given the mathematical assumption that there are now only ten or more distinct spatial directions, let's consider the second type of anomaly in the bosonic string. The second anomaly led to the prediction of a particle that was given the name "tachyon." The math paints the tachyon as similar to a Higgs boson particle, but with the curious feature that if one takes the square of its mass, one obtains a *negative number!* The only known solution to rid string theory of this problem is through the introduction of a new symmetry, one we have met before—supersymmetry.*

A theory that cannot describe fermions would not be able to describe our universe. Seeking to overcome this mathematical inconsistency in the bosonic string, physicists sought to include fermions by using supersymmetry. This led to even more string type theories and, eventually, to five superstring theories in all. The seeds that would enable understanding the landscape of string theory were planted much earlier, around the birth of string theory in 1971.

During 1970–71 Claud Lovelace discovered that the bosonic string is consistent with Einstein's special relativity with only twenty-five spatial dimensions. However, this theory has a class of problem, divergences, in which calculations run off uncontrollably. (The origins of this are unclear.) In the infinities of these mathematics, the tachyon of the bosonic string is revealed.

Shortly thereafter, Pierre Ramond added fermions to the model to create a two-dimensional supersymmetry. This had the effect of "taming the tachyon monster," making possible a full description of a supersymmetric string without need of the tachyon. Let's look more deeply at this monster and how it was quelled. In doing this, we will see the first intersections of ideas developed in

---

*The authors want to note that the concept of the tachyon has found its way into much science fiction over the past years and, in misunderstanding the nature and consequences of the tachyon, the phrase has become a kind of hollow techno-babble. (Nonetheless, it surely is entertaining for a physicist to encounter it, if not in its intended way!).

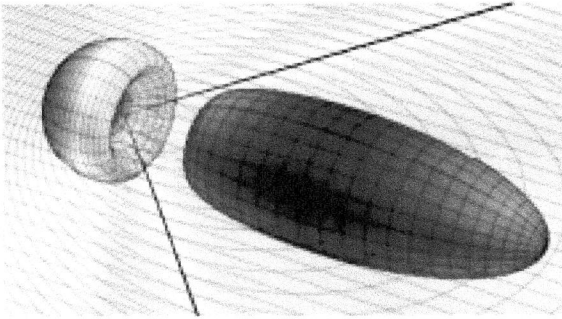

Figure 11.4   The Tachyon Monster

earlier chapters, and how their convergence leads us to a more reliable string theory.

But, for now, imagine a world where everything travels faster than the speed of light, where things are either very heavy or very light, and are unrecognizably different from the world we live in. Imagine the mathematics to be off the wall, that calamity happens all the time, and accidents of all sorts cause devastation. This is the world of the tachyon monster.

An artist's concept of what it might look like (were it to exist) is shown in Figure 11.4. It is a weird configuration, showing a leading shape with a wake behind it that is followed by a bullet-shaped object flying backward. The figure depicts the two-component object to be in motion, always at speeds greater than the speed of light. In special relativity, this would imply that it travels backward in time, violating the notions of cause and effect. Traveling faster than light has a similar effect in the universe because when traveling on Earth faster than sound, a shock-front develops. The first component in this visualization is a bow wake—a shock—shown on the left side of the figure. Because its speed is so great, the medium in which it is traveling forms a wake followed by an aft wake that looks like a bullet traveling backward.

Tachyons lead to other problems as well, such as defining what is probable. How would you answer the question, "What is the probability that something—*anything*—will happen today?" Because at least one thing happens every day—your cells divide, you take a breath, an apple falls from a tree, atoms jiggle somewhere—you would likely, and sensibly, reply one hundred percent. What would it mean to reply one hundred *and ten* percent. That doesn't make sense. You can't have more than all there is. You can't have more of a chance of something happening than *all* of the chances of that something happening. This, too, is an anomaly.

The tachyon, having faster-than-light speed and imaginary mass results in quantum mechanical probability calculations that violate totality and predict probabilities that exceed one hundred percent. This is a very disturbing situation—one that threatens the foundations of quantum mechanics. You can understand why this is a monster. The mathematics of strings, as a means to contemplate reality, is in jeopardy, unless the tachyon and other anomalies are eliminated.

Anomalies are nature's way of telling theoretical physicists that they have written elegant gobbledygook using her own beautiful language, mathematics. The real world has no numbers to identify measurable properties that imply imaginary mass. This was the downfall of the bosonic string as the correct idea to describe nature.

We earlier introduced the example of the Einstein hypotenuse to help you visualize invariant structures. The rest-mass of a subatomic particle is noted to be one such invariant. The tachyon monster can be thought of as having a hypotenuse whose final value, when squared, is a negative number. That is not possible in the real world! It was clear that something was wrong—not with the whole of the concept—just its mathematical workings. We have to capture this monster and tie it up.

The solution, it was found, could be achieved by referring to the spin properties of the particles described in the standard model. The bosons in the standard model all have integer-numbered spin (0 or 1). The fermions, however, have half-integer spin (½).

We referred earlier to the spin characteristics of bosons and fermions, but little was said of the property of spin except in reference to its rate. The bosonic string, the first generation string, was not able to interact with the electron and other matter particles. This is because until 1971, bosonic string mathematics had no way to produce those tonal vibrations that would create matter particles with their half-units of spin. This was the limitation of that first bosonic model, with its tachyon monster, making it clear that these problems had to be addressed.

To Spin a String

In 1971, three physicists—Andre Neveu, Pierre Ramond, and John Schwarz—brought a new slant to spin property and strings. Their innovation permitted the spin-rate of the electron (and all other fermions) to be incorporated into the mathematics that describes string characteristics. With this in place, matter particles could be included in the spectrum of string vibration modes. String theory was now deemed to have advanced from the first generation to its second generation. But, this still did not tie up the tachyon. More would need to be done to save string theory.

We've discussed quantum spin earlier. We have also discussed supersymmetry and developed independent notions of string theory in the early 1970s, which is a symmetry of space-time that unites the fermion and boson. When string theory incorporated supersymmetry into its mathematical framework, the monster was tamed. The anomalies that previously had appeared in the math—negative-mass, lowest-energy states of the theory—were canceled out

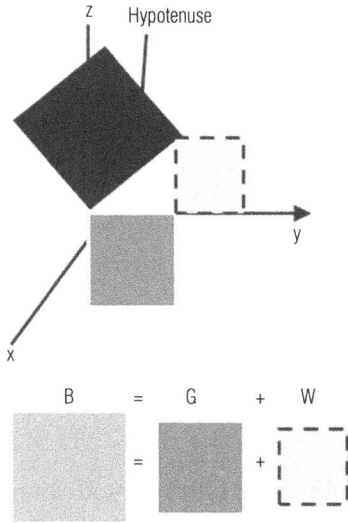

Figure 11.5 The Two-Dimensional Einstein Hypotenuse

by having equal numbers of fermions and bosons be describable by the behavior of strings. The trusty tool that we set up earlier, the Einstein hypotenuse, provides a means of understanding how supersymmetry tamed the monster.

This mathematical analogy describes a new form of symmetry that makes it possible to add the bosonic and fermionic string vibrations to each other, causing the anomaly to cancel from the theory. This new symmetry can be understood by referring to Figure 11.5 (a variation of which was first seen in Figure 9.4) The only difference between the two images is that, in this chapter, the shading of the various blocks has been changed. A form of symmetry will be described that *makes one function equal to another function.* In this figure you will note that the black hypotenuse is equal to the sum of the other two areas. This can be taken as a form of symmetry. As was described earlier, when the angle of the hypotenuse is varied up and down, the size of the other two areas varies with the change, always maintaining the sum of the two areas to be equal to that of the black area. Further, the area of the black hypotenuse remains constant; it does not change in size. It is the invariant.

Now, in the new symmetry, imagine that the gray area represents the bosons of a spinning string and the white area the fermions. If, for every fermion, there exists a boson, and vice versa, it is possible to trade the size of a gray area for the size of a white area while maintaining the analog of the size of the black hypotenuse area. This extends our Einstein hypotenuse to incorporate the concept called supersymmetry. It requires that a mathematical solution describing any fermion (e.g., the electron) can be traded for other solutions that describe any boson (e.g., the Z boson).

How does the cancellation then arise? We have to introduce a relatively wild concept, Grassman numbers. By associating Grassman numbers with the areas of the two hypotenuses, the cancellation occurs. This cancellation creates a kind of "Pandora's Box" from which the tachyon anomaly, our mythical monster, cannot escape.

You are, no doubt, familiar with how numbers behave when moving them around on paper using mathematical operations. Consider the humble multiplication table. You know that

$$2 \times 1 = 2$$
$$2 \times 2 = 4$$
$$2 \times 3 = 6$$
and so forth.

You first learned to memorize these tables, but, later, as your brain skills improved, you found that there are rules, tricks, and shortcuts for doing these operations more quickly, using even larger numbers.

One of the shortcuts one learns rapidly is to not be tricked by the question, "If 2 x 1 = 2, what is 1 x 2?" The order of the operations in multiplication doesn't matter, (the commutative law) creating overlaps among the times-tables of various numbers. Of course, in 1 x 2 = 2, the right-hand side of the equation doesn't change because you have simply swapped the order of the numerals on the left hand side. This we learned in youth and apply without thought today.

This is not so with Grassman numbers. Removing ourselves now, far from the automatic, the rules state that, if you can conceive of something and are able to write down sensible mathematics that do not violate known axioms or theorems of mathematics, it's fair game in mathematics. But, just because you can create sensible mathematics, this doesn't guarantee that it will agree with or have physical application. That said, we will find great utility for Grassman numbers in string theory and the Einstein hypotenuse thought experiment. Let's look more closely.

Grassman Numbers

Hermann Grassmann (1809–1877) conceived of a novel kind of algebra in 1844. Elements of that algebra are Grassman numbers. They behave according to this relationship:

$$a \times b + b \times a = 0;$$

if a = b, then a number times itself is zero. Grassmann numbers behave in a manner such that the sign of the resultant behavior changes depending on the order in which the multiplication is performed. This is odd behavior when one is familiar with commutative law, but so long as the rules of mathematics are conformed to this is quite acceptable. Do we see "Grassman-like" behavior in the universe? We do. The behavior of known fermions is very much like the behavior of Grassmann numbers.

For example, suppose we describe two electrons that collide and bounce off each other, and we view it in the context of the kind of string theory developed in the early 1970s. That theory states that the order in which the electrons collide affects the outcome of the collision. In that early string theory, electrons sometimes scatter due to their interaction with tachyons. Let us label the two electrons as "e1" and "e2." There are two different orders in which the scattering can occurs. Mathematically, a scattering such as this (even in the standard model) is described by a Grassman multiplication operation:

1. e1 x e2,
2. e2 x e1.

In this theory, if the two orders are compared, one finds that the result of e1 x e2 = −e2 x e1. This behavior is called non-commutative multiplication, because the order in which the computation is done matters. This strange behavior occurs for fermions that bounce off of tachyons or any other object.

We find this behavior in basic quantum mechanics. Consider a state of nature that contains two electrons described by a single wave function. If you impishly swap the two electron's position in space, you will find that the wave function notices this—it picks up a minus sign. This is an example of a non-commutative phenomenon. Grassman numbers are one of a class of mathematical methods that describe such behavior.

You might expect that, as it is in everyday mathematics, the principle of commutative multiplication should apply– you know, the rule that is used as a shortcut in multiplication tables:

$$a \times b = b \times a.$$

For bosons, this turns out to be true. One finds that it makes no difference in what order the bosons engage each other: there is no change in sign in when bosons encounter bosons. Nature doesn't seem to care if you swap them around, and they behave according to commutative law.

Let us delve into this idea of "non-commutativity" from another direction. Sometimes the order in which actions are performed matters. To see this we can apply the equation above to putting on one's shoes and socks. Suppose we let the symbol "a" stand for putting on one's socks and the symbol "b" for putting on one's shoes. Finally the symbol "x" represents the word "then."

If we follow these rules, the left hand side of the equation says, "Put on your socks, then put on your shoes." The other side says, "Put on your shoes, then put on your socks." In the real world, following the left side's instruction leads to a result that is very different from following the right side's instruction. It can then be deduced that the equal sign *cannot,* under these circumstances, be valid.

If we assign a plus one for a set of clean socks seen on a person who has walked through a muddy patch and use a minus sign for dirty socks seen on a person who has walked through the same muddy patch, we could then write an equation that describes our observations and it would look just like one at the beginning of this discussion! It would follow Grassmann's rule.

Grassmann simply introduced the idea that there could exist mathematical numbers under which the order of multiplication matters just as in our real world example.

It should be noted that the minus signs discussed by Grassmann and in our example about socks, has nothing to do with the minus sign seen in Einstein's Hypotenuse. Instead, the property of the minus sign in Grassmann numbers refers to the measurable properties of electrons and quarks. There is a sense in which all of these are "Grassmann-like" when you multiply them. Such particles are called "fermions."

On the other hand, photons and other force-carrying particles behave like ordinary numbers when you multiply them. Particles with this behavior are called "bosons."

These are also more akin to the numbers we use to measure space and time. Multiplication using these numbers obeys the ordinary "commutative" rules under which order does not matter.

Two interesting questions arise from consideration of this thinking:

"Why do the numbers that describe space and time coordinates obey only the rules associated with bosons?"

"Is it possible that there exists other than our space and time in which coordinates obey the rules associated with fermions?"

This second question opens the way to a bizarre mathematical structure called "superspace" and leads to supersymmetry. We will discuss this more later.

When supersymmetry was added to string theory, we left the era of plain vanilla string theory to enter the era of supersymmetric string theory, or superstring theory, for short. Supersymmetry makes it mathematically possible to tame the tachyon. Strings that spin are adjusted to have an *equal* number of bosons and fermions. This observation spurred an attempt to fully combine supersymmetry with Einstein's theory of general relativity. This combined theory is called supergravity. There exist a number of different versions of this construction, but there is a maximally extended version, very difficult, that required years of effort.

In 1978, French physicists Eugene Cremmer, Bernard Julia, and Joel Sherk found that this could be done somewhat easily—*if* one introduces not just one extra spatial direction, but seven! You see now how the extra-dimensional ideas of Kaluza and Klein, will come down from a dusty shelf and find new life in string theory's urgent need to add additional spatial dimensions to the universe.

Supersymmetry opened the door to describing bosons and fermions together within one theory. That is where this physics continues to stand today although the underlying ideas have continued to evolve, as we shall see, and have gone in many different directions.

But what if we don't actually need all of those extra dimensions to accommodate a string theory's ability to explain the cosmos? So far, a four-dimensional theory of strings has not caught the attention of a majority of string theory researchers. The reason seems to be that the current mathematics in such a theory is very complex and is not completely able to describe the physics of strings (with the exception of a few cases). However, since the 1980s, a group of string theorists bent on showing that a four-dimensional string theory is possible has emerged. We will discuss their work later.

Strings and gravity

We have made the repeated assertion that the idea of unification is fundamentally embodied in string theory—that it is one of the few frameworks so-far discovered that can not only unify the standard model forces, but *all* known forces can be unified. (After all, supersymmetry enables that to happen when one adds it to the standard model, and supersymmetry is a crucial part of string theory when one tames the tachyon), If *all* is the operative term, then that means it must include gravity. Let's look at this a little more closely.

As presented earlier, fundamental strings are like a clarinet or a harp in their ability to produce the kinds of overtones needed to generate fermions and bosons. Like a harp, plucking the string produces tones (vibrations), where each tone represents a unique matter or force particle. Depending on properties such as length and tension, the string can vibrate at different frequencies and produce varying modes. Given that these frequencies and modes represent particles of matter and force-carrying particles, the properties of the particles will depend on the properties of the string.

So, if the tension and type of the string is all that is needed for particles to emerge from this idea, what particles might one expect from string theory? One of the early-on discoveries was that there are modes of vibration in string theory that resemble a quantum of gravity, the graviton. This was expected (from earlier efforts to write down a quantum gravity theory) to be a boson (like the other force particles) having spin-2 (twice the spin of the standard model vector bosons), zero mass, and other unique features expressed in its quantum numbers. String theory did offer a means to generate just such a particle.

Thus, for the first time, the gravity force can be described by a quantum statement. Historically, this type of particle could not be described in quantum physics as a force particle that was unified with matter particles. In 1974,

Tamiaki Yoneya discovered that all then-known string theories include a massless, spin-2 particle that obeys the correct equations so as to be identified as a graviton. John Schwarz and Joel Scherk came to the same conclusion and made a bold leap to assert that string theory is a theory of gravity, not a theory of hadrons (particles held together by the strong force) as it had first been portrayed. It was then that string theory made the jump from a theory of particles found in colliders, as described in earlier decades, to a potential theory of everything.

Superstring theory, like all string theories, describes a set of mathematical vibrations. The key difference between the original string theory (i.e. the bosonic string) and superstring theories, is that the former describes only vibrations that behave like bosons, while the latter describes both bosons and fermions. Since electrons and quarks are fermions (with their Grassmann-like multiplication rules), the original string theory could not describe our cosmos as fermions are observable in our universe.

Recall that superstring theory is currently prognosticatory mathematics and will remain that until there occurs an observation in nature to confirm at least one of its predictions unambiguously. However, we can continue to examine these mathematics, comparing them with the mathematical descriptions of properties and objects we have already proven in nature. In the case of an obvious contradiction at this level, we could rule it out as a viable description. That is precisely what happened to the bosonic string. The mathematics of the bosonic string ruled out the mathematical description of electrons and therefore, we discarded the bosonic string!

The Laser Interferometer Gravitational-Wave Observatory (LIGO) observed waves of gravity in 2016 for the first time in human history. Such waves were predicted by the mathematics of Einstein's theory of general relativity in 1916. Comparing the mathematics of general relativity to the mathematics used to describe electronic concepts, one finds that the graviton must have a spin that is four times greater than that of the electron. LIGO is the only likely type of experimental set-up that will study this in the future.

However, the mathematics of superstring theory precisely includes both the mathematical descriptions of the electron as well as the mathematics of the graviton. They are both part of an all-encompassing mathematics that Einstein called a "unified field theory."

Revolutions and Evolution

The incorporation of supersymmetry into string theory marked the beginning of the first revolution in string theory, the superstring.

Superstrings were shown to require only ten dimensions (9+1) to obtain consistent mathematical behavior. Our universe, however, appears to most readily recognize (3+1) dimensions, a challenge to superstring theory, then, was the reduction of dimensionality in order to arrive at predictions relevant to the observable universe. It was thought at first, that to permit the extra six dimensions one could reckon them to be *very* small (like the Kaluza-Klein construct) and, therefore, not "see-able" in this universe using practical instrumentation (meaning devices that present technology allows to be built).

However, until an experiment is concluded that definitively reveals the character of extra dimensions, the debate about the reality and details of those dimensions in superstring theory remains academic. Debates ensue such that one school of thought adheres to the Kaluza-Klein view of extra dimensions, while another viewpoint is that, as there is no evidence for them, it is wise to adhere to what is known from the standard model—the real world is four-dimensional. If a ten-dimensional world exists, the additional six dimensions must be proven through experimentation.

What about the issues of dark matter and the accelerated expansion of the cosmos, which could be described as a kind of negative pressure or dark energy? Do these point the way toward resolving the problem of dimensionality in superstring theories while maintaining the successful standard model view of the universe? And, if supersymmetry is important to saving aspects of the standard model and string theory, where is it hiding? These and more questions led physicists to consider other methods of modeling and several types of "braneworld" scenarios that allow for extra dimensions and the possibility of broken supersymmetry, all while maintaining a 3+1 dimensional brane (our D3-brane) whereon the standard model appears to rule.

Dimensionality was not the only curse that plagued string theory after the first revolution. There was also a "problem of riches:" there were too many viable string theories. As mathematical approaches matured, theorists created five separate string theories. These are called the Type-I open, Type-I closed, Type-IIA, Type-IIB, and heterotic superstrings. Which one is the correct description?

Revolutions dissolve without dissidents. While much of the string theory community has moved in the direction of accepting the presence of multiple extra dimensions and are focusing on understanding the framework that underpins each of the many string theories (adding now, M-theory), some have chosen to move in a different direction, separating themselves from the mainstream. As we begin soon to focus more on the ways in which theorists have sought to embrace dimensionality, let us keep in mind the many roads to the frontiers of physics—not all of them head in the same direction!

Chapter 12

The Smallest Shadow
Part II: Current String Theory

During the search in the 1970s for a unitary view of the small and large worlds of our universe, the tachyon monster emerged to plague the string physics community. The development of superstring theory tamed the monster and opened the door to describing a universe having both fermion and boson particles, like the one described in the standard model. This marked the first revolution in the theory, and physicists clamored to work on string theory.

The next phase was also plagued by problems, that there had to be as few as at least ten spatial dimensions in the universe for them to work, and five different string theories emerged. We will look at how these problems were dealt with in the decades that followed, and see where mainstream string theory is today. In addition, we will explore some of the dissidence within the string theory movement, physicists (including SJG, one of the authors of this book) who embrace string theory, but reject the need for so many dimensions.

It helps if we look at some of the historical markers on the path from the first superstring theory into the 1980s, when the first of the modern set of solutions to these problems came out of this work in the shape of extra dimensions.

In the 1980s, physicist Edward (Ed) Witten showed that some of the string theories of that time would not accommodate certain fermions (such as the elusive neutrino), thereby causing violations of the conservation laws. This caused him to conclude that Type I string theories were inconsistent. In 1984, physicist Daniel Friedan showed that the equations of motion in string theory, generalizations of the Einstein equations of general relativity, emerge from the basic mathematical properties that are required to understand how the theory is mathematically well-behaved. John Schwarz and Michael Green then discover a property known as T-duality and construct two different superstring theories—Type IIA and IIB—that are related to each other by T-duality. This becomes the key idea that underlies the next phase of evolution. (They also construct Type I string theories with open strings.)

Then, in the 1980s and early 1990s, physicists Warren Siegel and Joseph Polchinski discover that string theory requires not only extra dimensions,

Superstring Theory	Symmetry Group	Number of Supersymmetries
Type-I	SO(32)	1
Type-IIA	U(1)	2
Type-IIB	none	2
Heterotic O	SO(32)	1
Heterotic E	$E_8 \times E_8$	1

Figure 12.1 Shows the primary distinguishing characteristics among the five superstring theories.

but higher-dimensional objects within those dimensions. These are called D-branes (which we encountered earlier). It quickly became clear that branes, not just strings, form the matter content of string theory. Even more stunningly, a connection is found between these objects and a mystery that was predicted by general relativity—strings and branes are a type of black hole. (We'll explore this more later.)

Five distinct versions of string theories were believed to exist. They possess specific properties that differ slightly from each other. They are classified by the following descriptions and are summarized in Figure 12.1. They have mathematical behaviors among each other that are based on the notion of dualities (a duality is shared mathematics among two or more theories) that suggest unexpected relationships that connect the constructs of the five theories. As proposed by Stephen Hawking, the 11-dimensional supergravity theory, which is not a string theory, was also seen to join them via one of these duality relations. Using other dualities, it was argued that all of the models are related; that is, they are different aspects of the same underlying theory.

In 1995, Ed Witten, at the Annual Conference of String Theorists at the University of Southern California, gave a seminal talk showing that it was possible to unite the five string theories. This marked the birth of an 11-dimensional theory that he calls M-theory. Not long after this, in 1996, physicist Cumrun Vafa (who had earlier, with Andrew Strominger, showed that string theory predicts the thermal properties of black holes, themselves first proposed by Stephen Hawking in the 1970s), introduces a possibility he calls F-theory, that appears to have its origins as a 12-dimensional theory. You can see that the problem of dimensionality was not improving during this era, but as we will see later, the handling of those dimensions has vastly matured.

As string theory evolved, a more complete understanding of M-theory was identified as needing to be tackled in order to provide a unified description of the different types of strings. M-theory, and its intellectual relative, F-Theory, continues as a work in progress that will require further development before the physics of strings is understood.

Space Within Space

The contributors to string theory research number in the hundreds, going as far back as 1969. Many theorists earned their Ph.D. degrees engaged in research on this subject. Thus, this branch of research has most often been advanced by the research work performed by the youngest members of the community, including graduate students. This is not unusual. In fact, unique work by Ph.D. candidates is a requirement for achieving the doctoral degree. Young scientists and new discoveries go hand-in-hand.

Lisa Randall, whom we met earlier, is one such physicist whose early work as a young physicist left an indelible imprint on the modern framework of string theory with the much-venerated Randall-Sundrum model, created together with Raman Sundrum in 1999. The Randall-Sundrum model is not a string theory. It is a hyperspace theory (one having more than three spatial dimensions) pertaining to interactions in the universe.

Brian Greene was another young physicist who carried forward the work of understanding superstring theory. Born in 1963, he attended Stuyvesant High School in New York City with Lisa Randall. He received his Ph.D. in physics as a Rhodes Scholar at Oxford in 1987. He is a professor of physics at Columbia University and co-founder and director of the university's Institute for Strings, Cosmology and Astroparticle Physics (ISCAP). Greene promotes educating the lay public on string theory and other aspects of the universe, having authored four books on science.

After his graduation from Harvard in 1984, Greene's research concentrated on string theory. He (and others) worked extensively on Calabi-Yau manifolds, tiny six-dimensional shapes that appear curled up at every point in space that preserve the forces of nature, but are otherwise inactive participants in physics. With a group of other physicists, Greene explored the consequences of applying these objects to the questions of string theory.

Calabi-Yau manifolds describe the mathematics of the six-dimensions that are compactified into elaborate shapes similar to origami. They lead to equations that are consistent with SUSY being observable in nature and are believed to play a role in the interaction of force and matter in the universe. Because they follow the paradigm laid out by Oskar Klein, compacted extra dimensions, they do not interfere with the normal workings of the universe because they are very tiny. Calabi-Yau manifolds are analogous to the pinholes that were punched into a smooth floor discussed earlier. These "holes" have intricate origami-like shapes and occur at every point in space. They, being tiny, permit matter and force particles to interact properly and consistently in all three spatial dimensions of the real world,

Let's take a closer look at how superstrings and Calabi-Yau manifolds work together to help us understand the quantum world we have been exploring.

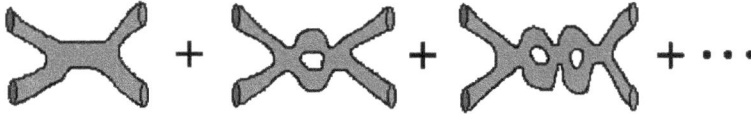

Figure 12.2   Perturbations of String Interactions (Adding Fluctuations)

Superstring Theory and Calabi-Yau Manifolds

In 1926, Oskar Klein's extended version of Kaluza's idea created a difficult situation. Instead of assuming the independence of the extra dimension, he assumed it to be compactified within the space of the four-dimensional model. He conceived of an approach in which the fifth dimension would have the topology of a circle, with a radius on the scale of the Planck length. Five-dimensional space-time, then, has the topology of a five-dimensional circle having a fifth coordinate that is a periodic circular string having characteristics such as those shown in Figure 12.2. Periodicity describes the number of times the string encircles the rod. It is equal to or greater than zero. A number of versions of this periodic type of string characteristic are shown in Figure 12.3. In the normal perception of space-time one would not be able to see this extra dimension.

When the dimension is compactified, its size is very small. The dimension can have structure, despite its size. For example, it might be as small as a very fine piece of string (a hair or thread), or it could be hollow, like a hose. It can take on many configurations. In each case the object will display a different form of behavior. We show in Figure 12.4 three types of surfaces having very small characteristics of dimensions. Each of these is a possible shape for extended dimensions.

Figure 12.4(a) shows, very close up, an ant walking on a curled-up dimension that looks like a rope, a string, or a straw. Standing far enough away, one cannot discern that the dimension is curled up as it appears to be a solid line with no third physical dimension. Segment 10.4(b) is an even closer view of the line thus registering on the eye a more apparent extension in an additional (third, in this case) dimension. At such close range it looks like a garden hose. You can see that there are three dimensions; R_X and R_Y are the two dimensions inside the hose. The third dimension is its length, L. Note that R_X and R_Y have

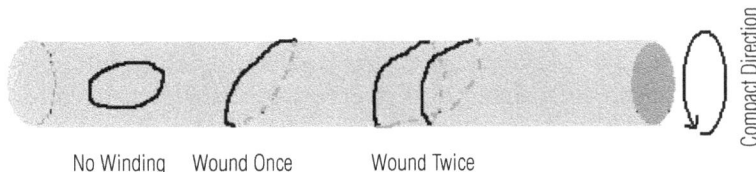

No Winding   Wound Once   Wound Twice   Compact Direction

Figure 12.3   Compactified Periodic Fifth Dimension

(a) Curled Up Dimension

(b) Curled-Up Dimension at Close Range

(c) Calabi-Yau Manifold (6 Dimensions)

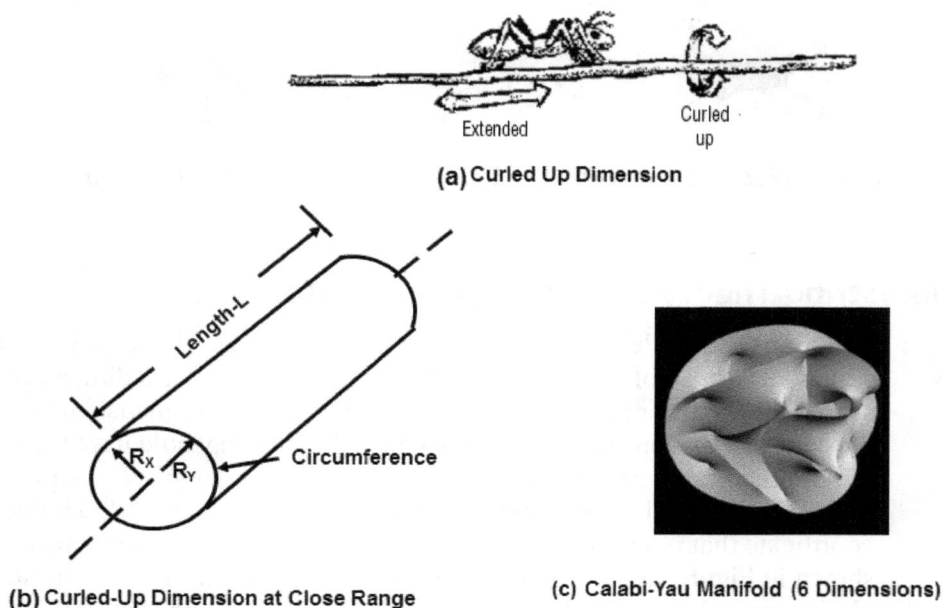

Figure 12.4   Kaluza-Klein Spaces with Compactified Dimensions

a limited extent in the inside dimensions of the hose, but L can have an infinite extent.

In Figure 12.4(c) we see a compactified, six-sided surface representing a Calabi-Yau manifold, named after its two creators, mathematicians Eugenio Calabi and Shing Tung Yau. Physicists became quite interested in this type of manifold as an object for describing the compactified six dimensions of a string over and above the (3+1) dimensions we can actually experience. To an observer viewing from higher dimensions, the physical effect that the manifold makes on the universe is negligible when it is considered in (3+1) dimensions. This surface shrinks the additional six dimensions into a very small space that resides at every point in the (3+1) dimensions of the universe. The universe can then be thought to appear like the hyperspace surface shown by the visualization in Figure 12.5. Instead of a single circular dimension occurring at each point in space, the six-dimension Calabi-Yau surface is placed there. To be explicit, the illustration in 12.4(c) is a projection of the 6-dimensional Calabi-Yau manifold into the three dimensions illustrated by the figure. This is but a 3-D shadow of a 6-D object, but this shadow is rich in interesting structure, a reflection of the even richer 6-D structure of its original form.

Kaluza and Klein's work resulted in a universe as visualized in Figure 12.5, a concept that unifies the gravity field of Einstein with the electromagnetic field of Maxwell. The compactified fifth dimension allows unification to occur

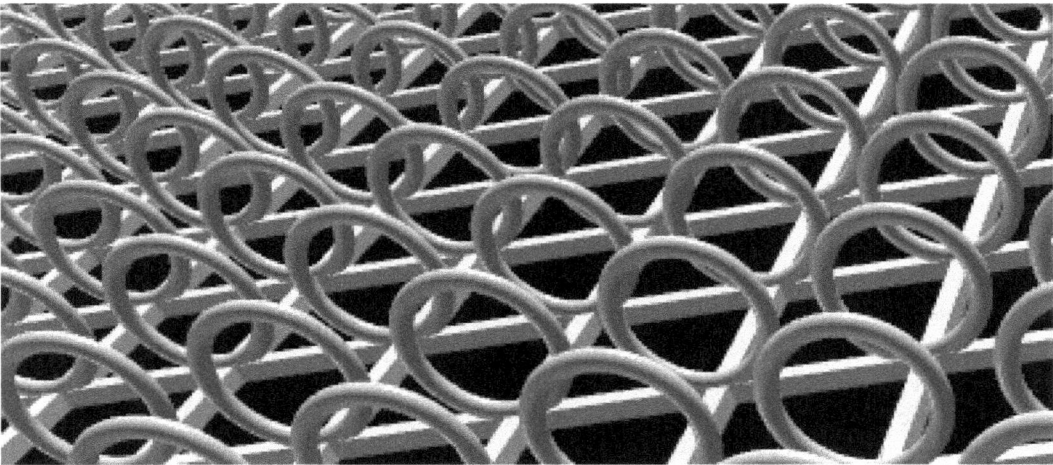

Figure 12.5   A visualization of a 2-dimensional space with an extra (third) dimension that is curled up at each point in the original space.

between the two forces. At the same time, this model did not upset the known 4-dimensional physics of electromagnetism and gravity thanks to the small size of the extra dimension. Once extra dimensions are admitted into a theory, the Kaluza-Klein approach demonstrates that unification is a likely possibility as a result of the extra mathematical freedom.

Kaluza-Klein space will be used frequently in the discussions that follow. It introduces and describes the multi-dimensional concept of hyperspace. The circles represent the fifth dimension compactified at every point in space. The grid on which the circles appear shows two dimensions (forward-backward and right-left); the third dimension (up) comes out of the page. Keep in mind that the dimension of time is not shown. The curled-up six dimensions may stay hidden from direct probing while still having consequences for the observable universe. Or, it could be that these dimensions are sufficiently large to be detected by a collider.

Given so many variables, it is necessary, based on the manner in which science is conducted, that physicists find experiments to assess or constrain the claims of string theory science. Experimentation and theory have always gone hand-in-hand in physics. In this case, string theorists are mathematically (i.e., theoretically) proposing objects of such tiny spatial extent that particle colliders are presently not capable of measuring them because these devices do not operate at high enough energies. As we noted earlier, nature is capable of accelerating subatomic particles to much higher energies than humans can; however, even these energies pale in comparison to the energy that is believed to be needed to directly probe Planck-scale dimensions.

While the scales of length in string theory are much too small to be measured directly in modern particle experiments, there are aspects of string theory that may be measurable using today's technology or with technology to come in the near future. Supersymmetry is essential to the functioning of string theories. Perhaps the LHC will detect direct evidence for superpartners of the standard model particles. If so, this would certainly be a victory for the idea of supersymmetry; but, alas, it would not make the string theory picture any clearer, as SUSY is permitted to exist independent of string theory. Nonetheless, since string theory demands that supersymmetry must exist, it would bolster the cause of this means of unification of all forces in nature.

Based on these challenges, physicists are presently focused on those physical properties that are predicted by large extra dimensions (either within string theory or by employing alternative methods). That is the intent of the research that Randall, Sundrum, and many others undertook. The detection of extra dimensions at the LHC would provide a moment of relief for long-struggling string theorists, as string theory demands extra dimensions without giving a clear picture about how those dimensions would manifest themselves. Perhaps detecting something such as braneworlds will light the way! Still, it is important to keep in mind that the idea of extra dimensions came independently of string theory, decades before it was first conceived; evidence for extra dimensions is not definitive evidence for string theory.

Calabi-Yau Spaces

Let us look more closely at small, compactified extra dimensions represented by Calabi-Yau spaces. The mathematicians and physicists who worked with these objects realized that Calabi-Yau manifolds are more like coffee cups with handles than they are like spheres. A handle is attached to a cup leaving a hole through which one can place one's finger. For a very long time, mathematicians and scientists have known how to write equations describing an idealized coffee cup. Such equations describe holes in the surface configuration of a compactified space that can affect a string's vibration pattern. It is easy to expect that the vibrational pattern of a ceramic ball, when it is struck, will be very different from that of a teacup being struck, even if the two are made from exactly the same mass and volume of ceramic material. This concept implies that the form a manifold takes will affect a string's performance, and the mathematics confirms this intuition. Figure 12.6 depicts such surfaces. An arrangement of Calabi-Yau spaces located at each point in a 3-dimensional space is shown. These small surfaces determine the force structure of the space while they do not disturb its four-dimensional behavior. The question is, does this concept work?

Two dimensions of the three-dimensional universe are shown by the illustration—plus, at each point in this two-dimensional picture there is a six-dimensional Calabi-Yau manifold compactified to a very small surface. The third dimension can be assumed to be perpendicular to these two dimensions such

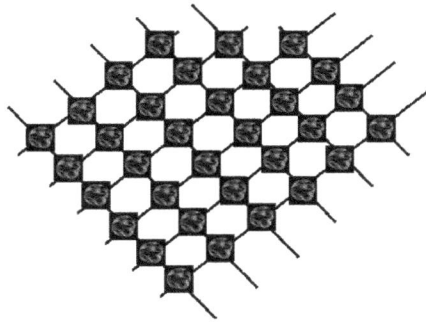

Figure 12.6: 3-Dimensional Space Using Compactified Calabi-Yau Manifolds

that it comes out of the page. The third dimension would contain an identical configuration of surfaces. This structure goes much of the way toward answering one of particle physicists' most intriguing questions: why are there three families of elementary particles—up/down quarks (electron and electron-neutrino), charm/strange quarks (muon and muon-neutrino), and top/bottom quarks (tau and tau-neutrino)? Why not one, or four, or any other number? The answer proposed by string theorists follows.

The universe, as viewed by an observer, would behave as though it had only four dimensions. The low-energy behavior of the higher dimensions would not be observed due to the inability to sense them. Nonetheless, the properties of the Calabi-Yau manifold have important implications for low-energy physics. The types of particles observed, their masses, quantum numbers, and the number of generations, constitute some of these properties. A major problem has been that there are many types of Calabi-Yau manifolds (thousands upon thousands) and no way to know which is the correct one to use. The mathematics of superstring theory began by describing a universe with ten-dimensional space-time (one time-like direction and nine space-like directions). By assuming that the shape of the extra directions are mathematically similar to surfaces like coffee cups possessing handles, string theory was found to possess the possibilities for a four-dimensional representation of physics.

A long-standing hope of string theorists is that a detailed knowledge of full superstring theory (M-theory, the underlying framework that leads to the known superstring theories) will provide an explanation of how and why the universe flowed from the ten-dimensional physics thought to exist during the high energy phase of the Big Bang, to the low energy four-dimensional physics that is observed today. This is a rich area of exploration in modern string theory. Can there be a singular, unique Calabi-Yau manifold that makes this work? Maybe there is a mathematical theory or framework that shows how all possible Calabi-Yau manifolds are related to each other in some simple way—so that it doesn't matter which one is chosen.

Or, maybe the notion of requiring extra dimensions is the problem in the first place. This is where some dissidence crops up. While supersymmetry plays nicely with ten-dimensional superstring theory, one of the authors of this book (SJG) proposes that supersymmetry also enables the possibility for four-dimensional, not extra-dimensional, string theory.

The Dance of Theory and Experiment

Theory and experiment in physics go hand-in-hand. The latter prevents the former from being a branch of philosophy (from which physics sprang as it evolved through the proofs provided by experimentation). Experiments confirm to physicists the real-world behavior of particles and forces. Using string theory, physicists employ mathematics to explore objects so tiny that experimentation is currently unable to test for their existence; so tiny that naively scaling present accelerator technology, one would require a collider that is at least the size of the Milky Way galaxy to detect them.

Physicists, instead of seeking to prove such small dimensions, are now focusing on those physical properties that are shown by string theory to be within the current range of experimentation. In his book, *The Elegant Universe*, Brian Greene presents his viewpoint on experimentation. He relates that many physicists believe that it is important to use experimentation as a means to underpin string theory. Such activity would strengthen the confidence that their achievements have merit and that their pathway is focused. This has been the way science has always worked. Through trial and error, experimental physicists have moved theoretical work along, demonstrating the detailed behaviors of particles and forces by using scientific approach to observe and explain investigated phenomena. Historically, theory has evolved from practical demonstration or it has been driven extinct.

Some now argue, according to Greene, that this may be the time for theoretical work to take the lead in defining theory so that experiments can be developed to assess the theoretical claims. This takes progress out of the realm of intuition into a mode in which it steers experimentation to yield indisputable outcomes. If not, we may continue to move haphazardly into an abyss of uncertainty. The result would be to achieve solutions in a more efficient manner; targeting with a rifle rather than shot-gunning.

To a degree not generally perceived by non-physicists, the reversal of the roles of theory and experiment *has* been occurring for a few decades. Almost all major discoveries in particle theory since the 1960s have had mathematical avatars that came before the actual discovery. The charm quark, first observed in 1974, is often described to have been a surprise discovery, except that physicists Sheldon Glashow, John Iliopoulis, and Luciano Miani had argued in 1970 that such a quark had to exist. It was found four years later. The recent detection of the Higgs boson is another example of a discovery preceded by the mathematical framework that predicted it.

On the other hand, it is wise to recall that, just when you think you have covered everything, nature has a way of throwing a curveball. When hearing the report of the discovery of the muon, the heavy cousin of the electron whose existence was not anticipated (because it was not needed to explain anything in the atom), the physicist Israel I. Rabi (1898–1988) is reported to have queried, "Who ordered that?" The tau lepton, the next-heaviest cousin of the electron, was discovered in 1977 by a team of physicists led by Martin Perl (1927–2014). Its existence was not anticipated by a theoretical need. While the existence of a fifth and sixth quark (the bottom and top quarks) was theoretically necessitated by the observation of certain symmetry violations in nature, there was no prior need from mathematics for another generation of heavy leptons. Similarly, the observation in 1998 that neutrinos have a tiny amount of mass was not anticipated. (We will return to this issue later.)

It is clear that a great deal of theoretical work is needed in string theory, and that this work is best guided by considering experimental results, but it is also wise to be open to the possibility that any future experimental measurement can yield a surprise, one that could topple a mountain of well-held theoretical physics.

Let's look to where the exploration of extra dimensional shapes and our understood reality can intersect—a place in string theory where, maybe, theory and experiment can help each other out.

The existence of generations of matter, each object heavier than the last, are described in the standard model, and the known number of generations, three, (see the discussion in Chapter 4 beginning on page 77) is the minimum number that a universe like ours would have to possess to allow for certain important symmetry violations to exist. However, the first cause of generations lies outside the standard model. Calabi-Yau manifolds, those compactified extra dimensions with holes, offer a striking way in which generations of matter could have emerged. Calabi-Yau spaces contain holes of various numbers of dimensions, each of which can produce a group of low-energy vibration patterns. Multiple holes can, therefore, cause multiple groups or families of vibration patterns. This conclusion goes far in answering a most intriguing question: Why are there three families of elementary particles? Why not one, or four, or any other number?

The fact that Calabi-Yau manifolds offer a physical means for matter generations to exist is an intriguing feature of the theory. Recall that Calabi-Yau manifolds are far more numerous than physicists had at first believed. When the idea was proposed, it was thought that there might be a few hundred or even as many as a few thousand such constructions. If the appropriate criteria for selecting a specific shape were to be found, this would provide a powerful validation of string theory. Logically, since strings vibrate through all dimensions, the shape of the curled-up strings will affect their vibration and thus, the properties of the elementary particles, making them observable.

Andrew Strominger and Edward Witten have shown in a Calabi-Yau manifold model that the mass of a particle depends on the manner in which the intersections of the holes and their positions appear relative to each other. This, of course, would have similar consequences for all particle properties. This idea has been extended from the notion of counting and restraining Calabi-Yau manifolds to a more general construct called "the landscape." Some of the best recent estimates argue that the number of such possibilities may be around 10^{500}—ten multiplied by itself 500 times!

How does one narrow this ever-expanding group of choices? How did nature choose? Or, maybe no choice was needed to be made. Let's look at how one might go about selecting a specific Calabi-Yau space and explore the question.

Choosing Calabi-Yau Spaces

String theory is written using perturbation theory, the same foundation that underlies the successful quantum field theories represented in the standard model. The idea is simple—start with the largest effect that could explain a physical process, and then add to it the next largest effect, and the next largest after that, until you've added up all possible effects from largest to smallest. It is this idea, embraced within an area of mathematics, that makes theories like QED so precisely predictive. Is more precision wanted? This is accomplished by adding additional small terms to the assessment of the problem.

However, in the case of string theory, this approach does not account for the criteria required for selecting a Calabi-Yau shape from the myriad of choices. To take the first step in a calculation, one must know what space to begin in to be able to identify the largest effect. Current research has refocused to find new non-perturbative methods that might yield more exact equations and more accurate criteria for selecting the shape.

Although there are currently no defined criteria for selecting a Calabi-Yau manifold from among the myriad choices, there are ways to narrow the field. At present it appears best to start with a Calabi-Yau that has three holes, thereby yielding three families as suggested by the standard model and historical experimental evidence. The mathematics of a Calabi-Yau manifold can be smoothly deformed to yield an infinite number of topologically equivalent shapes. Since the holes in a Calabi-Yau shape can be made to change in shape, size, and position due to these deformations, many Calabi-Yau manifolds can be eliminated. (Remember that a string's vibration which is dependent on the properties of its holes is what determines a particle's properties.)

Topological equivalency is the recognition that a coffee cup, with its one hole in the handle (a single perforation of the ceramic material), and a donut you might eat with the coffee, are mathematically (topologically) the same object— each has one hole. If ceramic were made malleable, like clay, it would be possible to deform a coffee cup into a donut shape without tearing the material—

they are topologically equivalent. If two objects are equivalent, it's a powerful reason for grouping them together and treating them identically. If a promising shape is topologically identical to a shape previously discarded because it fails to explain our universe, we can eliminate it.

Although Calabi-Yau compactification was greeted with enormous enthusiasm at the time of its discovery, in the time since then, physicists have realized that there exists a large number of alternative approaches. The names of some of these include "Gepner Models," "Orbifolds," "Asymmetrical Orbifolds," "Non-geometrical," and "G-Structures," among others. This list is not stagnant and has, for every few years since 1984, grown. There is no reason to believe this will stop. Among physicists who support the idea of "extra dimensions" or "hidden dimensions" in the context of a string theory-like approach to describing the physics of our universe there may be a holy grail found that provides an accurate description of our universe. This consideration of how generations of fundamental particles can emerge from the structure of extra dimensions (e.g. Calabi-Yau manifolds) is a current example of a continuing work-in-progress in string theory. The reader will appreciate that string theory, while progressing relatively rapidly, continues in its early stages of discovery as each problem is addressed while developing a mathematical formalization, observing experimental evidence of behavior, and the logical processes needed to describe interactions. The important question is: how does this science apply to furthering our understanding of the universe?

String Theory, M-Theory, and Hyperspace

After the type-I bosonic string was developed, a flurry of activity ensued to develop that theory, evolving into five string types. It was put forth that these theories are each aspects of an underlying and as-yet-unknown, more comprehensive theory. Witten suggested that these individual theories infer the existence of an underlying theory that, as does a foamy sea, hides what is beneath. He called this hidden theory M-theory. The letter "M" might stand for Master, Mathematical, Matrix, Mother, Membrane, Mystery, Magic, you choose it; but what was suggested by Witten himself was, *Murky!* Murky is plausible because the current understanding of the five superstring theories continues to be unclear as it is unknown how things will evolve.

As M-theory was being explored, the community continued to seek a breakthrough that would unify the forces of nature. Despite finding a greater set of challenges at each stage of its development than those overcome in the previous stage, string theory continues to be held by a majority of physicists to have merit as a means for studying particle interaction and other behaviors. Developing this idea, as had the development of all of the ideas that competed to explain the particle zoo in the 1960s, is a crucial step toward a fully developed theoretical framework that is ready for as many surprises as nature may hold.

Some string physicists prefer ten dimensions; M-Theory enthusiasts claim eleven; there is a small group who support F-Theory that hints at twelve. One of the authors, SJG, is self-described as a dimensional refusenik who is working to develop mathematics that proposes just four dimensions of space-time.

During the late 1980s, three different groups of physicists presented approaches that did not rely on "hidden" or "extra" dimensions. Among these, during 1988 and 1989, one of the authors (SJG), working with Warren Siegel, described an approach that was precisely built upon the mathematics of the standard model. The vast majority of today's physicists live and work in the extra dimensions of hyperspace, at least figuratively. That being said, the 4-D approach has not been investigated extensively and may offer a pathway—one that is a bit daunting mathematically—that avoids the criticisms of M-theory that have been posed in recent decades. We'll return to those criticisms, but for now let us consider a 4-D approach to superstring theory.

Let us recognize and respect our universe as it has so far been revealed by experiment: a 4-dimensional space-time containing forces and matter described by the standard model and its symmetry groups, a space-time that bends and curves in response to particles and forces while directing those particles and forces on trajectories prompted by its bending. There is nothing in this that violates any principles we know. If superstring theory is the correct description of nature, how do we make it comport with this long-observed set of facts?

One way might be to imagine that the universe is not simply a 4-D space-time, but a 4-D space-time in which the symmetry groups of the standard model themselves describe a kind of trajectory in the universe, like fibers stretching out from a 4-D world-sheet on which are embedded the matter and forces we think of as "the universe." Each symmetry group defines a fiber, and together you can think of these fibers as forming a bundle. This is not a bundle of fibers in the physical sense of a steel cable, braided from many individual strands of steel wire, rooted in a metal plate; rather, this is a mathematical description of the relation between these gauge groups and the 4-D space-time in which are played out the forces described by these groups.

How do superstrings fit into this picture? Our conceived fibers are not limited only to the symmetry groups found in the standard model. While it is the case that those symmetry groups are established as the backbone of the quantum realm, there could be other gauge groups that describe aspects of nature not realized in the standard model. Instead, those symmetries might describe the behaviors of superstrings. The presence of such symmetries has not yet been directly detected in nature. Indeed, in extending our mental picture of a bunch of mathematical strands rooted in 4-D space-time, we can imagine a fiber bundle of many symmetries, perhaps some (or all) of them contained in the table of superstring theories we presented earlier!

There is nothing in this view known to violate any principles established about nature, nor that renders superstring theory untenable; e.g., by reintroducing, say, the tachyon, which doomed early bosonic string theories. Supersymmetry is also preserved in this view. Strings can be open or closed; the graviton still emerges from this picture, and the standard model continues to be a reliable description of nature. However, this presently is a less-explored, dissident realm of superstring theory having challenging mathematical puzzles to solve. There is also no observation as yet that discriminates between this idea and the extra-dimensional view offered by M-theory and F-theory. There is only the absence of any detection of extra dimensions.

That cautioned, we are wise to keep in mind a core mantra of the scientific method: that the absence of evidence is not the same as the evidence of absence. Only a positive observation, positing the favor of one idea to counter another, will discriminate among things that seem otherwise equally valid under the "anything goes within the rules" of pure mathematics. (And, of course, both could be ruled out in the process!)

M-theory continues to be worked on as it is contributing to our further understanding. At the time of its discovery, it sought to promise to become the next great leap in creating a comprehensive description of string theory but has not yet done so. To restate, think of M-theory as a planet on which five islands appear in a global sea, each having characteristics at its shores (boundaries) that are similar to each other. With M-theory one becomes aware that the continents all exist on one planet. (See Figure 12.7.)

There are calculations in some of the five theories that are very difficult to perform. However, because of duality, the solution for these calculations can be found by utilizing one of the other theories, as it is just the same as looking at the sky from a different continent!

With all this said, a *complete* theory continues to elude researchers who have characterized M-theory as possessing boundaries described by the five superstring theories (as if it were a pie having five slices). Each theory is perceived at the boundaries of its operating characteristics, with little understanding of

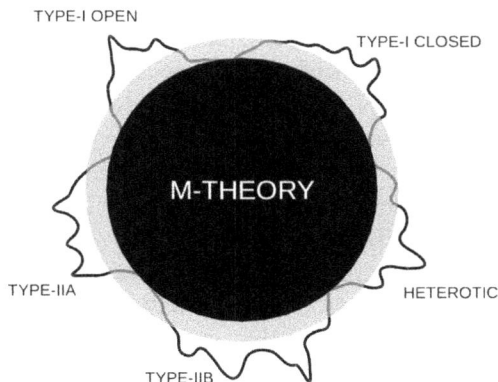

Figure 12.7 String theories are to M-Theory like isolated continents are to an imagined planet; their interconnections lie hidden under an ocean.

the pie's inside filling. For now, it is difficult to proffer a description that cuts more conclusively into M-theory. And, as we mentioned earlier, M-theory is not alone; there is F-Theory.

Paris Fires Upon Achilles

There is an "Achilles' heel" in the application of the compactification concept hiding those extra dimensions that is required by most of the efforts pursued by string theorists. This weakness is a scientific one that was noted by physicist Sheldon Glashow (of standard model fame) that goes as follows: there exists no systematic way to classify all possible compactification schemes. Most work on string theory ignores this point, but why should one care if there are not guidelines for ways to carry out compactification, as is pointed out by our earlier discussion about choosing Calabi-Yau spaces?

It matters because there result implications as to how to confirm string theory as an accurate describer of nature. Let us imagine that there are only ninety-nine known ways to compactify a given version of a string theory. (Of course, the actual number of compactification methods is tremendously large.) One would devise experiments to test whether any of these ninety-nine approaches accurately describes our universe. If none is found to be satisfactory, we would conclude that string theory is not viable in the description of nature. But, suppose that there is a one-hundredth compactification scheme out there that has not been recognized by any mathematician or theoretical physicist. It would then occur that the universe *is* described by a string theory, but no one has proven it (or, more correctly, *disproven* it) because there was no systematic way to arrive at all possible compactification schemes.

This may seem fanciful, but in reading the history of string theory it is noticed that new compactification schemes are presented with regularity. Although the Calabi-Yau method was the first to be proposed, there are now a great many. There are now so many that it is sometimes referred to as a landscape of possibilities. Until one can be certain that all possible routes to compactification have been exhausted, Glashow's challenge remains in effect. String theory might turn out to be a theory of *any*thing, rather than a theory of *every*thing.

With no end in sight, new features of string theory are discovered from time to time as with the discovery of fluxes (so named because the mathematics resembles that needed to explain the magnetic field lines from a bar magnet penetrating a metal conductor, or the "flux" of such lines). These are "yellow caution lights" that suggest that one should proceed carefully in the attempt to find the correct string theory to describe our universe.

Most of these developments are largely unknown to non-physicists. The systemization of extra-dimensional compactification mathematics is the critical mathematical challenge to the proposition that string theory is capable of providing testable predictions. There has so far emerged no systematic way to deduce all possible ways to compactify string theory. If observations are inconsistent with the predictions of then-known techniques for compactification, it is always possible to argue that there exists some other *unknown* method of compactification that *would* be in agreement. This is the essential technical point reflected by Glashow's remark that string theory is "permanently safe" from falsification. His criticism is valid so long as there does not exist a framework to deduce a "theory of compactifications" of the extra dimensions.

We cannot tell how the universe works without understanding the moments following the big bang and tying them to the fundamental structure of the universe. It is the only way we can analyze how it works and where it is likely to go. The standard model falls short of merging gravity with the other forces of nature and, even where it does apply, there are problems. The origins of dark energy and dark matter are unknown, and their influence on the cosmology of the universe is not understood. Will the universe inflate forever, grow cold, and die? Will inflation stop and reverse itself? Will some new action reveal itself? Until we know more about the workings of the dark side of the cosmos, we cannot make reliable predictions.

As a means for unifying forces of nature, physicists remain divided on the subject of extra dimensions. A major problem of string theory is that its mathematical results cannot be tested in the laboratory. Consequently, mathematics alone cannot solve the mysteries that are the pursuit of unification. There is the need for greater observation and more experimentation, but the technology to perform these activities is not always well defined by the ongoing work of string theory.

Even then there remains Glashow's challenge—without meeting that challenge, the way in which string theory can be utilized as a practical paradigm for future progress is not entirely clear.

Against Extra Dimensions

No tale of revolution is complete without dissidents, and one of the authors of this book (SJG), is a dissident about a core-claimed necessity of string theories—extra dimensions. Do we really need them? Perhaps the manner that string theory spun away from the first string framework set physicists off on the wrong road when it was required that there be twenty-five spatial dimensions.

This extra-dimensional debate continues on within a small community of string theorists who actively challenge the present ten-dimensional theory. This group pursues mathematics of four-dimensional string theory because it represents what is seen and experienced in our real world. Jim Gates, working in collaboration with theoretical physicist Warren Siegel, published a series of papers in the 1980s demonstrating that a four-dimensional string theory could be viable and should not arbitrarily be ruled out. Appearing in various scientific papers during the late 1990s, the basic mechanism that a four-dimensional description of strings requires was provided. This was corroborated independently by two other groups who employed methods different from each other. So, logically, while string theory admits descriptions that *permit* extra directions, it can be shown that it does not *require* more than three spatial and one time dimension to describe the behavior of our universe. Although the mathematics is more complex, it does work. So, there is room in the mathematics for a comfortable place having just our lovingly familiar four dimensions.

Getting back to topology, it is the area of mathematics that is concerned not with the detailed shape of objects, but their general surface and what properties remain fixed (or not) when that shape is altered. As we noted, a cup having a handle and a donut are considered to have the same topology. The importance of topology and its effects on the physical world cannot be overstated. The Nobel Prize in Physics was given in 2016 to theoretical physicists David Thouless, F. Duncan Haldane, and J. Michael Kosterlitz for their mathematical work showing how the topology of thin (plane) atomic structures can lead to many kinds of physical behaviors, potentially to include room-temperature superconductivity, a holy grail of technology.

In some of Gates's work in the 1990s, looking at how to define calculus in the presence of gravity *and* anticommuting numbers (the Grassman numbers we discussed earlier), it was shown that topology is important. So it is possible that quantities in mathematical models of physical reality that appear to be directions are not actually directions except when they relate to topology that is changeable.

The four-dimensional approach taken in the work of Gates and Siegel was inspired by looking at the symmetries of the original heterotic string having ten directions and 496 charges. The key point was how best to describe them in a way so that they aligned precisely with the concept of phases known already to appear in the standard model. This work can also be modified to accommodate numbers beyond 496 charges in such a way that the mathematical variables associated with them do not permit changes of topology and is consistent with the four directions of space and time. There are other string variables that permit changes of topology, but there are only four such variables, precisely the number needed to describe a four dimensional universe.

The statement declares that if directions exist whose properties describe topology change, they can be said to have an extra-dimensional characteristic. In the case of rotations, however, these do not appear to be caused by added dimensions. Rather, they appear to be rotational directions at different orientations in a four-dimensional geometrical space within which rotations can take place. It would appear that these rotations are more like fixed angular changes rather than continuously varying amounts. Such supersymmetric constructions do not lead to additional dimensions in their space. However, it must be first proven that supersymmetry occurs in our reality. It should also be understood that there has not occurred widespread interest in nor investigation of the Gates-Siegel approach. It occupies a dissident position within the community. However, a dissident position may make sense as it makes the case for a four-dimensional solution and is not subject to Glashow's challenge as are higher dimensional viewpoints.

Wrap-up

This chapter has presented two key characteristics of string theory—curled-up extra dimensions and hints of superstring theory: M-theory. The curled-up dimensions make it possible to satisfy the extra-dimensional mathematics of 10-dimensional string theory. This extra-dimensional construct conjectures six dimensions that are compactified into a very small surface (manifold) called a Calabi-Yau space.

This six-dimensional space has a great many possible configurations and is under study to determine how it might be correctly used and how to choose the correct configuration. At the present time there is no experimental evidence to demonstrate the existence of a 10- or 11-dimensional universe. We must continue to consider a four-dimensional string theory because that possibility exists in nature. On the theory side of this framework, there is no clear guidance about how to know if one has captured all possible ways of compactifying extra dimensions. Without addressing this and other challenges, the theory runs the risk of being "permanently safe" from falsification, a poor position for an idea that wants to be seriously considered as a scientific one.

We have progressed from the type-I bosonic string of the late 1960s to the supersymmetrical string, eliminating anomalies that existed in the earlier theory. The tachyon was successfully dealt with in the late 1970s and early 1980s. Supersymmetric string theory has emerged as the only string theory not possessing mathematical inconsistencies and is, at present, the dominant theory. There are five string types having similarities, similarities that Edward Witten observes and states implies the existence of an underlying M-theory, an eleven-dimensional universe in which the weak and strong forces and gravity are unified. Extra dimensions may be large or small. That there are so many questions left open, this is either a crisis or an opportunity, but the allure of string theory to unite all of the known forces and particles under a single framework is undeniable. The view of the cosmos as composed of many-dimensional objects and a rich variety of interactions and phenomena is grand.

But, maybe it is needlessly grand. We've put forward the ideas of a small group of physicists who embrace the string paradigm, but argue that there is nothing in nature, nor in the requirements of string theory itself, that makes it necessary that those extra dimensions be there. The mathematics starts out more difficult than one might like it to be at the beginning, but when has difficulty ever been an impediment to forwarding scientific principle?

What is unknown is whether this smallest shadow of reality, the superstring and its constellation of related ideas, is itself a trick of the light or embodies the true form that casts all other shadows.

## Chapter 13

A Shadow Where No Light Shines

On this odyssey to discover the workings of the universe, we must travel many paths. It is remarkable how pieces of knowledge gained by many researchers in mathematics, physics, and other subjects (that are not necessarily related to each other) have, together, over time, yielded important insights into the cosmos. Unrelated results influenced the following of new directions that otherwise might individually have gone undiscovered. String theory is a good example. While it has provided many new paths to discovery, its application to black hole behavior is of particular significance. Also, a recent discovery may put us on the way to experimentally probing space-time in ways that were not possible in past decades. A convergence of theory and experiment may occur thanks to one of the most unusual and mysterious actors in the cosmos, the black hole.

Curiosity about black holes stems from a realization that its mathematics are of the type typical of the physics of the universe, requiring both quantum mechanics and the mathematics of Einstein. Understanding the conjunction of these principles would help to understand the universe better, and could serve to validate string theory, as well.

The concept of the black hole was born even before Einstein's theory of general relativity. It was first proposed by Rev. John Michell (1724–1793). In addition to being a clergyman, Michell was an English natural philosopher and geologist. His interests included astronomy, optics, and gravitation. He was both a theorist and an experimenter. In 1783 he sought to discover the effect of gravity on light in a paper he wrote to the Royal Society: "If the semi-diameter of a sphere of the same density as the sun were to exceed that of the sun in the proportion of 500 to 1, a body falling from an infinite height towards it would have acquired at its surface greater velocity than that of light, and consequently supposing light to be attracted by the same force in proportion to its vis inertiae [inertial mass], with other bodies, all light emitted from such a body would be made to return towards it by its own proper gravity."

In other words, a star of the same density as that of the sun, but whose radius was five hundred times larger, would not be able to emit light because the star's own gravitational force would capture its own light! This is a remarkable insight given that Newton's *Principia* was written just ninety-six years earlier and Einstein's theory of general relativity would be completed one-hundred and thirty-three years after Michell's suggestion. Michell had conceived of "dark stars," an intellectual first given the then-recent foundation of the laws of mechanics. He also the fathered the idea of the binary star, a system that played a key role in the discovery of the accelerated expansion of the cosmos and plays an equally key role in a subject to be discussed later in this chapter, the collision of two black holes. Michell was a remarkable thinker.

An understanding of black holes lies at the intersection of cosmological and quantum theories, an understanding that did not arise without a struggle. When the quantum mechanics of black hole behavior was being developed, serious contradictions appeared. A black hole must be described by accounting for the behavior of both its singularity, the remnants of a dead star at its heart, and its event horizon, a virtual surface we perceive of as surrounding the black hole, beneath which surface even light, the fastest thing in the cosmos, cannot escape the singularity.

During the 1970s, many physicists researched and portrayed the black hole and the physics surrounding its behavior. We'll meet some of them, including one notable physicist, Stephen Hawking. Among other findings, Hawking's studies showed that, unexpectedly, energy could radiate from a black hole, developing an equation defining the temperature at which this radiation would occur.

The Black Hole Connection

Black holes are very mysterious celestial objects. We are just beginning to understand what they are and how they work.

It was Karl Schwarzschild (1873-1916), an astronomer and physicist, who first solved Einstein's equations for the case of a star collapsing to a very small radius, and thus preventing its light from escaping its own gravitational field. Schwarzschild's short life of forty-three years was filled with many accolades and accomplishments. He wrote his first two papers on celestial mechanics at the age of sixteen. He defined the Schwarzschild radius, the radius below which, if a body of mass, M, is shrunk, the ensuing warping of space-time became so extreme that not even light could escape. This radius is given by the equation:

$$r_S = 2GM/c^2$$

where r_s is the Schwarzschild radius, c is the speed of light, G is Newton's gravitational constant, and M is the mass of the body.

For instance, our sun has a Schwarzchild radius of 3 km. If one were to make our star into a black hole, its present radius of 700,000 km would have to be compressed inside of a sphere having a radius of 3 km. The sun is expected to end its days as a white dwarf star, having a radius of about 1% of its current size, or about 7.000 km. This is still very safely above the Schwarzchild radius; so exhale, the earth will not fall into a black hole.

The Schwarzchild radius also marks a black hole's event horizon. Much like the horizon at sea, an event horizon is the illusion of a boundary caused by curvature. The earth curves away in the distance causing the ocean to appear to end at some distant point. It is why sailors in the past feared sailing off of the edge of the world, even though the earth has been known to be round for almost 2,500 years. Similarly, space-time around a black hole is curved such that there is a point beyond which light can no longer escape, creating an event horizon and the illusion that there is an "end" to space (and time) at that location. For now, no event that occurs inside an event horizon can be known to people outside an event horizon.

Stars are born, they live for a while, they die, and, if conditions are right, they leave a black hole behind. However, there is another potential source of the creation of black holes; the big bang itself. Shortly after the big bang, the conditions were ripe for the creation of black holes. These are called primordial black holes to distinguish them from those that arise from the deaths of stars. Primordial black holes, together with dark matter, must have been the anchors for galaxy formation preparatory to the first generation of stars that began to shine with their own light.

The formation of a black hole depends on both the mass of and the radius of the star. After much searching of the heavens, a body of data has been compiled that suggests that black holes can be found in many places—in interstellar space, for sure, and at the centers of galaxies. This last statement may be true for *all* galaxies.

Black holes may be to galaxies what plate tectonics is to continents. The continents of earth are anchored to "plates," whose relative motion is a cause of earthquakes. It seems possible that galaxies are similarly anchored to black holes. Our own Milky Way galaxy possesses such a black hole "plate" called Sagittarius A* (pronounced "Sagittarius A-star"). It possesses the mass of about four million suns.

But the supermassive black hole at the center of our own galaxy is neither the only, nor the first, black hole of which humans became aware. The very first black hole candidate was catalogued in 1964, but only later (in 1971) was it realized to be a candidate for a black hole. Decades of experimental work was required to fully satisfy the scientific community that this was definitively a black hole, but by the 1990s this was well established. Let's look more closely at the black hole called "Cygnus X-1," to help understand how one can detect an object from which no light can escape.

Cygnus X-1 is an x-ray-emitting object in the constellation Cygnus, the swan. Earlier, we mentioned binary stars, partner stars that dance around each other for hundreds of millions, or even billions, of years. Cygnus X-1 is part of a binary system. The partner star, named HDE 226868 is visible. It is a blue supergiant star. Blue supergiants are at the top of the stellar "Main Sequence," a nomenclature plot of brightness and color. This type of star lives fast and dies young as they deplete their hydrogen core quickly and move to a late stage of hydrogen fusion causing them to become supergiants. HDE 226868 is about 400,000 times brighter and has a mass about twenty-five times our sun. She is probably only a few million years old. When HDE 226868 was born from a condensing cloud of interstellar hydrogen gas, the first modern humans had probably only just parted ways with our upright-walking ape cousins.

HDE 226868 is seen to "dance" with its partner. Measurements of this dance have concluded that it takes about 5.6 days to complete one mutual revolution. Compare that to the earth, which requires 365.25 days to complete one revolution with our dance partner, the sun! Factor into this calculation that HDE 226868 has a mass that is almost eight million times greater than that of the earth. What fearsome gravitational partner could make a supergiant blue star dance so quickly while locked in the strong embrace of its gravity?

If you look with your eyes, you cannot see an answer to this question, as no visible light comes from the partner. How strange! If one uses the measured information about HDE 226868 to estimate the mass of its partner, one concludes the mass to be about fifteen times that of the sun. We know many stars with masses in that range. They shine brightly in the sky and are easily seen with the naked eye. However, no instrument to aid the eye will reveal HDE 226868's unseen partner.

It was the detection of Cygnus X-1 using x-ray light that provided the first evidence of its existence. The partner star, HDE 226868, was established as its partner much later than that. From careful measurement we know that the pair lies about 6,070 light-years from Earth. We presently see this system as it existed 6,070 years ago, when its visible light and the x-rays we observe today began to move toward us on their long journey.

Cygnus X-1 was the first object to qualify as an excellent candidate for being a black hole. Its mass is huge, yet it is invisible to the naked eye while it emits x-rays. This is the same way we know that there is a black hole at the center of our own galaxy, Sagittarius A*. We know this not by direct observation, but because we can see stars very close to it that cycle at incredible speed around nothing that appears to be there.

But how do we know that the unseen partner in Cygnus X-1 is a black hole? To comprehend that, we need to know the proper conditions for the formation of a black hole. The conditions for the collapse of a shining star to a black hole is called Chandrasekhar's Limit, named after Subrahmanyan Chandrasekhar (1910–1995). Chandrasekhar was born in Lahore in British India, before it was partitioned into India and Pakistan. He earned his Ph.D. in 1933 from Trinity College at the University of Cambridge and is famous for his ground-breaking work in the evolution of stars.

Chandrasekhar discovered an important limit in the mass of a star. While he was not the first to do so, his work was independent and far more precise than the work that had preceded his. He was only nineteen years old.

In 1935, just after earning his Ph.D. and seeking to have his results accepted by the scientific community, he found himself in contentious disagreement with the world's then-most eminent cosmologist, Sir Arthur Eddington. The Chandrasekhar limit he proposed was a statement about the maximum extent in mass and radius that can be possessed by a white dwarf star. Chandrasekhar argued that stars above this limit (1.4 times the mass of the sun) would collapse in runaway fashion to a black hole. Eddington, while he was certainly aware of the possibility for black holes to exist, rejected this idea and, being such a force of authority, many listened to him. But Chandrasekhar's argument was ultimately seen to be correct.

In further investigations over time, we now know that this limit is not the limit at which a star will collapse to a black hole. The further development of quantum mechanics revealed that the nuclear matter in the collapsing core must first almost entirely convert to neutrons, forming a neutron star. Because neutrons are fermions, and fermions are forbidden by their spin-1/2 nature to occupy the same point in space together, they resist further collapse *unless* the stellar core is about three times the mass of the sun. Physicist Robert Oppenheimer, and others, using the concepts that Chandrasekhar had developed, combined with the state-of-the-art quantum physics of 1939, refined this picture and found that this slightly higher mass was one at which collapse is unstoppable. Even the fermion nature of neutrons is not enough to save this dying star. Stars above the Chandrasehkar limit will result in a neutron star. Stars above the Tolman-Oppenheimer-Volkoff (TOV) limit (a limit on neutron stars analogous to Chandrasekhar's white dwarf limit) will form a black hole. You can see why Cygnus X-1 is an excellent candidate to qualify as a black hole—it possesses a mass about five times above the TOV Limit.

A black hole is a highly compressed state of subatomic matter. The location in space where the original matter resulting from the death of a star resides is known as the singularity of that black hole. It is predicted to occupy a volume smaller than the nucleus of an atom. It was John Michell, with all of his imagination and armed with the power of Newton's mechanics, who conceived of dark stars larger than our own sun and having the immense gravity necessary to stop its own light from escaping. General relativity and quantum mechanics go further, to allow for something far, far stranger. A black hole is truly a dark atom, one whose mass is greater than our sun but whose size is smaller than the nucleus of a single typical atom. Where typical atoms reveal their presence by emitting light, black holes do not.

We see why a unified field theory becomes a necessary tool for probing the universe. If we are ever to fully understand the black hole (and the birth of the universe), we need a framework that comfortably marries gravity and the quantum realm.

Some Things Black Holes Do

Black holes are like the beast, Charybdis, of Greek myth. This ship-eating whirlpool, depicted in Homer's *The Odyssey*, was a danger that Odysseus, the protagonist, was forced to sail near to if he was to continue his heroic journey. Charybdis didn't come to him, Odysseus had to go to it. The danger of a black hole to matter is not that a black hole encounters matter because collisions between stars, alive or dead, are rare. Rather, it is when matter comes to a black hole that things get strange.

Because a black hole is an extreme warping of space-time, space and time behave oddly as one draws near to a black hole. Time runs slower and slower as objects come closer and closer to the singularity. Because time measurements are relative to something, by slower we mean slower relative to the clocks held by an observer far from the black hole. Space-time is so stretched that time is stretched, too. While people close to a black hole do not notice this effect, far away observers will perceive astronauts who are growing closer to the singularity to move more and more slowly as they approach it.

While this may sound inexplicable to you, this effect is present in our everyday lives. If you have a mobile device with you, take it out. Does it have the ability to locate your position on the earth using GPS, the Global Positioning System? If so, choose an app that has a map that shows your location. That a GPS device linked to the GPS system can so accurately place you as a marker on a street map is thanks to time dilation. How is this done?

You and your phone are standing in a place on the surface of the earth. The satellites that calculate GPS location information are orbiting the earth twice a day far above you, about 20,000 kilometers (almost 12,500 miles) from the earth's surface. The earth's gravitational field at the surface is much stronger than it is 20,000 km from the surface. As a result, time runs ever-so-slightly more slowly down here than it does when compared to the orbit. Astronauts do not notice this effect, but the atomic clocks on GPS satellites can!

This effect must be corrected for. If not, the clocks on the GPS satellites will, more and more, lose synchronicity with their twin clocks on earth. This would result in your position becoming more and more inaccurate each day! Software and hardware are designed to correct for this, keeping the GPS system accurate. Without these general relativistic corrections, the GPS system would misplace positions by about 10–11km (about 6 miles) during the course of a day.

Let's look at some other aspects of black holes. They are often accompanied by a halo of ordinary stars and other matter (such as gas) that orbits them just as planets and asteroids orbit our sun. This halo-like structure is called the "accretion disk," but, unlike our solar system, their activities can be violent and chaotic. Sometimes material of the accretion disk crosses the event horizon and is sucked into the singularity. For now, we cannot know what happens to matter that sails over this horizon. We can only observe the matter left behind, spun into a hot frenzy on its journey toward the point-of-no-return.

It has been observed that black holes are sometimes dormant for periods of time. Dormancy means that they do not appear to suck in matter. Sometimes stars and gases that are in close enough proximity to a black hole are *not* swallowed by that black hole. Then, suddenly, a black hole begins again to take in matter. This is due to the orbital mechanics of the material that circles the black hole. It is similar to a bunch of demolition cars driving around a track. The matter around the black holes is like the cars. When two or more of them bump into each other, one or more might be slowed down enough so that the gravity of the black hole is sufficient to draw them in. Or, another way of thinking about it is that the orbiting material, in rubbing against itself, causes friction that slows the matter down so that it is drawn into the black hole.

However the slowing down occurs, once it does, matter is drawn into the black hole and astronomers have then observed that the region around the black hole emits radiation in copious amounts. This radiation is in the form of high-speed jets that likely come from the poles of the accretion disk. (If the disk represents the equatorial line in our picture, the jets are emitted from the north and south poles above and below the disk). The reasons for this are not clear, and are an area of active investigation. One would expect to turn to Einstein's general relativity and the standard model for answers, and, for the region around the black hole where gravity warps space and time but is still weaker than the forces in the standard model, this works well.

But, if we want to try to understand whether the black hole *itself* is capable of emitting radiation, we need to look deeper—to investigate where no reliable theory yet exists. Black holes were the subject of intense study in the 1970s, not only because they were simply interesting, but because the equations that govern their behavior are identical to equations that describe the universe, requiring solutions both to Einstein's equations as well as quantum mechanics. Intense efforts to characterize black hole physics inspired and were led by Stephen Hawking. One surprising result of this study was that the mathematical laws that describe black holes are identical to the laws that describe energy, heat, and entropy. These latter laws, the laws of thermodynamics (the scientific description of the character and behavior of heat energy), are one of the great triumphs of eighteenth- and nineteenth-century physics. The equations that relate to thermodynamics apply equally to black holes with just a little switching of symbols!

Let us meet some of the physicists who played notable roles during this period, and explore their ideas. Stephen Hawking was born in Oxford, England. His early academic life was fraught with fits and starts, and by his own account he did not present as academically adept or gifted when he was in school. He studied physics and chemistry, switching to cosmology in graduate school where he struggled as well because he felt insufficiently prepared for his area of research, general relativity and cosmology and because he was diagnosed with amyotrophic lateral sclerosis (ALS), a motor neurone disease known also in the United States as "Lou Gehrig's Disease." Nonetheless, Hawking recovered from these early setbacks, encouraged by his Ph.D. advisor, and completed his thesis on subject material related to the big bang and the birth of the universe from a singularity-like state, much like the state found in the center of a black hole.

Another important player in this story is Jacob Bekenstein (1947–2015). He was born in Mexico City, the child of Polish immigrants. He located to the United States when he was young and eventually gained U.S. citizenship. He earned his Ph.D. from Princeton University in 1972, working as a student with the physicist John Archibald Wheeler (1911–2008). Wheeler is one of the fathers of modern general relativity, having written a book on the subject with two other co-authors. He is better known outside of the physics community for his succinct description of general relativity, which can be paraphrased as "Space and time tell energy and matter how to move, and energy and matter tell space and time how to bend." Bekenstein also worked on general relativity with a focus on black holes and the states of matter and how they affect space-time, as well as how cherished notions of energy, matter, and information are related to these concepts.

Bekenstein and Hawking in the mid-1970s each made the startling proposal that black holes, due to the laws of quantum theory, must emit heat! Although general relativity would have them simply to be black holes that can emit nothing at all, if one looks at the quantum realm in the area just at the event horizon of a black hole, one finds that things are more complicated. Bekenstein predicted, based on his own work, that black holes have temperature (implying that they are capable of radiating energy, much like a hot cup of coffee radiates heat energy) and that they can be described as having entropy, the physical measure of the disorder of a system. Temperature and entropy are strongly related. When both of these features are present, it means that the laws of thermodynamics and black holes can be thought to be interconnected.

However, some of Hawking's early work on black holes suggested that what Bekenstein concluded was not correct, and so it was that the two found their work to be a bit at odds. But Hawking went further to consider what might happen, from the quantum perspective, at the event horizon of a black hole. In doing this, he concluded that it is possible for black holes to radiate energy. This would give them both temperature and entropy, contradicting his own earlier work.

To reach this new conclusion, Hawking had to make a leap of faith without scientifically rigorous investigation. He did this by making a guess about some of the terms in an equation as there was no way to mathematically substantiate them. It was found later, in studies led by Andrew Strominger and Cumrun

Vafa in 1996, that string theory provided mathematical evidence that Hawking had gotten it right! Thus, one can unify the two theories, thermodynamics and black hole physics, when defining a black hole's behavior. This gave new motivation to physicists by providing a reason to pursue the physics of black holes in order to help to move string theory forward.

Until string theory was applied, the black hole was very difficult to analyze. Because string theory unifies gravity with quantum mechanics, it is an important tool that combines the positivist's knowledge of the macro-world with that of the quantum physicist.

The singularity at the center of a black hole is the heart of the system. Its characteristics define the environment surrounding it. It is so small that it must be part of the quantum realm, but it is so heavy that it has macroscopic effects describable by general relativity. Still, we have no well-established fundamental theory that unites these two pictures. Singularities have characteristics that cannot be predicted on the basis of what is known about nature; that is, they cannot be described by laws known to physics prior to string theory. Note, too, an important characteristic of string theory is its apparent ability to unify gravity with the standard model particles and forces is now extended to the realm of heat, energy, and entropy, the realm of thermodynamics.

We believe it is important to note now the further unifying ability of superstring theory. Superstring theory has shown an ability to unify gravitation with the elements of the standard model as well as to unify the two with thermodynamics. None of the independent alternatives, such as branes or supersymmetry, have been able to successfully perform this triple play. (Incredibly, as we have noted, none of this was the vision for string theory when it was first created.) This poses that string theory with supersymmetry can be more credible from a physics standpoint, even though it presently lacks having experimental proof. Because it seems capable of fully analyzing a black hole, string theory physics may be useful in going farther in the description of the behavior of the universe. Hang on as we go deeper into the black hole.

Black Holes, Strings and a Paradox

We will use the intersection of black holes, thermodynamics, and string theory to illustrate how superstrings and their associated constellation of mathematical insights might help us to better understand the black hole. This is a story about a schism in part of the physics community, about what radiating black holes might do to the universe, and our understanding of the rules that govern it. This story was detailed in a book by Leonard Susskind dramatically entitled *The Black Hole War*.

The process of analyzing black holes uses string theory and a combination of Einstein's equations and quantum mechanics. For the latter, a majority of physicists agree that this can be done using string theory, because string theory can describe the graviton, making it capable of synthesizing both of these realms. Susskind describes how, in the unorthodox environment of a new-age gathering in 1983 in Werner Erhard's home that became known as Erhard Seminar Training (EST), Steven Hawking gave a presentation that reported on the status of his black hole research which rattled the foundations of the laws of physics.

Hawking's work had led him to a conclusion that was startling to other physicists: *that the the information pulled into a black hole is lost forever.* Matter and energy, along with all of its quantum numbers and force interactions, carries with it information about the state of the universe. What if those bits of information are sucked into a hot, black hole? At a certain temperature, Hawking asserts, Hawking radiation occurs. When this happens, the black hole evaporates, losing matter and, therefore, with it, its information. This evaporation, he continued, *causes the information to be lost forever.*

This is an astonishing assertion! The reason for this startling result, he claims, is that quantum mechanics cannot be reconciled with Einstein's theory of gravity in describing black hole behavior.

Hawking states that there is a quantum gravity contradiction that causes a breakdown in black hole theory. He suggests that quantum mechanics needs a correction. He claims that a black hole gives off radiation when it is hot, thus causing it to shrink. This radiation eventually causes the black hole to evaporate entirely. He further maintains that when this happens, all information collected in the black hole is lost forever (the so-called "Hawking information paradox"). This created the fear that, should this be true, the foundation of all physical sciences would be in jeopardy. Why? Because the basic laws of physics, including thermodynamics, would experience key violations should this happen.

Many eminent physicists were present at this meeting. After Hawking's assertion, the physicists split into two factions that were immediately at odds with each other; the cosmologists (led by Hawking) siding against the quantum physicists (Leonard Susskind, one of the "fathers" of string theory, and Gerard 'tHooft, Nobel Prize winner for his contributions to the foundation of electroweak theory, led the principal opposition).

Susskind and 'tHooft sided against Hawking's assertions. They had been involved in researching black holes in connection with gauge theory and string theory, respectively, and, believed this matter (or, rather, the *destruction* of matter) to be totally incorrect, strongly communicating their opposition to Hawking.

It is true that the black hole is a singularity, and whatever matter passes too close to its event horizon will be sucked in and should never be seen again. According to Claude E. Shannon (the father of information theory, 1916–2001), matter can be shown to be bits of information and would therefore be preserved. Information theory, which has deep connections to the laws of thermodynamics, suggests that even though the matter is lost across the event horizon, its information content should not be gone. It is simply locked into a box with a one-way door. That information will continue to exist because the first law of thermodynamics says that energy can neither be created nor destroyed.

The second law of thermodynamics describes entropy, stating that all matter within a closed system, where no external energy or force can act upon it, moves from an ordered state to a state of disorder. Entropy is the mathematical measure of that disorder, which, by the law of entropy cannot decrease, so it must increase.

There are many ways that the atoms in your body can be arranged. Each possible way of arranging those atoms is called a "micro-state." Consider the atoms in your hand. There are only a few ways, out of the many possible ways of arranging them, that the atoms in your hand can be arranged to make your hand a functioning hand. Because there are so few, out of the possible total of arrangements that "work" as a hand, we can say your hand possesses a state of low entropy. But should you leave that hand to itself for a long time—stop eating, stop drinking water, die, and decay—the atoms will begin to occupy the *more likely* arrangement of micro-states. Those atoms will move to configurations that vastly outnumber the ones that allow for a functioning human hand. We can then say that *those* atoms are in a state of high entropy, occupying more probable arrangements of micro-states. Only the action of an external force that would put those atoms back into micro-states that lead to a functioning hand can reverse this situation. This is the march of entropy, absent external forces.

Black Hole Theory

Hawking's and Bekenstein's work in the 1970s showed that black holes can have temperature and entropy, demonstrating a deep connection between black holes and the laws of thermodynamics. Indeed, one can relate physical properties of the black hole, such as the area represented by its event horizon (treating the event horizon as a thin spherical shell, with a specific radius around the singularity) to its thermodynamic properties, such as temperature. By combining quantum mechanics and thermodynamics at the event horizon, Hawking arrived at one of his most famous equations:

$$T_H = \hbar c^3 / (8\pi G\, M_{BH}\, k_B)$$

where T_H is Hawking temperature, $\hbar$ (h-bar) is Planck's constant, G is Newton's gravitational constant, M_{BH} is the black hole's mass, c is the speed of light, and k_B is Boltzmann's constant. (Some of the "greatest hits" among the constants of nature appear in this equation!) The black hole's mass would appear to be a measure of its temperature, the Hawking temperature, T_H.

Bekenstein showed that the area (A) of the black hole is related to its entropy (S_{BH}) in his own famous equation:

$$S_{BH} = (k_B A)/(4L_p^2)$$

where L_p is the Planck length, 10^{-33}cm. From these equations we begin to see predictions emerge. As the mass of a black hole declines, its temperature increases. It radiates more energy, which would, in turn, lower its temperature. Black holes might "explode"— radiate faster as they shrink until they disappear altogether. We also see that, if the area of a black hole declines as it radiates away, its entropy decreases too.

This latter formula is the Bekenstein-Hawking formula, and we'll return to it when we discuss gravitational waves. For now, what's important to realize is that while Hawking and Bekenstein were able to *derive* these relationships, they could not *prove* them. There was no mathematical formalism allowing a demonstration of the deep origins of these relationships. This left open the possibility of interpretation without proof, and in part this is what sparked the "black hole war." One can read implications from these equations, but without an underlying framework to tell you *why* they are of this form, you are making predictions on shaky ground.

Black hole entropy poses some interesting conundrums. To help one to understand them, a discussion of entropy in the context of the black hole is appropriate. In a typical real-world situation, a system has entropy when a large number of different micro-states can satisfy the same macroscopic condition. Recall the analogy of how to arrange atoms to make a functioning human hand. However, unlike your hand, a black hole is believed to be featureless. What, then, are the degrees of freedom that can give rise to black hole entropy?

What does a physicist mean by a degree of freedom? It means all the possible ways a system can carry energy. For example, a bowling ball can move within any of three independent directions in space, but it can also spin about any of three independent axes. Any fourth axis you might choose in either of these two cases can be written in 3-dimensional terms as regards the other three axes you have chosen. You have run out of freedom to allow the system ways in which to move. So we would say that it has 3+3 (totaling 6) degrees of freedom. How does one describe these for a black hole?

We need a framework on which gravity and quantum mechanics play nicely together. One of these is superstring theory. Instead of a pure superstring approach, theorists have taken higher-dimensional approaches that can ~~could~~ maintain supersymmetry. Calculating the entropies of hypothetical black holes yield results that agree with the expected Bekenstein entropy. Unfortunately, cases studied so far involve higher-dimensional spaces—D5-branes in nine-dimensional space, for example. They do not, as yet, apply to the familiar

Schwarzschild black holes that are observed in the real universe. The challenge of having many dimensions in string theory is again apparent here, in the calculations of black hole physics.

Inasmuch as string theory is a quantum theory, black hole behavior is an important approach to help physicists to understand how string theory might describe real-world behavior. By taking the position that black holes lose their information forever, Hawking was appealing to the physicists in attendance at the 1983 meeting to be certain that string interaction mathematics agreed with time-tested well-established physics. One principle of long standing is that all known physical processes are "time-reversible," a fact that has been proven across many years. When time is imagined to go backward from its present state, the conditions of a process or a system at its initial state can be retrieved. This allows that, for any earlier state, in any known process, one can retrieve detailed information about the state of that process; its atoms, their locations, how fast they were going, and so on.

To understand how this works, assume that you had put two ice cubes into a tub of water on Monday. The ice cubes will melt into the tub, disappear, and are not seen on Tuesday. On Tuesday, should another person want to know what the condition of the water was on Monday, that person should be able to measure the details of the constituents of the tub (the quantity of atoms, their speed, their temperature, and so forth). Using these results one could go backward in time to Monday to learn that, on that day, two ice cubes were put into the tub. In reality, this would be very difficult to accomplish, but in principle one should be able to do this.

Entropy completely defines the makeup and information content of a black hole. As matter in-falls on a black hole, its mass necessarily increases, requiring its area to increase and thus pushing out its event horizon. This implies that the event horizon characterizes all of the matter that has fallen into it, matter being synonymous with information. Thus, the surface area of the event horizon should contain information that can completely define the contents of the black hole. The event horizon is a 2-dimensional structure, the thin surface of a sphere. The black hole is a 3-dimensional structure, a sphere with internal structure. However, a 2-dimensional surface seems capable of giving us overall information about a 3-dimensional structure. Is this possible?

It is. There are such things in nature with which humans are already familiar: holograms. Holograms are 2-dimensional structures that, when light is shone upon them, allow for the eye to perceive the original 3-dimensional object used to make the hologram. Turn the hologram in your hand, and the 3-dimensional image also turns, allowing you to see features that weren't present to the eye a moment before.

The event horizon is like a hologram of the black hole, a surface that possesses all the information about the micro-states within the black hole, sufficient as to reconstruct that black hole. Let's look at this more closely.

The Holographic Principle

The holographic principle can be better understood by reviewing how a physical hologram is created. The characteristics of a hologram, what it is and how it is obtained, will help to describe the principles behind it. The holographic process was discovered by electrical engineer and physicist Dennis Gabor (1900–1979) in 1947, for which he received a Nobel Prize in physics in 1971.

Gabor developed the theory of holography while working to improve the resolution of the electron microscope, a device that uses the short matter-wave-length of electrons to make images of atomic surfaces that light cannot resolve. A hologram is an image made by exposing film (or some other memory medium or device) to the interference patterns created when two laser light beams are made to shine on an object. When two beams shine on object, their light waves "interfere." This interference is similar to the behavior of water waves as they pass through a board with two holes, as illustrated in Figure 13.1. The wave to the left of the board is created, say, by moving your finger up and down in the water. This movement induces ripples (fluctuations) in the water that travel outward in concentric circles.

When they strike the board, they pass through the two holes, spawning two, new circular waves that emanate from the holes. These waves spread out causing them to intersect with each other. As they intersect, they either reinforce or cancel each other (interfere with) dependent on their amplitude and phase at the point of intersection. To the degree of their differences, when they are *in* phase, their amplitudes (wave heights and falls) add together; when out of phase, their values are subtracted. For example, when one intersecting crest is positive, while the other is negative (they are *out* of phase), they cancel each other to the degree of the mathematical difference in the value of their positive and negative amplitudes.

The technique for creating a hologram is shown in Figure 13.2 In this diagram, a laser transmits a light beam to a beam splitter, which is an optical device that breaks the beam into two separate beams.

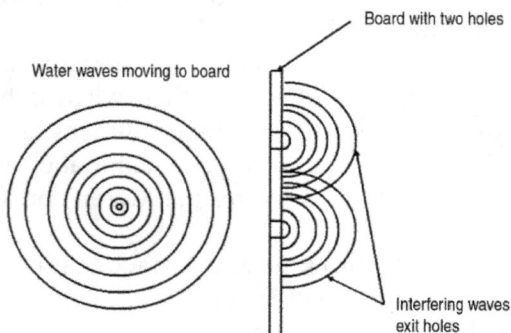

Board with two holes

Water waves moving to board

Interfering waves
exit holes

Figure 13.1 Water waves passing through a board with two holes.

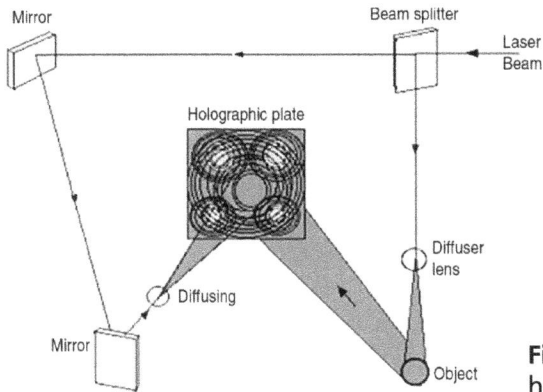

Figure 13.2   Creating a holographic image.

One beam passes through the either semi-transparent or dually reflective mirrored glass of the beam splitter, traveling thence to a fully reflective mirror. The other beam reflects off the beam splitter, directing it toward the object through a diffuser. The beam is then reflected toward the photographic memory (plate). The direct light, after passing straight through the splitter goes through two reflected diversions passing through a second diffuser before it, too, illuminates the photographic plate.

The interference of these two waves creates the hologram. The plate stores the resulting holographic image (the interference pattern) of the object. This process can also be completed using a computer's memory as though it were the photographic plate. The stored hologram preserves the amplitude and phase relationships of the interference pattern. This recording of the amplitude and phase pattern can be used to re-create a three-dimensional image of the original object.

Most people think of a hologram as a three-dimensional image of an object, but it is, itself, but a two-dimensional image, from which the three-dimensional image of the original object can be resurrected. An example of such an image is shown in Figure 13.3. The amplitudes and phases of the interference pattern on the hologram completely describe the image. It is called the holographic image. Not at all a pictorial representation of the object, note that the image has ripples

Figure 13.3

and undulations in it. These are a result of the interference of the two laser beams as they impinged on the photographic plate; one beam from the mirrored path, the other from the object. When the two beams overlap, their waves interfere with each other, creating this type of pattern. The beam from the object contains information from the object that is not in the beam that is reflected directly. The interference pattern contains the modulated information caused by one beam impinged upon the other, thus conveying the object's appearance.

The important information to take from this is that a complete three-dimensional image is captured on a two-dimensional photographic plate. In fact, any part of this interference image can be cut away multiple times using scissors and the remainder will continue to contain all of the information sufficient to reproduce the entire image of the object. Once again, this means that a three-dimensional image is preserved on an image plane having only two dimensions, one less dimension than actually describes the image.

It was work by Gerard 'tHooft, whom we mentioned earlier, who concluded that the quantum fluctuations at the event horizon contain all of the information necessary to describe the contents of the black hole. The contents, described by its entropy, are not lost, even as the black hole evaporates. 'tHooft and Susskind found that the information that falls into a black hole is preserved on its event horizon, a phenomenon they define as the holographic principle.

This principle establishes a new characteristic of nature that needs to be understood. The holographic principle can be extended to describe the entire universe. In such a context, the information contained within the universe is fully preserved at the boundary surface of the universe.

A critical insight for the physics of the holographic principle as it pertains to strings is that quantum fluctuations (which are like the light wave information encoded in photographic holographic patterns) at the surface (event horizon) of a black hole must also describe the space-time (the image of the black hole) within it. Thus, holography is a technique that allows a low-dimensioned system (the event horizon) to, in turn, describe everything that falls through it (which can be anything at all). Therefore, using the parallel concept of the universe's boundary as equivalent to its event horizon, black hole theory is a theory of everything! In this manner, string theory demonstrates that the behavioral characteristics, not only of a black hole, but the totality of the universe, can be described by the same mathematics.

The holographic principle shows that the contents of a black hole will be completely described on the surface of its event horizon, a two-dimensional figure. But, what does this mean more specifically? How, exactly, does the holographic principle apply to this issue?

The Black Hole Paradox and anti-de Sitter Space

Leonard Susskind generalized the holographic construct by asserting that the information content of a space in any defined region is stored at the boundary of that space. It doesn't just have to apply only to black holes. This means that it should be possible to obtain the description of a room in which one sits, a room that may contain tables, chairs, bookshelves, curtains, and windows, merely by studying the pattern of information from light on the walls of the room.

In many ways, this feels like a reminder moment calling back our use of the allegory of the cave. We have long watched shadows dance on the cave wall, but perhaps by thinking *holographically* we can learn about the higher-structured objects that cast those shadows without seeing them directly. Carrying this idea to its completion, Susskind showed that the contents of the entire universe are stored at its boundary surface.

This conclusion is fascinating, and will make greater sense after a discussion of an important concept known as anti-de Sitter space (AdS). In this discussion we show that this spatial description offers the ability to relate a higher dimensional space (4+1), called AdS, to a lower dimensional physical theory, one that humans can understand, (3+1) dimensional space (our comfortable home, the residence of the standard model). Indeed, the mathematics of quantum mechanics and the general laws of physics can be well described for this condition even if the universe is modeled by a higher-dimensional theory. Therefore, the discussion that follows is presented as a two-dimensional description to make it easier to understand. In this picture, the "elusive" higher dimension is an unseen third dimension. AdS mathematics has had many applications beyond the present discussion. It has been used successfully to demonstrate physical reality for the case of a quantum-critical state of electrons leading to high-temperature superconductivity. We know, therefore, that the mathematics of AdS works in the real world.

According to the field equations of general relativity, the shape of space-time is, in part, what determines the fate of the cosmos. The interplay of that shape and the energy-mass density of the universe allows the universe to expand, collapse, or remain still and unchanging. The case that we are interested in is the one that Edwin Hubble observed in his telescope in 1929. He noted that all distant galaxies were receding from earth in all directions, as if we were dots on the surface of a balloon that is inflating over time. As inflation continues, the spots on the balloon move away from each other in a manner that corresponds to the inflation pressure. Our universe is, thus, expanding.

The shape of space-time plays a role in this. We know from observations during experiments like the Planck satellite that the shape of space-time is very nearly flat, but it isn't mandated that it must be this way. In the early days of general relativity it was important to explore many possible space-time shapes to understand what kind of a universe results from that shape. For instance, should you want to define an expanding universe, one way to do this is to have a non-flat space-time. As was discussed in chapter three, this circumstance can take on a number of shapes. We will refer to one of these shapes as a manikin shape, such as that shown in Figure 13.4.

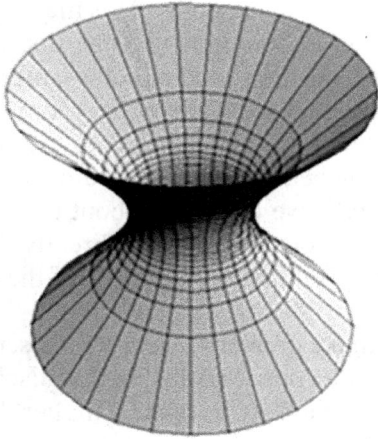

Figure 13.4

Saddle-like shapes, possessing negative curvature, are key to a model of space-time known as anti-de Sitter Space (AdS). It takes its name from the physicist Willem de Sitter (1872–1934), who discovered this curved surface when solving Einstein's field equations for positive curvature. We will see that these spaces have special mathematical properties that make them behave like our universe. Understanding *anti*-de Sitter space requires that we first learn about de Sitter space.

An example of a de Sitter space is the mercator projection method of mapping the surface of the earth (see Figure 13.5). If a paper cylinder were wrapped around the earth's sphere just touching its equator and the earth's contents then

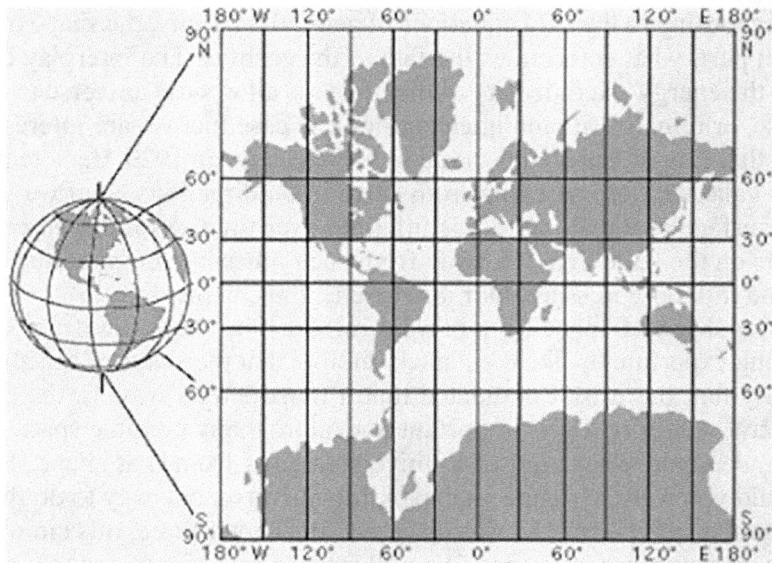

Figure 13.5 A Mercator Projection

projected onto the cylinder, the map on the right of the figure is what is produced when the cylinder is unrolled. Earth's surface is projected onto the cylinder producing a flat, two-dimensional map displaying all of the details that appear on the spherical surface.

The important feature of a mercator projection is that all latitude and longitude lines are straight and perpendicular to each other. This projection is like a surface with positive curvature, similar to a de Sitter space. The curvature produces a flat, two-dimensional surface like a mercator projection whereon the features will be distorted at its extremities. A sphere, like a mercator projection, has a two-dimensional surface on which latitude and longitude are its two dimensions. When a mathematician or a physicist says that a sphere is two-dimensional they mean that it requires only two numbers to describe a location. Although the earth is three dimensional, one may state the location of any point on its surface by giving only the latitude and longitude. By this definition the sphere is a two-dimensional surface.

Look at Greenland (between north latitudes 60° and 90°, to the right of longitude line 60° W at the top of the map, just left of its center). As portrayed on this map, Greenland is as large as Africa, although everyone knows that this is not true. This distortion occurs due to the steep curvature at the top of the earth's sphere, distorting the images north and south of the 60-degree parallels. Notice, also, that the 90th parallel is separated considerably from the sixtieth parallel, all other latitude lines being almost equally spaced. It is this distortion that results from projecting the sphere onto a flat plane, as though there was a positive pressure distorting the earth's shape to make it fit onto the rectangular surface.

This same condition exists south of the equator below the 60th degree parallel. This is the property of a positive curvature. Note too, for this case, that the sum of the angles of a triangle on a mercator projection add up to 180 degrees, but, on a sphere, due to the curvature of the sphere, the angles of the triangle would total greater than 180 degrees.

Let us now compare this to an AdS (anti-de Sitter) space. Recall again the saddle curvature in Figure 13.4. Its curvature is negative. Adding the angles of a triangle on that curved surface, they sum to less than 180 degrees. Figure 13.6 shows another version of a saddle and manikin that is comparable to the mercator projection for a *negative* curvature. It shows a stretched surface that is an *anti-mercator* surface and can therefore be called an AdS Space. This surface is equivalent to the saddle figure and the manikin in Figure 13.4, except that it has been stretched to lie flat like a drum's head.

In this flat surface depiction an important thing happens! Notice that the top and bottom dimensions are normally sized, but the smaller, constricted dimensions of the waist are squeezed down as if by a belt that cinches around the waist of a manikin. However, if we stretch the manikin space like a drum skin and rearrange it, we can make it look like the Escher figure, *Circle Limit IV*

Figure 13.6

shown in Figure 13.6. Look closely at this figure, because this two-dimensional portrayal depicts what the space-time of a black hole might look like (exclusive of the angels and devils, of course).

The angels and devils touch each other across the entire surface, but the surface is shaped to look flat so that the waist is on the periphery of the surface and "normal size" is depicted at its center. The figures on the surface do, however, depict the distortions of space-time. The figure shows an endless array of devils and angels, each touching the other, out to the boundary of the surface. As the figures progress outward from the center, their sizes become smaller. This is due to the distortion or warping of AdS produced by the geometry.

Susskind shows that this surface has negative curvature. The angels and devils get smaller and smaller as they progress from the inside at the center of the disc to the outside edge, where they appear like fractals. Because the curvature is negative it distorts the images in a negative direction, just the opposite from that of de Sitter space and the mercator projection.

A surface governed by AdS geometry results in a pressure that causes expansion in space-time (much like the surface of a balloon responding to the pressure from lung-supplied air). In the language of general relativity, AdS is a symmetric vacuum solution of Einstein's field equation (EFE) with a repulsive cosmological constant (Lamda, Λ) corresponding to a negative vacuum energy density having positive pressure. In short, this definition means that the Einstein equation, having a negative or repulsive energy field just like that observed by Hubble, behaves like the AdS surface.

The reason why AdS space is so important to the holographic principle is that its curvature results in expansions of space-time, very much like that observed in the real universe. Moreover, this negative curvature of the AdS has characteristics which make it very attractive for use in the analysis of black holes. It allows one to work with behaviors like those of Hubble's observations to conclude that all galaxies in the universe are receding in all directions. An additional discovery makes AdS even more palatable, the discovery made by physicist Juan Maldacena in 1997 that demonstrates the holographic principle to be real and viable.

Juan Maldacena is an Argentinean physicist at Princeton University. He proposed that there is a duality, which he called the AdS/CFT (anti-de Sitter Space/Conformal Field Theory correspondence). A conformal field theory is one where the field theory's rules are invariant under a special class of transformations: those that change the space but preserve, locally, angles in the space. For instance, imagine warping a square in such a way that the sides are no longer straight in the original space, but the corners all still meet at 90-degrees. This is a conformal transformation, and it can be abstracted into a quantum field theory context. The Ads/CFT theory demonstrates that an anti-de Sitter space in three dimensions can be described by an equivalent conformal quantum field theory in two dimensions at the boundary of this space. This behavior is schematically shown in Figure 13.7.

At the left of the figure is a conformal field theory such as QCD (the gauge theory of the strong nuclear interaction). To the right of the gauge boundary is the anti-de Sitter space that is like our universe (how this is so will be described later as being like the braneworld model of a five-dimensional universe). In this case, AdS describes the quantum gravity physics of a (4+1) space. Thus, the CFT correspondence describes a (3+1) field theory while the AdS is a (4+1) theory (or it could be yet another theory with even higher dimensionality—we merely take this one as an example). The CFT looks like the hologram of the AdS. The AdS, therefore, is a hyperspace theory that can be related to a CFT; it is like a hologram of a four-dimensional space. This is a very neat and very elegant realization, because it can be used to simplify the mathematics of the problem by working in a lower dimensional mathematical domain that is better known.

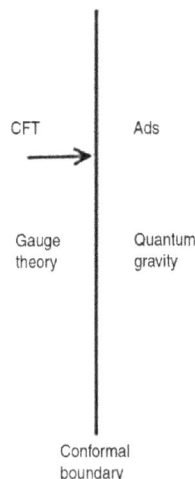

CFT

Ads

Gauge
theory

Quantum
gravity

Conformal
boundary

Figure 13.7

Although a four-dimensional theory has been demonstrated, string theory makes use of at least ten dimensions to describe its physics. Using AdS/CFT correspondence it is easier to develop solutions and relate them to a lower dimensional problem because it simplifies the mathematics and can be related to a theory that is well known. And, most importantly, the holographic principle, which can resolve the contradictions of the quantum physics of black hole theory, can also help to resolve this very difficult multi-dimensional behavior. By taking advantage of AdS/CFT correspondence with a conformal field theory on the boundary between them simplifies the analysis and seems to resolve the black hole paradox.

The beauty of AdS theory is that a 3-dimensional space (as in an AdS universe, or an AdS description of a 3-dimensional black hole) can be completely described by 2-dimensional conformal theory.

The present belief predominant among physicists is that information is preserved on the surface of the event horizon of the black hole and that Hawking radiation is not precisely thermal (which would result in all information being lost) but receives quantum corrections that lock in information. The AdS/CFT correspondence would then further prove that black holes do not lose their information because the conformal field theory assures proper behavior.

The theory also shows that there is an equivalence between string theory defined in one space (a sphere, for example), and a quantum field theory without gravity defined on the conformal boundary of this sphere, one or more dimensions less than that of AdS space. The name suggests that the first space is the product of anti-de Sitter space with a closed manifold-like space (a sphere, for example), and the quantum field theory is a conformal field theory (CFT) like QCD. An example of this duality (and named for him) is Maldacena's proposed correspondence between type IIB string theory on a product of a five dimensional AdS (a 5-dimensional sphere) and a supersymmetric four-dimensional Yang-Mills gauge boundary (which is a conformal field theory on the 4-dimensional boundary). It is thus far the most successful realization of the holographic principle.

Look again at the angels and devils in the AdS Space of Figure 13.6. The images become smaller as one goes from the center of the figure to its outside boundary. This negative curvature is akin to a repulsive gravitational field. The gravitational behavior of this surface is repulsive so that an object placed at the outer boundary of this surface will be pushed toward its center, unlike the opposite effect on the mercator surface we considered earlier. This curvature has the effect of preventing the surface's outer edge from touching the walls of any container that might hold it (such as a black hole). Thus, within an AdS space, even black holes can be tamed!

This is where anti-de Sitter space (AdS) shines! When the event horizon of a black hole grows to the point where it is nearly touching the extent of the Escher disk (see Figure 13.8) the black hole reaches a limit at the diameter of

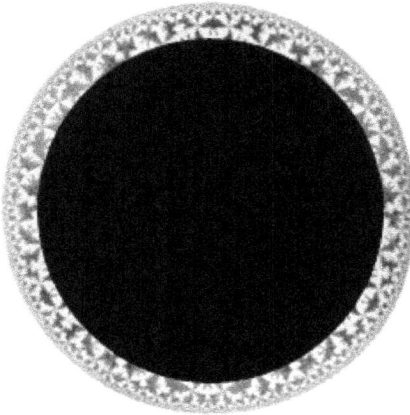

Figure 13.8   A Black Hole in anti-de Sitter Space

the Escher surface. The repulsive negative curvature of the space, according to Susskind, inhibits its ability to touch anything.

This is because the spherical wall of AdS exerts a powerful force, an irresistible repulsion, on everything and anything that approaches it, even a black hole. There is a tremendous repulsion that forces this space to shrink from the surface of the space-time cylinder and the event horizon around a black hole (Figure 13.8) such that no contact with it is possible. How fortuitous that is! The warping of AdS space is not different from the warping of time.

Look at the space-time continuum of Figure 13.9. Time is moving upward in the vertical direction. The graphic illustrates the Escher AdS disk space, propagating in space-time in the vertical time direction. As it does this, it creates a world map. One could take a slice of this figure anywhere in time and this slice would be an AdS space (a disk) at a particular time. Think, then, of AdS as a thin slice of space-time. When these slices are stacked together they

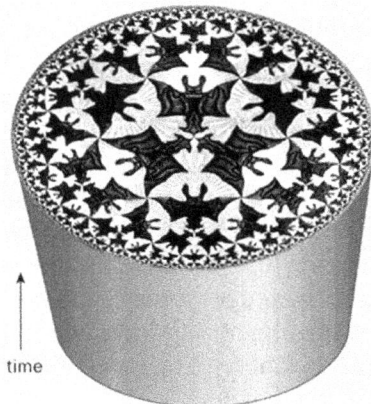

time

Figure 13.9   Space-Time Continuum

constitute a space-time continuum. As the disk moves in time it creates the cylindrical tube shown in the figure. Suppose that the angels and the demons all have clocks. The clocks in the center of a space-time slice would run at normal speed but the clocks located at different places as one moves out from the center of the disk, along its radius to its outer edge, would run faster and faster, speeding up to twice their speed for each ring of figures, all the way out to the edge of the surface.

The vertical surface of Figure 13.9 is the outside of a cylinder produced by the movement of the Escher disk through time. AdS space-time is confined inside this cylinder, not touching its surface because the repulsion caused by its negative curvature causes it to shrink away from the sides of the space-time continuum. If you take a slice out of the space-time continuum, it looks like the Escher disk (again, without the angels and devils, of course!).

Now, imagine there is a black hole in the space-time continuum of Figure 13.9; it might look like the surface shown in Figure 13.8 (which is a slice of the continuum), its event horizon being the boundary of the black hole. When material falls into the space-time continuum, the entropy at its singularity increases and so does that of its event horizon. Then, as the entering material is captured, you would see the black hole's event horizon become larger and larger. The black hole gets larger, too, causing its diameter to increase, approaching the edge of the AdS disk. The black hole increases in size, causing it to approach the boundary of the space-time continuum. The AdS surface, however, resists this condition from ever being reached, thus preventing the boundary of the black hole from touching the AdS boundary.

Resolving the black hole information paradox

In previous sections we explored some of the early ideas about black holes and thermodynamics and the refinement of those ideas in the context of higher-dimensional theories, like superstring theories. The original work, discovering relationships without regard to the underlying theory that predicts them, resulted in the contradiction between the information-preserving roots of the work and the consequences of a black hole that evaporates and loses the record of all of its information in the process. What is the state of the resolution of this problem?

The black hole is a three-dimensional sphere (with a two-dimensional surface area). When material falls below the event horizon, it falls into the black hole, reaching its unimaginably small singularity (we cannot illustrate its size in scales familiar to us). When this happens the event horizon entropy increases, causing the black hole to grow. As the entropy of the black hole increases, its size increases. The event horizon's surface, the edge of the black hole, retains the contents of all things that fall into it according to Susskind and 'tHooft (the event horizon surface is a hologram of the black hole). The contents of the event horizon describes everything that is in the black hole.

Hawking did not all-at-once nor completely retire from the battlefield of the "Black Hole War." Instead, he turned the tables to use the ideas of string theory to amplify the role of hidden dimensions. Perhaps the information that boils away as the black hole evaporates is, indeed, lost only from our 3+1 dimensions. But, viewed from all of the dimensions of the cosmos, that information is preserved. This would necessitate the existence of parallel universes, a landscape of spaces wherein we occupy but one of those small spaces. This is an idea we'll return to later.

String theory, and the cosmology that accompanies it, is mathematical conjecture, with no proofs of these theories at this time. The debates within the cosmology and string theory communities that we have talked about merely provide an intersection where these ideas converge in an effort to make sense of an extreme and not yet fully explained phenomenon in our universe: the black hole.

Because these discussions are based on mathematics only, the black hole war rages on, with no end in sight. Susskind asserts that he has completely proven that no information is lost when the black hole evaporates while Hawking allows that this may be true, but implies that the problem is more difficult to solve.

The holographic principle is a fascinating synthesis of the quantum mechanics concerning black hole theory and the survival of information trapped inside a black hole. It remains to be proven whether such information is actually retained and is retrievable even in the presence of Hawking radiation-driven evaporation of the black hole. At present there is no experiment that can be performed to offer proof of this premise. However, the mathematics of string theory has shown that this is feasible, proving that quantum mechanics is viable and does perform as physicists say it should. Whether the mathematics is correct and verifiable, no one knows today.

One important concept on which the current assessment of the problem hinges is the mathematics of the AdS/CFT correspondence. Evidence that the AdS/CFT correspondence works has been demonstrated mathematically. There have been reports that AdS/CFT supports a transition from a superfluid to an insulator. In these reports, the superfluid begins by packing together trillions of atoms at a temperature near absolute zero into a space illuminated by multiple laser beams. Initially, the atoms move collectively, like a fluid, but a fluid presenting no friction. By definition, a fluid that possesses no friction is a "superfluid." A superfluid can exhibit bizarre behavior such as crawling upward along the walls of a vessel in which it is contained. Liquified helium gas was the first example to be found in the occurrence of superfluidity. As the intensity of the lasers shining onto the ultra-cold atoms is increased, the helium transitions from superfluid to insulator.

Realize that the phenomenon of superfluidity and similar spectacular "finds" are far outside the realm of particle physics. So, too, is the assertion that the universe can be defined at its boundary by a hologram, although it is a fascinating conclusion that must be pursued. If true, we look to it to be tested and the facts concerning it revealed as the process of discovery continues.

String theory is one of the more advanced pathways being followed in this search for the origins of the universe. We learned that to understand string theory it is necessary to allow extra dimensions, although there is also a strong basis for assuming that string theory may need only the conventional (3+1) dimensions. Anti-de Sitter space replicates the observed behavior of our universe, although measurements indicate that our universe is actually best described by a flat space-time whose destiny is controlled, at present, by dark matter and dark energy. However, the AdS approach might work very well to help us understand physics in a black hole, imagining a multi-dimensional space that is related to a conformal boundary of a space of one less dimension than AdS space itself. Thus the holographic principle is preserved, producing a result that is a hologram of the contents of the AdS.

The description of the holographic principle is an example of contemporary theories and debates that abound. There is much to be done to assemble the facts that pertain to a description of the universe. Today's explanations for the origin of the universe are mostly unverified mathematical frameworks yearning for an experimental test to verify or refute their claims.

Before we close this chapter, let's look at a spectacular event in 2016 that might be a door to testing the early notions of black hole physics—gravitational waves, and the possibility for testing the Bekenstein-Hawking theorems.

When Black Holes Collide

On September 14, 2015, things got very exciting for the members of the LIGO Scientific Collaboration. LIGO is an interferometer, using twin 2 km-long arms that host laser beams to look for miniscule distortions of space-time gravitational waves. LIGO was first conceived of in the 1960s and became a project commissioned by the National Science Foundation in the 1980s that first operated in 2002. It has been on a path of operation and upgrade ever since. The LIGO instruments were just completing an engineering shakedown test-run in September, 2015, when the event occurred.

Like any good scientific group should be, the members of the LIGO collaboration were initially skeptical of what they observed on September 14th. They spent months analyzing the data and testing alternative hypotheses, but in the end the only interpretation of the data that survived the testing was that they had witnessed two black holes collide and merge. This was the first time anyone had ever observed this phenomenon and it was the first definitive detection of gravitational waves, the ripples of space-time that were expressly predicted by the general theory of relativity.

What had LIGO observed? The interferometer measured distortions of space-time on Earth that first struck the instrument in Livingston, Lousiana, and then seven-thousandths of a second later, struck the independent twin facility in Hanford, Washington. When the numbers are crunched, one finds that the time lag between the two signals is consistent with a phenomenon that traveled the distance between the two sites at the speed of light, just as Einstein's theory of space-time predicted would be the case for gravitational waves.

The distortions that they saw corresponded to a stretching and squashing of space-time that was but one-thousandth of a billionth of a billionth of a meter—roughly a million times smaller than the size of a single proton. The actual pattern of distortion in space-time is similar to the amplitude pattern of air distortion that a chirping bird makes when it emits a single "*chirp!*" In fact, the event has been nicknamed "*the chirp*"—a name that belies the fact that it resulted from the collision of two enormous space-time monsters.

The most reliable interpretation of the data is that the chirps at each site were the result of two objects, orbiting each other closer and closer and closer. What LIGO "heard" in space-time was the final moment as the two objects came so close that their boundaries touched and they merged. Think of two rain drops sliding along a car windshield, coming closer and closer. At some point, those shining round surfaces, the result of the surface tension of water, touch one another and the drops become one. So it was with the two objects that, circling each other in their final moments collided at their boundaries and became one.

And what were these objects? The data are consistent with one of them having a mass of (36 ± 5) times the mass of our sun, and the other having a mass of (29 ± 4) times the mass of our sun. From the way the wave pattern changes as the chirp "rings down," it is possible to measure the mass of the resulting merged object as (62 ± 4) times the mass of our sun. The astute reader will do the math and note that naïve addition of the original masses leads to an expected total of 65 solar masses . . . so where are the missing three solar masses?

We detected them. Or, rather, we detected a small part of the three solar masses of energy because they were carried away as gravitational waves emitted at the moment of the merger. Just as waves spread out in a pond into which a rock is thrown, the total energy of the event is wholly there, but the waves spread that energy more and more thinly over space as they expand away from the event. By the time this mind-boggling amount of energy from the collision reached earth 1.3 billion light years later, there remained only enough energy to warp space-time by a very tiny amount. Nonetheless, LIGO can detect it!

Recall that the Chandrasehkar limit for neutron star formation is 1.4 solar masses, and the Tolman-Oppenheimer-Volkoff Limit for black hole formation is just 3 solar masses. The only things that could explain such heavy objects are black holes. The data tells us that the distance to this merger is 1.3 billion light-years. When the merger originally occurred, Earth was inhabited by basic multicellular life such as simple bacteria. On September 14, 2015, the ripples of space-time, traveling at the speed of light, finally reached our planet, 1.3 billion years after the event had occurred.

When this discovery, named GW150914, was announced on February 11, 2016 (it took that long to process and verify the data), the scientific and popular press was taken by storm. For the first time, humans had put an ear against the floor of the cosmos—space-time itself—and heard the thud of distant giants tumbling around in the dark. For the first time, we had received information that may have come straight from the event horizons of two black holes. In just one day, an entirely new form of astronomy—gravitational wave astronomy—burst into existence. What Galileo Galilei did for the world in creating astronomy by turning a telescope to the night sky, LIGO has done for us by turning our ears to space-time to listen for the distant messages of cataclysmic events.

The LIGO scientific collaboration announced a second detection on June 15, 2016. This second event had occurred months earlier, on December 26, 2015. Again, careful analysis was done to rule out alternative hypotheses. The conclusion was that this was yet another black hole merger, named GW151226. This merger involved lower-mass black holes, approximately fourteen and seven solar masses, but definitive nonetheless. The occurrence of binary black hole mergers may be common in the universe, and if they are, LIGO expects that it will detect hundreds of them per year as the collaboration continues to upgrade and develop its instrumentation. Other instruments like Virgo, GEO 600, and TAMA 300 should expand the sensitivity of these searches, allowing for the creation of a global-scale gravitational wave observation network. The entire earth will be tuned in and listening for these events.

Wrap Up

Black holes are the result of a stellar remnant of mass in excess of three times that of our sun, where the runaway gravitational collapse of a star cannot be stopped by known quantum effects resulting in a singularity that strongly warps space-time around it—a black hole. Physicists like Schwarzchild, Chandrasehkar, and Oppenheimer had early-on described the conditions of this phenomenon. The discovery and subsequent characterization of Cygnus X-1 gave us tools to better understand how to measure black holes using material outside their event horizons—the point of no return beyond which no information escapes the black hole.

However, as Bekenstein and Hawking conjectured in the 1970s, perhaps information does escape the event horizon of a black hole in the form of Hawking radiation. Further, the physical scale of the black hole, its mass, and the surface area of its event horizon, might define the thermodynamic properties of the black hole, its temperature, and entropy. But in doing so, Hawking suggested that this might doom information that enters a black hole. This led to a paradox in the laws of physics that theorists like Hawking, Susskind, and 'tHooft have been struggling to understand by using tools like superstring theory and extra dimensions.

These efforts have led to the suggestion that a black hole—and perhaps our entire universe—can be understood by the information encoded in quantum states on the surface of the black hole. This holographic principle tells us that the higher-dimensional space of a black hole might be summarized entirely by the state of its event horizon, a 2-dimensional surface. This leads to an intriguing set of new ideas about relating higher-dimensional space-time models to lower-dimensional field theories (the AdS/CFT correspondence). Perhaps, in thinking about the black hole, physicists have discovered a means to resolve the problem of dimensionality in superstring theory and M-theory.

Data, however, might be nipping at the heels of these ideas. For the first time, physicists have detected gravitational waves by observing colliding black holes. We may now have access, via gravitational wave chirps in space-time, to information from that mysterious boundary of the black hole, its event horizon.

Chapter 14

What is a Universe?

When you think about the whole of existence; i.e., the "universe," what comes to mind? Some people think about the fields or the forests that lie at the town line. To them, the whole of the meaningful world may lie within those lines, the borders of a town in which they may have spent their entire lives. To others, the whole of existence may be the few city blocks of their neighborhood, just sufficient to describe where they live and work and shop and play. The whole of existence can be intensely personal, for it is these personal things that occupy our daily thought and interest and help to make our lives worth living. There are still others who are more global in their thinking and consider the Earth, with its diversity of land, people, resources, and ideas to be their universe. Still others reach for the stars, some in reality, most in their imagination, and want to understand where they really are in the cosmos. To them, the universe is more than just an earth that circles a mid-sized main sequence star—it is all the planets, all the stars, all the galaxies, and all the spaces between these things that are the universe.

When Nicolas Copernicus (1473–1543) advanced the idea that the earth is a planet circling around the sun, this was, indeed, a rethinking of what it meant to consider the whole of existence. Accepting this idea meant, to many, that we were not in a *special,* privileged place. This idea, however, also places us in a grander universe, one where we and other planets dance in endless orbits around a central star, worlds waiting to be explored.

When Edwin Hubble (1889–1953) provided evidence that there are stars located beyond our own Milky Way galaxy, in what appeared to be *other* galaxies, it truly boggled minds. We now know that our universe has more than one-hundred billion galaxies. This is an awesome and stunning revelation! Rather than interpreting this in a way that makes us feel less special, let us instead interpret this humbling revelation to be confirming of just how special it is that a species like our own, trapped on the surface of a small planet around a mid-sized star, has come so far in understanding how vast the universe truly is.

But even more wonders exist in the understanding of our place in the universe. The Milky Way galaxy is but one of fifty-four or so galaxies known as "The Local Group." There is a similar but much larger cluster of galaxies called the "Virgo Cluster" nearby. For a long time it was assumed that our local group

268

was part of the Virgo Cluster. However, it was recently realized that our local group is part of an immense "supercluster" containing at least one hundred galaxy clusters, each composed of hundreds or thousands of galaxies.

So, how do we know that our local group is not part of the Virgo Cluster? One way to know whether a planet or a planetary-like object is part of a solar system is to determine whether it orbits a sun. We can resolve this question about the trajectory that our local group follows. It turns out that the answer is that the local group is moving toward a region of space called the "great attractor." Our local group is just one of thousands of such galaxies moving toward the great attractor. The great attractor is to galactic clusters as the sun is to planetary objects in the solar system. If you happen to think that this is the biggest structure in the universe, it's not.

The Virgo Supercluster is but one lobe of an even greater structure, a cluster of superclusters. This immense structure, home to 100,000 galaxies, has been given the lovely sounding name, "Laniakea Supercluster," derived from the Hawaiian phrase for "immense heaven."

Long before we knew about superclusters, or clusters, or began to understand galaxies, one of the minds working hard to understand the early work of finding our location in the universe belonged to Albert Einstein. In 1916 he completed the theoretical magnum opus that he had worked on since 1907, the general theory of relativity that we discussed earlier. After completing it, Einstein realized that these equations could be used to describe the history of the entire physical universe. At that time, the world's most respected cosmologists thought the universe was static and eternal. In order to get his equations to agree with this commonly held belief, Einstein had to introduce a new term into his equation. He needed, he thought, only to put in a constant which he named the "cosmological constant" and the toy universe he imagined in general relativity would match what "everyone knew was true"—that the universe was static and unchanging, neither growing larger nor growing smaller. Today, we would say that he had added a "fudge factor" to his equations. It turns out that the conventional wisdom motivating his choice was wrong.

So, after 1929, when Hubble showed that the universe was expanding, Einstein had to go back and erase the fudge factor, calling it the biggest blunder of his career. While we know from observational evidence that, in fact, the universe grows larger over time, there may still be a need for a cosmological constant—one that, rather than hold everything still in the cosmos, makes it grow larger over time at a faster and faster rate.

The earlier chapters of this book describe aspects of the universe as its fundamental building blocks—the forces that bind together those building blocks; the ideas of quantum physics and space-time that are the basis of successful explanatory frameworks for matter and forces, including the standard model and general relativity; and the speculative ideas that attempt to unite all known matter and forces under a single theory of nature, such as supersymmetry, extra dimensions, and string theory. We have reviewed the history of these discoveries and the ideas that explain them, as well as some of the people who contributed to these discoveries. The laws and the phenomena that govern each area have been evaluated and were found to provide insight into the vastness of the subject and begin to comprehend its complexities.

A common refrain in the history of physics is that, just when researchers thought they had developed a complete understanding of the universe, they discovered they were merely jousting with shadows. The body of what they knew was not found to be incorrect, but incomplete, requiring that each new discovery beyond the then-comfortable framework of their knowledge required new understanding. The collection of data drove the development of new frameworks, new frameworks drove the search for previously unknown phenomena, and new data revealed new unknowns. As scientists, we learn again and again that when you shine light on the shadows, you sometimes discover new obstacles that cast their own shadows on the wall.

As physicists cast light into the darkness, technology progressed considerably, adding to our knowledge and interest and providing impetus to knowledge-building via new scientific observations that gave us a more accurate glimpse of reality. The way that science grows is through the increasing accuracy and precision of measurement, thus enabling a route to better understanding.

For example, astronomical instruments have shown that the universe has a space-time geometry that is essentially flat, revealing two new characters that are the major players in the cosmology of the universe, dark matter and dark energy. As recently as 1998 these were totally unknown!

A series of observations since 1998 has further changed our understanding of the largest structures of the universe. Before then, if you asked a physicist what the universe was made from, they would have told you "normal matter." standard model stuff that composes you and me, the earth, the planets, and the stars. It's the stuff we experiment with in labs.

Soon after 1998, the light in the universe observed in a variety of instruments, including devices like the Hubble Space Telescope and satellites like the Wilkinson Microwave Anisotropy Probe (WMAP) and PLANCK, led to a convergence of data that revealed that normal matter may account for only four to five percent of the energy density of the universe. Dark matter and dark energy compose the remaining 95%! We went looking for the details of the universe we thought we already knew, only to find we really didn't know the universe as well as we thought. This is how science makes progress.

As recently as February, 2017, a new planetary system was found containing at least seven planets that revolve around a single sun, those planets of similar size to the earth. This finding, Trappist-1, more than any other so far, contains environmental potential that could sustain life of the kind we know.

This process of discovery, followed by a cycle of technological evolution, followed in turn by new discoveries is well understood in science. It is a process that takes place continuously in all areas of research. It is why science is an exciting *art*. We call it art in this description because not everything in science achieves repeatability under the same conditions; but, over time, each scientific gain proceeds forward from ever-improved knowledge about the laws that govern its topics of concern. This process that scientists are engaged in can be called "the search for the universe." There is no ultimate certainty in science except that—at all times—we must remain uncertain and be prepared to be disabused of cherished, long-held notions about nature. There are areas of science where what is known at a given time is so meager as to be described as crossing the boundary into art.

What is our universe?
How does it really work?

Our species has expended enormous effort, money, and time across thousands of years seeking to answer what become more and more *basic* questions. It might appear as though we are back where we started, as the sum total of our knowledge can describe only about five percent of what it is that composes our universe. The scientist, faced with such a humbling revelation, is defined, not by a frustration with how little they might actually know, but by an *excitement* to meet the challenge to learn more; to develop new technology to empower the next phase of discovery; to push boldly and bravely into the new frontier, thrusting forward the light of science in an effort to cast out the shadows.

At the frontiers of knowledge there is opportunity to bring together old data and explanatory frameworks with new ideas and technology that lead to new experiments to create new data to be defined by computational simulation and groundbreaking theoretical ideas. Physicists who collaborate together perform experiments that cross the traditional boundaries of theory and experiment, seeking to make sense of the things that, for now, make no sense. When successful, emerging ideas are put to the test by predicting new observations, the best ideas will survive those tests, leading us to a deeper understanding of observations that, for now, are merely confusing shadows dancing on the wall.

Let's take a look at how some of the ideas that we've discussed might come together with state-of-the-art computational ideas to lead us forward to a greater understanding of the universe. The goal, as always, is to seek an understanding of everything, not just the world up to the town line, not just the world up to the few blocks surrounding our neighborhood, not just our planet, or our solar system, or the Milky Way, or the local group, or the Laniakea Supercluster, but *everything*—that which is seen, and all that is unseen.

The Computational Cosmos

The frontier ideas discussed earlier: braneworlds, supersymmetry, and superstring/M-Theory, can be used to build mathematically practical models that could be useful in helping us to better understand how the universe works. Withstanding that final test, making accurate predictions while standing up to falsification, is not yet within the grasp of all of these models. Nonetheless, we will discuss this process here because it is universally applicable to new ideas. Using these mathematical postulates, it is possible to compare mathematical behavior with real behaviors as they are witnessed in the world.

We have powerful computers to help us explore models so that they can be tested with ever-greater accuracy and detail to determine how a model universe described by mathematical equations behaves and what are its characteristics. However, because no one knows which models will be successful, each model having different attributes, capabilities, or characteristics must be tested. Most of them will be found to be shadow-forms that do not accurately depict the observed behaviors we encounter in the real world.

While these new models can be mathematically comparable to what actually exists in the universe, they accurately describe only broad outlines of aspects of the standard model, not all of it. As will become apparent as we move along, these models begin as simple descriptions of the universe. They develop deeper refinement as physicists learn more about the characteristics of the particles, strings and forces, branes, bulk, and how these and other descriptors of our universe interact or manifest. The models we will discuss describe interactions with branes and the bulk space, their consequences on particle collider experiments, and how well these models jibe with observed results in fitting as the standard model describes how they should fit, or what is observed in astrophysics and cosmology. This process allows physicists to assess the validity or the viability of such models to describe our universe.

The standard procedure for testing theoretical frameworks is to first construct the simplest possible model. If that fails, one either changes the simple model or adds complexity and repeats the process of projecting what the universe, predicted by that model, would look like. The more that physicists do to improve the theoretical understanding of objects like extra dimensions, supersymmetric particles and forces, strings and superstrings, branes, and braneworlds, the more refined the models become and the more accurate the core ideas of those models become. Physicists can then compare the predictions of these models to astronomical data, particle collider data, or other observational evidence. Using this procedure, predictions can be tested using running experiments or existent experimental data to assess their application to nature.

The role of the computer and computer simulation is increasingly important. Computer simulations have become an essential tool for both theoretical and experimental physicists, because they allow for physics equations and pre-existing knowledge of the universe to be projected forward into new models or imagined within the context of new experiments. Computers can take pure mathematics and convert it into observable phenomena that can be measured by physical experiment. They also allow for analysis through crunching of the huge quantities of data that are produced by many modern experiments.

Computers provide for the possibility of predicting the outcome of an experiment by inputting all that is already known about a given set of conditions. When the experiment is completed, the output of the physical data can be compared to what the computer expected from them. If the data deviates from the prediction, we may have found a new behavior of nature in the laboratory that could point toward a correct and complete theory of nature. Or it might mean that the experimentalists did not understand or anticipate all of the ways in which known physics could manifest in their experiment. Having multiple independent experiments running at the same time can rule out the latter, giving confidence to the former possibility. Physicists routinely use mathematics and computation together in modern practice. The best modern physicists know how to combine mathematics, computation, hardware, and electronics to imagine the next steps needed to test frontier ideas.

Modern theory uses this newly established tradition to make it possible to suggest physical interactions, describing them mathematically, at ever-more-distant venues long before physical experiment is available to demonstrate their accuracy. This approach makes it possible to plan prudent experimentation that is both fruitful and cost effective.

This combination of the mathematical and the physical focuses the physics community's attention on future experimental efforts that could yield unique answers to specific questions. To do so requires an unprecedented expansion of this approach to present a unique opportunity to science. While much has been said about individual studies of the science of the universe (as in cosmology, quantum physics, and other disciplines), little has been said about how these individual scientific pursuits fit into the unified description of the universe, because, until now, this has not been a plausible pursuit.

High performance computers, consisting of many processors operating in concert (parallel computing), large volumes of memory for storing steps of calculations, and large volumes of disk storage to hold "big data," are in regular use at various university and research laboratory centers. These systems train generations of new physics graduate students and enable the pursuit of research programs by post-doctoral researchers, faculty, research staff, and engineers. Investigators from various aspects of the fields of study use computers to explore new ideas about the universe. The equipment offers advanced processing technology that includes a large array of individual processors that, together, produce tremendous processing power capable of handling a diverse arrangement of mathematical models.

Particle collider physicists, such as those who work at the LHC, use advanced computer techniques to model the behavior of the hundreds of millions of channels of data output by their experiments. That data is analyzed in real time, and the data and simulations are made available on-demand, globally, to thousands of experimental collaborators.

Computer modeling is very important to the detailed study and simulation of universe behavior at all levels. It is a unique way, an increasingly powerful third way along with theory and experiment, to study the universe and to provide guidance in determining testing selection before committing a project to more expensive collider or satellite experiments.

Model-Building, the Key to Experimentation

Model-building is an art and a very necessary part in building a scientific theory of the universe. This unique specialization of science has been in existence for more than fifty years. A researcher develops mathematical equations to describe a process or theory and then writes a computer program to run a simulation that performs the functions described by those equations. The simulation produces results to demonstrate what took place in response to the forces, particles, and other stimuli that were imposed on the problem.

Many manufacturing and scientific fields use model-building to help researchers understand their data, particularly before an actual experiment is performed, because it reduces the cost of subsequent research and allows certain freedoms of analysis and experimentation while exercising control over the experiment in a way that unguided laboratory experimentation cannot always achieve. However, model-building is not a panacea. The equations governing a theory may not yet accurately describe that theory (producing the shadows we see in our view of reality). The researcher may have to resort to a simplification of the original mathematical models in order to program the situation to stay within the limitations of the computer and its software capabilities.

While engineering and science are different from each other, they share some attributes. When the desired knowledge cannot be obtained by an analytical technique, the researcher must resort to a real-world experiment. The aerospace field uses this technique extensively. Detailed studies and analyses using model-building are employed in the development of aircraft, missiles, and spacecraft.

Blind trial-and-error experimentation is costly. From a financial standpoint, scientists and engineers don't have the luxury of being able to try every new idea when planning a new vessel, or instrument, or project. The complexity of these systems requires model-building and simulation in order to successfully perform detailed analysis and design. Modern computation offers the ability to pose variations and narrow the field of interesting possibilities before embarking on hardware development or construction. Models of aerodynamic, thermal, and stress behaviors under dynamic conditions are developed and used to assess design proposals. Studies of the moon and Mars have included simulations of their geological and gravitational characteristics to help understand what astronauts will encounter during landings and explorations.

For instance, aircraft, missiles, and spacecraft are first designed on the computer. The design is implemented right down to the smallest detail before any metal is cut. The Boeing 777 is an example of this approach. Boeing used 350 design personnel in several geographic locations, creating a huge distributed processing system to design the airframe, propulsion, and avionics subsystems. This distributed-processor technique was set up to allow engineers in several locations across many companies to each design their particular parts of the system, the results of which were merged at Boeing to achieve the overall detailed global design. This saved many millions of dollars. This same approach was used by physicists and engineers at the Large Hadron Collider to design the original accelerator and experiments, and to plan the necessary upgrades for the projects that will allow them to continue taking in useful data through the 2020s and into the 2030s.

Model-building is a necessary in-line technology and has become a major branch of physics, and science more generally. It is regularly applied to such occurrences in the universe as particle–force interaction and can be used to accomplish many other things. Among them is the evaluation of interactions to obtain specific results that help demonstrate and quantify corresponding mathematical theory. Once mathematics has shown that a specific behavior results from a given interaction, real-world experiments can be defined that assess this prediction; model-building can also help to modify the mathematics so that it agrees with observed data.

Let's build a model

So you want to put humans on the moon? That seems simple enough. Sit them in a rocket, aim it at the moon, and launch. Oh, but it's far from that simple. The moon is moving and it rotates. So does the Earth. Not only that. While you don't need to know where the moon is now, you do need to know where it will be in the future so that you can plan a launch trajectory that ends at the right time in the right place on the moon. The moon "feels" the earth through the warping of space-time (gravity) and vice versa. You'll need to take that into account. The moon and the earth are not the only bodies in the solar system, we may need to consider that. This is getting complicated! Let's build a model.

Model-building is accomplished by defining the mathematical relationships that describe a behavior that may result from a specific set of interactions. For example, suppose that one wishes to learn how the moon interacts with the sun and the earth under a specific set of circumstances. This is called a "multibody" problem in which gravity and, possibly, electromagnetic phenomena (e.g. the sun's large and strong magnetic field) may cause interactions among the bodies that influence the behavior and orbital characteristics of interest to the researcher. Moreover, other bodies or objects can influence the behaviors and interactions of the earth–moon system.

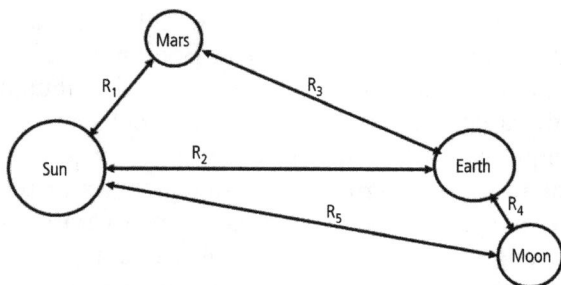

Figure 14.1 A model of the sun–Mars–earth–moon system showing a number of possible gravitational interactions that should be considered among the four bodies.

The equations that describe the gravitational interactions among these astronomical bodies are exact and have been known for centuries. This is illustrated in Figure 14.1. A student who has completed a first course in introductory physics will know how to write down the equations for this picture, but the problem is not in writing them down, the problem is in *solving* the equations—finding the outcomes of the interactions at every moment after these astronomical bodies are set in motion. It is most expeditious to solve these equations within a computer in which we have set the initial conditions for the astronomical bodies and then solving the equations in steps of time, one after the other.

What makes this particular problem so complicated? The observable interactions involve the physical relationships in the multi-body problem. For example, the sun, earth, moon, and Mars are attracted to each other by the gravity forces acting between each possible pair of the bodies. The moon and earth pull on each other; the earth and Mars pull on each other; the moon and Mars pull on each other. Each object has an effect on every other object. Gravity alone, however, won't give you the full picture of the motions of these bodies. Because of their magnetic fields (the sun and earth possess them, while the moon and Mars do not) and their rotational properties, they may also be acted upon by the electromagnetic forces between the bodies. All of these forces interact to affect motion.

If one wishes to consider the full inner solar system, it gets exponentially more complex. For example, Mars, Venus, and Mercury pass in close proximity to the earth and the moon at various times. As they pass close to each other, the effects of their passing alters the individual orbits and other properties of each object. Depending on distance, the proximity of these bodies can cause electromagnetic effects and undulations of their orbits.

Starting with the full complexity of a problem is usually considered wasteful of time. Instead, one can make choices to seek to simplify the problem, simplifying the model in the process. Complexity can be built up from there. For instance, suppose we decided that Mars' distance and mass will likely have the most significant effects on our study. Therefore, our mathematical model need not include the effects of other nearby planets, that may be close, but due to their smaller masses, greater distances, or weaker magnetic fields, can be neglected. We will make the assumption that only Mars' effects need be of interest. (We might not even have the technical ability needed to detect the effects of the other bodies!)

This illustrates the importance of measurement in science. If a model produces an effect that is beyond the ability of the measuring or computing devices, then there is very likely no need to include that effect to explain observations. Good, well-established science is about what can be measured, while keeping in mind what assumptions have been made that might, upon improvement of measurement techniques, need to be reassessed. Using a model to speculate on outcomes is a crucial part of science, but one must always be aware of whether or not speculations are measurable or, if not, that they are at least based on reasonable assumptions about what can or should not be neglected in the problem.

Let's revisit the idea of simulating our multi-body solar system to project the motions of four astronomical bodies. We show in Figure 14.1 a sun–Mars–earth–moon model, inferring the possible interactions among them. The bodies of interest are situated in their orbits relative to the sun. There are two parameters of interest for each body, parameters that can either be set for calculations or obtained from the calculation of time and distance. For example, we can define the range vectors (distances and directions) between each of the astronomical bodies ($R_1,...R_5$) as the vector distances between these objects. As objects grow closer or farther apart due to forces and motion, these range vectors change. The gravity and electromagnetic behaviors, as a function of these range vectors, are stored as code in the computer. "Code" is simply a way of expressing mathematics inside a computer.

We have to "initialize" the simulation, that is, set the astronomical bodies in some starting configuration based on real observations of their positions and speeds at those positions. This moment is "time zero," the starting moment in time for our computation. Almanac data or astronomical measurement databases found online will provide the required inputs. We then step time ahead, each time by a small amount, allowing the bodies to move from their original positions with the speed and direction taken initially from the data. The computer then applies the forces and updates the speed and direction of motion of the bodies. This gives us a new set of initial conditions for use in the next time step, and the process is repeated over and over. This is why computers are so useful for this process. They are outstanding at high-speed, repetitive computation.

Every time step is charted and the result is time-dependent predictions of the motions in the sun–earth–moon–Mars system. We then compare these motions to almanac or database information that can be found on line. For example, if we began our simulation using data at January 1, 1990, we advance the time steps in the simulation until June 1, 1990. We then compare the numbers we derived in the model regarding the position of the moon relative to earth to real data. We will find that our simulation was based on a model that worked and accurately predicted this . . . or not, in which case we must revisit the computation. Did we make a bad assumption? Is there a mistake in the mathematics? Is there a mistake in the code we wrote?

Perhaps we should have factored in the gravity or the electromagnetic effects of Venus, Mercury, and Jupiter. Perhaps we should have accounted for the non-spherical shapes of the earth and the moon in the simulation. We need to adjust the assumptions of our original model, adding the next-most-potentially important feature back into the original model, to see whether the simulation is then sufficient to reproduce published data on astronomical motion.

Let us suppose that we have correctly determined that the system composed of the four bodies shown in Figure 14.1 are the key contributors in affecting the earth–moon system even though other bodies do cause some of the effects that are observed. We have reached a point where the predictions agree with existing data and we no longer need to add more features to the model. The model, encoded in computer software, is now ready to make predictions about the future, forecasting times that have not yet happened and for which we have no data. This software could be used to plan a mission to the moon, projecting the earth's and the moon's relative positions that will inform future launch times, launch locations, and flight trajectories that will result in landing on, and then returning from, the moon.

This method of multi-body modeling was developed for the APOLLO Project in the 1960s as a design tool in the development and fulfillment of the needs of trajectory calculation to anticipate trajectory effects on the guidance system and computer programming, rocket design, propulsion system design, communications, and so forth. Such models were built for other space projects such as the Mars missions and planetary fly-by missions. For APOLLO, the sun–earth–moon system alone was found adequate to execute the required navigation function.

But what if we want to simulate an entire universe? The problem we've just described seems, already, very complex! Lucky for us, the universe has an extremely useful feature. It is smooth and homogeneous on large scales, meaning that if we think big enough, the universe looks the same in all directions to any observer. The visible universe is about ninety billion light-years in diameter. We don't have to concern ourselves with every star and every planet contained in that vast volume, because on scales greater than about 650 million light-years (about 1% of the diameter of the observable universe) one cannot tell up from down, front from back, or left from right in the universe.

To simulate an entire universe, one can begin with the assumption that all that matters are the major players present in the universe—matter, with its various forces, radiation (light), the geometry of space-time, and any vacuum energy ("cosmological constant" or a similar player) that may be present. So, what is important in these considerations of our own universe? What must a simulation of the cosmos need to represent the equivalent of the astronomical bodies in our sun–Mars–earth–moon example?

So, what's in a universe?

Before we look at speculative ways to explain what we know about the universe, and perhaps to predict things we don't yet know (or which may turn out not to be true), let us understand a bit better about what is known about the composition of our universe . . . and a little about how that is known.

The universe plays out on a stage that, itself, executes a major role in how the universe behaves and evolves. That stage is space-time. We know from observational evidence—perhaps some of the earliest observations our species have ever made without appreciating the depth of those observations—that there are three spatial dimensions (up-down, front-back, left-right) and one temporal dimension. Space-time is a 3+1 dimensional framework that, as we learned from the general theory of relativity and its immediate predecessor ideas, is itself a character of the cosmos. The shape of space-time can affect the evolution of the universe, and, for now, the shape of space-time is known to be quite flat.

Beyond the stage itself, we know that there are two major players in the universe that are extremely well-described by quantum field theory, the theories of the very small and the very fast. These were discussed in the early chapters of this book; they can be summarized in two major categories.

First, there is matter. Matter comes in two families, quarks and leptons. There are twelve identifiable matter particles, each having its own set of quantum numbers that multiplies the unique states of these twelve basic building blocks. The atom is a kind of focal point where these two families meet. The quarks compose the nucleus of the atom (up and down quarks) and the leptons compose the cloud that dances about the nucleus (the electrons).

Second, there are forces. There are three known forces that have quantum field theoretical descriptions that are summarized in the standard model. These are the electromagnetic, strong nuclear, and weak nuclear forces. Again, the atom is a meeting ground where matter and forces unite in a single event. The electrons orbit around the nucleus thanks to electromagnetism. Everything we know about electronics, chemistry and, ultimately, biology and medicine, is explained by the quantum behavior of electrons and electromagnetism in the atom.

In the nucleus, the strong nuclear force overwhelms the mutual repulsion of protons, and the lack of electromagnetic attraction of neutrons, to form a tightly packed, sticky mass that is stable for very long periods of time. Well, mostly stable. The weak nuclear force, like an impish cousin who can't play nice when adults aren't looking, makes trouble in the nucleus, causing instability and nuclear decay in some atoms. But, we should give thanks to both the strong and weak nuclear forces as the strong nuclear force provides certainty and stability, while the instability of the weak nuclear force explains why the sun burns and how the earth has remained warm for billions of years longer than can be attributed to the heat of its formation.

Finally, standing off to the side of the standard model, but no less important in the life of the cosmos, is gravity. Gravity is best described as the deformation of space-time that is caused by energy and matter. The deformation tells matter and energy how to move. Gravity literally shapes the cosmos on its grandest scales, from the formation of stars to the vast structures of galaxies, galactic clusters, and superclusters.

As we have learned in the past twenty years, there is more to the universe than the matter, forces and gravity that we have spent thousands of years analyzing and describing with greater and greater accuracy. The matter and forces described by the standard model make up only about 5% of the "stuff" in the universe. We have at least two other players to be introduced, each mysterious and each little understood.

First, there is dark matter. Dark matter is known only because of its gravitational influence on normal matter, causing things to hold together more than they would if only normal matter were present. The constituents of dark matter are almost completely unknown. The standard model has only one character that could act like dark matter, the ghostly neutrino particle ("ghostly" because it passes through matter so effortlessly). Much evidence has been gathered to determine whether neutrinos can explain dark matter, and the answer is that they cannot. Dark matter appears to lie entirely outside the scope of the standard model.

Second, there is the accelerated expansion of the universe. This has been confirmed by multiple independent assessments of the universe. Experiments have revealed the fingerprints of a force that presses on space-time, acting against the clumping of matter. We call the origin of this force "dark energy," although we do not know what composes this energy, what explains this force, and whether it has any connection to things in the standard model. In fact, much of the evidence suggests that the standard model overdoes things when it comes to dark energy, so, again, we have something that appears to lie outside the scope of the standard model. Dark energy is an area of far frontier theoretical speculation. While it can be summarized by a cosmological constant, à la Einstein and his general theory of relativity, the cause and content of that constant is unknown.

We have many knowns, but we have many more unknowns. Together, these are the fuel for the next step, asking theoretical frameworks to describe and explain the unknowns, predict new behaviors or new phenomena, and propose to experiment, tests that can be done to assess the theoretical claims. Let us look at an example of how to model an entire universe, and to predict the consequences of such a model should it really describe nature.

Computing Braneworld

We've made a point in this chapter about connecting theoretical ideas to computational methods. Let us close this discussion by doing the same for KK particles and the possibility for the existence of branes and braneworlds and that the universe *is* a braneworld. Physicists who want to search for these particles at the LHC don't know (before they are discovered) what their masses will be. To simulate what these particles will look like at the LHC, one first has to select some expected test sizes for the extra dimensions. Once that is done, software is required that can simulate the interactions of proton constituents, quarks and gluons, that could lead to KK particles being produced. This is usually provided by a software framework. Present-day theoretical and experimental particle physicists do a lot of software development as part of their day-to-day activities. Sometimes this is done in one-off exercises to meet some specific goal while sometimes done as part of a software collaboration, with a few or dozens of physicists located at one institution or spread across the globe. The Internet and the collaborative tools built atop it make this possible.

One popular framework is MadGraph5_aMC@NLO. This software is available free on the Internet and is used by major experiments such as those at the LHC to simulate fundamental physics processes. These fundamental simulations translate the mathematics of a developed theoretical physics idea into computer code. The code produces a list of subatomic particles resulting from an interaction. These particles are then provided to the next step of the simulation, decaying the particles and allowing their decay products to then have simulated interactions within the particle detector.

For example, the Madgraph5_aMC@NLO framework contains a model, the RS (Randall-Sundrum) model, that is an implementation of the extra-dimensional propagation of particles produced in a standard model interaction. The effects of the wave functions moving in extra dimensions is then simulated for use in studies by experimental and theoretical physicists. Such a simulation can be executed with just a few lines of code in the Madgraph5_aMC@NLO framework (the lines that begin with "#" are comments to the human using the code and are therefore not commands to the computer):

```
# Load the RS model
import model RS
# Specify the contents of a proton ("p") -
#   ... quarks and anti-quarks and gluons
define p = g u c d s u~ c~ d~ s~
# now generate some Z bosons by colliding
# protons and using the RS model:
generate p p > Z Z
```

The simplicity of these few steps are shown in compiled computer language (reductions to stand for considerable numbers of expanded groupings of more fundamental underlying machine language) that belies the complexity of the actual machine implementation that is done out-of-sight of the day-to-day users of the code by processing the mathematics that is needed to simulate proton interactions, applying the RS model, and then obtaining the output (in this case, a pair of Z bosons). Physicists can then compare the behavior of the Z bosons from this process to the behavior of Z bosons produced using standard model means. If there are differences in this, the beginning of an analysis, it might, in real data, reveal the presence of extra dimensions through the fingerprints of new, heavy particles that decay down to Z bosons.

Wrap Up

We have asked the question, what is a universe? Is the universe all that we see, including dark matter and dark energy, or is there more out there that helps us understand both the seen and unseen (e.g., strings or branes)? We have applied one idea about how to model an entire universe, the braneworld model, with its extra dimensions of space, and seen that this has consequences for the observable universe. It might explain the relative weakness of gravity as compared to the standard model forces, but in doing so it requires the fixing of the three spatial dimensions in which the standard model plays out to lie in a higher dimensional space. Our three dimensions (plus time) form a brane, a multi-dimensional surface embedded in a higher dimensional space. The space between us, and other possible branes, is the bulk. If the universe actually has four spatial dimensions and one time dimension, the bulk is distinguished by an extra spatial dimension.

Such models can be represented in computer code and through modern computing methods (Internet-based software collaboration and high-performance computing) that allows for the shared development of publicly available code and for the simulation of universes wherein extra dimensions affect what is seen at particle colliders like the LHC. We have looked at one such framework that does this for LHC experiments, Madgraph5_aMC@NLO and its "RS" model. By simulating a universe having reach into extra dimensions, we can figure out what the effects would be on things we can already produce and see in particle colliders, like the Z bosons of the standard model. By comparing what is predicted from those models to real collider data, we can rule out the existence of extra dimensions within some space of the reach of our measurements, or perhaps see the first signs of extra dimensions.

Simulation and modeling are powerful tools. They allow engineers to try out new technology designs without wasting millions or billions of dollars on failed prototypes. They allow engineers and scientists to narrow the continuum of potentially good designs to the few that are worthy of exploration with precious research and development budgets. They allow physicists to apply the mathematics of exotic universes to what can be seen in laboratory experiments. They help us understand the fingerprints of a more complex universe, one that might have new particles or extra dimensions, and to then apply the age-old test for truth—to ask of the data: "Is this correct?"

## Chapter 15

What the Heck's the Higgs? —Part II

In the first part of this book we explored the history of science and, specifically, physics and astronomy. We looked at the cosmic and the subatomic realms. We came to see how conditions within the standard model pointed the way to the discovery of the Higgs particle. We explored the consequences of the standard model, and found that, while it has had much success, it leaves a number of important questions unanswered. In addition, it fails to explain a number of important observations. What is the nature of dark matter? Why is the cosmos experiencing accelerated expansion? Here the standard model either falls silent or, worse yet, offers nonsensical answers.

We have met the Higgs boson. We learned something of the story of its conception, the struggle for this idea to be tested, and the eventual success of the model as an explanation for the origin of fundamental mass. We struggled with supersymmetry and extra dimensions, owing to untestable string theory, and we introduced the idea of a system-modelling approach to help understand how these "components" might work in our universe. You might feel as though this is the end of a great story. However, as we will explore in this chapter, the discovery of the Higgs and the confirmation of its role in nature is but the opening chapter of a much larger narrative *and*—this story is full of peril and mystery!

We will see how the Higgs boson is incapable of explaining all of the mass that is present in the standard model. We will see how the measured mass of the Higgs boson suggests that we live in a universe set on the edge of a precipice that threatens, one day, perhaps, to end our existence. We will see how the Higgs mechanism connects to ideas beyond the standard model: extra dimensions; supersymmetry; and superstring theory. While exploring this newest frontier, the Higgs boson, we will grapple with a most engaging question: *What the Heck's the Higgs?*

The Little Ghosts

The Higgs boson, and its associated Brout-Englert-Higgs-Guralnik-Hagen-Kibble mechanism for realizing fundamental mass in a quantum field theory, is the secret to how particles acquire mass. Or is it? It seems to work exceptionally well for the electroweak bosons—the photon, and the W, and Z bosons. Early evidence suggests that it also explains the origin of mass for quarks and the electrically charged leptons (electron, muon, and tau), but there is another place in the standard model where mass has been found— a place where it was not expected to be. This turns out to be a sticky bit of the standard model, one that continues to vex theoretical physicists. Neutrinos, the little cousins of the electron, muon, and tau, also have mass . . . and it seems that the Higgs boson may have nothing at all to do with this.

To understand why this may be, we have to jump back in time to the time of the origins of the idea of the existence of a neutrino. The story begins during the same booming era in which the atom was being thoroughly scrutinized, the 1920s. The quantum theory was under development. Einstein had already wowed the physics community and the world with his theories of relativity.

Radioactive decay had been discovered in the late 1800s, and, in 1899, Ernest Rutherford (1871–1937), who would go on to discover the nucleus of the atom and the proton, was studying this newly discovered strange phenomenon. Seemingly without any prompting from an external entity, atoms would spontaneously emit energy. Rutherford classified the then-known kinds of radiation into two categories: alpha and beta radiation, named for the first two letters of the Greek alphabet.

It will be beta radiation that concerns us here. Beta radiation is distinguished by the fact that the radiated energy is carried by a specific subatomic particle. Research revealed that this particle was the electron, discovered by J. J. Thomson (1856–1940) in 1897. The electron that emitted beta radiation was no ordinary electron—it moved very fast and carried tremendous energy as a result. Beta radiation—fast electrons—will penetrate through millimeters of aluminum metal. It was Henri Becquerel (1852–1908), the discoverer of radioactivity, who found that beta radiation was the same as Thomson's then-recently discovered particle, the electron. In 1901, Ernest Rutherford and Frederick Soddy (1877–1956) learned that beta radiation was accompanied by atoms that changed their type, something that we later came to understand was a nuclear process—changing the number of protons in an atom.

When two physicists, Lise Meitner (1878–1968) and Otto Hahn (1879–1968), measured the energy spectrum of beta radiation in 1911, a real mystery began to take shape. Unlike other kinds of radiation (e.g., alpha radiation), in which the emitted particles carry very regular and specific energies every time they are emitted, beta radiation particles appear to carry a smoothly varying amount of energy, even when emitted from different atoms of the same atomic element!

A beta particle emitted from one atom could have a given energy level, while a beta particle emitted from an identical neighboring atom carried a different level. What could explain this unusual phenomenon?

Many other mysteries abounded about beta radiation, including the conservation of spin angular momentum in nuclear transmutations. In 1930, after decades of study of beta radiation and much theoretical labor to understand the problem, a physicist, Wolfgang Pauli (1900–1958) wrote a famous letter stating his speculation about the cause. Unable to attend a meeting of the Physical Institute at ETH Zurich, he wrote in a letter to the participants that he had hit upon a "desperate remedy" to the problem involving the prediction of a new and never-before-observed particle that he called the "neutron." Too afraid to publish this "desperate remedy," he sought input from his colleagues and provided his thoughts in the letter.

The particle we today call the neutron was discovered by James Chadwick in 1932 and he named it using the same word as had Pauli. But this constituent of the nucleus was later determined to be too heavy to explain Pauli's vision of neutrons. The physicist Enrico Fermi (1901–1954) renamed Pauli's hypothesis "neutrinos" (Italian for "little neutral one") so that they would remain a distinct category of as-yet-undiscovered (and possibly non-existent) particles.

The reason that this idea was not taken seriously was that the neutrino, if real, was so elusive as to be potentially undetectable. Fermi was the first to develop a theory of beta decay, attributing the emission of the beta particle to the following reaction, the decay of the newly discovered neutron (n^0):

$$n^0 \rightarrow p^+ \, e^- \, \nu$$

where p^+ is the proton, with its positive elementary charge, e^- is the electron, with its negative elementary charge, and ν (the Greek letter, pronounced "noo") is the neutrino, having a proposed zero electric charge. We can read the above equation like a sentence that tells us about a physical process in nature: The neutron (n) decays (the arrow) to a proton (p), electron (e), and neutrino (nu). Fermi's work appeared in 1934 and represented a kind of unification: the unification of the neutrino hypothesis with Paul Dirac's postulation of the anti-matter electron, the positron (which was, itself, only relatively recently discovered, in 1932). By putting it in a clear framework, Fermi had set the stage for making predictions. However, in an ironic twist, his paper was rejected from the most prestigious journal of the day, so he published it in an alternative journal. The idea did not immediately catch on (even though it turned out to be correct), so Fermi decided to switch from the pursuit of theoretical physics to experimental physics. This was likely a setback for theoretical physics but a huge gain for experimental physics.

The neutrino was predicted to be discoverable by inverting the neutron decay reaction by moving particles around in the equation. The rule is that when you move a particle from one side to another in a reaction, you have to also reverse quantum numbers such as charge and particle-ness. So a discovery reaction for the neutrino would look like this:

$$\nu + p^+ \rightarrow \bar{n}^0 + e^+$$

in which we interact an anti-neutrino (the anti-matter counterpart of the neutrino) with a proton, resulting in a neutron and a positron.

Many ideas for detecting neutrinos in this way were proposed, but it was the atomic age that enabled its discovery. The development of sustained nuclear reactors, a by-product of research conducted by Enrico Fermi when he switched to experimental physics, enabled this to happen (we now see how important it was that he switched fields!). In 1956, two experimental physicists set up a project next door to a nuclear reactor—hypothetically, a copious source of anti-neutrinos—and detected the process. These physicists were Clyde Cowan (1919–1974) and Frederick Reines (1918–1998). The key experimental signature of the above reaction, which they observed, involved a neutron produced during the capture of an antineutrino, as it was captured by a nucleus, causing a flash of light to be emitted. A positron, also produced in this reaction, will readily smack into an atomic electron, thus annihilating themselves, producing a pair of gamma rays. The production of characteristic light from neutron capture coincident with a pair of gamma rays was a unique signature of this process.

So the neutrino was determined to be a real particle of nature, twenty-six years after it was reluctantly hypothesized by Wolfgang Pauli! Eventually, it was discovered that there are three kinds of neutrino in nature: the electron-neutrino, which accompanies the beta radiation process; muon-neutrinos, which are produced when muons decay; and tau-neutrinos, which are produced in reactions accompanying the creation or decay of the tau lepton.

From studying the energy spectrum exhibited by beta particles in decay, it was inferred that the mass of the neutrino is exceptionally tiny, smaller than could be measured using experiments. It was therefore assumed that neutrinos of all kinds are mass-less. Like photons, they carry no electric charge and would travel at the speed of light as a result of their lack of mass. Unlike photons, neutrinos were determined (by experiment) to be fermions—spin-1/2 particles, just like quarks and other leptons. As pointed out earlier, we see that the building blocks of matter are fermions.

The assumption that neutrinos are massless turned out to be fatally flawed, but discovering this took decades. Neutrinos are very hard to capture and their properties, as a result, are difficult to observe. They interact *only* via the weak nuclear force—by the exchange of weak bosons with quarks and leptons—and, as a result, their interactions are short-ranged and thus rare. This is why a neutrino can pass through a light-year's worth of lead and not be guaranteed to strike an atom—the weak interaction, is very weak due to its short range. Neutrinos have to pass directly through a nucleus, or very close to an electron, to have any hope of interacting. Nuclei and electrons are such small targets, and neutrinos equally tiny bullets, that the chance of this happening is small.

But it does happen, and if you can produce lots of neutrinos you do have hope of detecting a few of them. This was the gamble that Reines and Cowan took with their experiment, and it paid off. There are other sources of neutrinos in the cosmos. One of our most copious sources is our sun that helped to reveal the error in the assumption of zero mass.

Hans Bethe, as well as a handful of other physicists, were the first to succeed in explaining where the energy of our sun comes from. Using then-state-of-the-art methods in nuclear physics and nuclear interactions, working with colleagues at a small workshop hosted by Carnegie Institute and George Washington University in 1938, Bethe took up the challenge of explaining solar energy. By the end of the workshop, they had a workable first model of the core of the sun and the nuclear processes that explain the tremendous energy output and high temperature of the sun. This theory would continue to be developed over decades, but it is a landmark in the connection between quantum physics and the theory of the nuclear forces, and astrophysics on the grand scale of the cosmos.

This nuclear theory of stellar energy production predicted, also, that the sun (and other stars) should emit copious numbers of neutrinos as a by-product of their nuclear reactions. Neutrinos would offer the first insight into the heart of the sun. Because neutrinos barely interact with matter, they should escape the core of the sun, where they are created, and fly almost unaffected by the sun until they reach a target on earth where they could be captured.

After the neutrino was discovered in 1956, an ambitious experiment was devised in the 1960s to measure the number of neutrinos coming from the sun. This would be determined by capturing solar neutrinos using a large batch of cleaning fluid placed deep in a mine underground to shield it from everyday radiation raining down to earth's surface from space. Called the "Homestake Experiment," named after the mine in South Dakota where it was located, it was led by an experimental chemist and physicist, Ray Davis, Jr. (1914–2006), and by a theoretical physicist, John Bahcall (1934–2005). They turned up a very strange result: every time they measured the flux (number) of neutrinos from the sun, they came up short of the predictions of the nuclear stellar model. They seemed to find only 1/3 of the predicted number of neutrinos.

This was referred to as the "solar neutrino problem." For decades, physicists argued whether it was a problem with the Homestake Experiment or a problem with the nuclear stellar theory (or both). It was a contentious issue and it wasn't resolved for thirty years when an experiment in Japan provided a new piece of the puzzle.

The Japanese experiment started small and got very, very big. It began in the 1980s as the "Kamiokande II" experiment. Like the Homestake Experiment, it was conducted deep underground in the Kamioka Mine in Japan. It used a tank of very pure water instead of cleaning fluid, and it could tell the direction from which neutrinos came from as they arrived at Earth. The objective was to count the number of neutrino interactions resulting from these particles emitted by the sun. Kamiokande II began taking data in 1985 and after a couple of years of shake-down it was able to show, not only that the sun beams neutrinos at us (the directional information told the physicists that these neutrinos, indeed, point back at the sun and not some other source), but also that the number of them was lower than was predicted: 1/3 of the expected number.

By the 1990s, the emphasis in the physics community was to better understand neutrinos to determine whether they, the neutrinos themselves, could be the reason for the "solar neutrino problem," the missing 2/3 of expected neutrinos. A much larger version of Kamiokande II, called "Super Kamiokande," was constructed in the Kamioka Mine. It contained 50,000 tons of pure water, observed by 11,200 photomultiplier tubes—reverse lightbulbs that *take in* light from neutrino interactions in the water, convert it to electricity, and allow humans to interpret the signals and analyze the interactions. Super-Kamiokande was the first experiment that was able to study neutrinos in the level of detail required to reveal what was actually going on. It seemed that neutrinos had the ability to morph from one kind of neutrino to another. This phenomenon is called "neutrino oscillation"—the changing of a neutrino's type from one, to another, and on to yet another.

It would not be until the Sudbury Neutrino Observatory, a large experiment in the Sudbury Mine in Canada, that we would have definitive evidence that neutrinos of a given type can disappear and reappear as a different type (that electron-neutrinos can morph into muon-neutrinos, and vice versa). When that evidence was gained in 2001, what was going on with neutrinos became clear to physicists—neutrinos must have mass.

Mass enables this kind of morphing to occur. It's a very strange phenomenon, but it is allowed by the rules of quantum mechanics. An analogy will better explain the concept and reveal the underlying activity that is happening. Imagine that you have a jar full of jelly beans. They are all red, and they taste like cherry with a hint of lime. You leave the jelly beans in the jar on the counter and go to work. When you get home, you look at your jar of jelly beans and see that there are two-thirds fewer red jelly beans in the jar. In their place, green jelly beans have appeared. How strange!

You take a bite of one of the new green jelly beans. It tastes like lime, with a hint of cherry. You want to describe what has happened to the jelly beans, and draw a conclusion. You can write an equation that explains the red jelly beans, in terms of cherry and lime, as follows:

$$[red] = A[cherry] + B[lime]$$

where the coefficients, A and B, indicate the degree to which you have to add cherry flavor (lots) to lime flavor (a little) to get a red jelly bean. You can also describe your green jelly beans in the same way:

$$[green] = B[cherry]+A[lime].$$

You realize that there are two "spaces" in which jelly beans can be described: taste ("cherry," "lime," and a combination of the two) and color ("red," "green," or a mix of the two). Your eyes are sensitive to the color space as they can easily discern red from green (unless you are color blind), but cannot "taste" a color. The tongue is sensitive to the flavor space, but blind to the color space.

In mathematics and physics, if a single system can be described by two (or more) valid spaces, then "mixing" between the states that define those spaces is permitted. This is what is going on in neutrinos.

There are two spaces—the space defined by the weak interactions and the space defined by mass. While they are distinct from each other, they can mix. When a neutrino travels through space, the states of the two spaces mix spontaneously, on their own. When you measure the neutrino's state with an instrument, you will observe it to collapse into one or the other of these states—the state it happens to be in at the moment you measure it. We only observe neutrinos through their weak interactions, so we are only directly sensitive to their weak space, but their mass space is also there, and varies the states of the weak interaction space.

Here is how this mixing in our jelly bean analogy is shown mathematically. Since you were able to write "color space" in terms of "taste space," you can also write a similar representation of taste space in terms of color space. Simply rearrange the equations to find:

$$[cherry] = (B/(A^2–B^2))[red] – (A/(A^2–B^2))[green]$$

$$[lime] = BA^2/(A^2–B^2))[green] – (AB^2/(A^2–B^2))[red].$$

Since there is nothing to prevent this relationship from being written down, it is mathematically possible. In our analogy, when you allowed the red jelly beans to remain in the jar for a day, the behind-the-scenes chemical process that makes the mostly cherry ones also red-colored, switched around the chemicals that makes them lime and cherry flavored, resulting in the jelly beans morphing from red to green. In the case of neutrinos, this means that mixing of the weak force states, due to changes in the mass-states, is physically possible. Models like this one have been written down for neutrinos, relating mass-space to weak interaction-space, and they appear to explain the mixing that has been observed in many independent experiments.

Neutrino mixing is only supposed to happen, however, if the neutrino type has mass. In the original formulation of the standard model, neutrinos were assigned to have no mass and, for the most part, this seemed to agree with what was observed. Does this mean that we should just toss neutrino mass into the standard model, assume that the Higgs boson is responsible for it, and move on? Things are not that simple.

Give Me a Hand

According to the data we have, left-handed people are a minority among human beings. About 90% of people in a randomly selected group favor their right hand, using it for writing, and manipulating objects. Left-handed people face the challenge of a world that is designed for right-handed people—from the use of scissors to the opening of a door.

Nature, too, seems to have a preference for "handedness" in its matter particles and the interactions that they are permitted to engage in. Physicists define fermion-handedness by looking at how a particle's internal spin angular momentum points with respect to its linear momentum. (See Figure 15.1.) When an electron travels through space, it is observed that its spin vector tends to point almost entirely in a direction that is opposite to its direction of linear motion. This is known as "left-handedness." On the other hand, positrons do the opposite. When they move in the same direction as electrons do, their spin vectors point mostly in the direction of its linear motion. This is known as "right-handedness."

This "lefty" vs. "righty" behavior among fundamental particles is a curiosity. It becomes more defined when you imagine a fermion having zero mass (as was assumed of neutrinos for many decades), all massless fermions are expected to be left-handed and all massless anti-fermions are expected to be right-handed. Because the universe is dominated by matter, not anti-matter, the universe is preferentially *left-handed*!

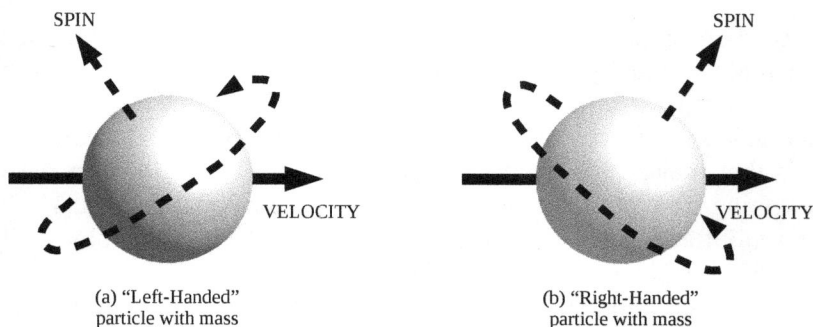

(a) "Left-Handed"
particle with mass

(b) "Right-Handed"
particle with mass

Figure 15.1   A depiction of "handedness" in nature. It changes depending on how the motion a particle relates to the direction its spin can be thought to point.

Even more interesting, the weak nuclear interaction appears to want to act only on fermions that are in a state of left-handedness and anti-fermions that are in a state of right-handedness. W bosons are blind to the existence of other fermions—it is as if someone has turned off a switch in the cosmos for one half of the weak interaction! While this may seem curious, it is an observed behavior of the weak interaction, a behavior encoded in the mathematics of the standard model.

When the Higgs boson interacts with a fermion, just as it does with an electron, it changes it from left-handed to right-handed. This is not a problem for the weak interaction in the standard model, because electrons have mass and there is always some part of their wave function having the correct handedness to interact with, for instance, a W boson.

Neutrinos were originally and incorrectly assumed to have no mass. That means they never interact directly with the Higgs boson. In that case, there is no risk of a change of handedness and no risk of them becoming isolated from weak interactions, the only kind of interaction they are allowed to have. However, now that we know that they do, in fact, possess mass, we have two problems: first, it is possible for them to have interactions with the Higgs boson, which complicates their behavior in the standard model; second, they must get their mass from somewhere, but the standard model, as written in the 1960s, neglected this possibility. This is quite a puzzle.

The standard model assumes massless neutrinos and, therefore, the left-handed nature of the weak interaction. Neutrinos and anti-neutrinos have a component of handedness becasue they are not entirely without mass. These issues complicate the standard model and even strongly imply that neutrinos have an aspect about themselves that lies outside the standard model.

One *could* just go ahead and modify the standard model by adding in mathematical terms to represent the interaction between the Higgs boson and the neutrino. Such interactions would, as hoped, result in mass. But this would lead to two problems for the standard model. First, the interaction strength required to obtain the tiny masses of neutrinos is at least a trillion times smaller than that required to obtain the masses of charged leptons: the electron, muon, and tau. Such a crazy hierarchy is not present in the quarks. This creates another of those troublesome hierarchy problems.

One can avoid this by postulating instead that neutrinos have a wavefunction of a kind not yet observed in nature. Massive fermions have wavefunctions that are of a class called "Dirac spinors," referring to the way their spin is described. But, if one wants to avoid the new hierarchy problem, one has to postulate that the neutrinos are not Dirac, but Majorana spinors. Ettore Majorana (1906–unknown) was a mathematician who made early contributions to atomic structure calculations. Majorana even derived his own relativistic theory of particles, one that would be in direct competition with those of Paul Dirac. Much of Majorana's work languished in obscurity, as it was published in his native Italian. He disappeared in 1938, and was never found. This is why it's not known when he died, or what happened to him after 1938.

The solutions to Majorana's relativistic equation are called "Majorana Fermions" and are built on different spinor representations than those found in Dirac's relativistic equation. It's Dirac's equation that lies at the heart of the standard model, but that doesn't preclude neutrinos from being special. In fact, if neutrinos are described by a Majorana spinor model, their mass can be explained by a mechanism independent of the Higgs boson. This would put them out of reach of the standard model.

Neutrino mixing, best explained by the existence of neutrino mass, poses a challenge to the assumptions of the standard model. Either the standard model is more flawed than we first assumed, having a new hierarchy problem in which something forces apart the Higgs interaction scales of charged and neutral lepton masses, or we have to immediately use the properties of neutrinos to light the way to the more inclusive theory of nature, one that explains both the standard model and the strange smallness of neutrino mass. Perhaps the neutrino is the vanguard laboratory observation of a new theory that lies beyond the standard model.

How might the Higgs play a role in this question? An interesting search at the LHC is for the invisible decay of the Higgs boson. For instance, were the following type of decay of the Higgs boson to be observed:

$$H^0 \rightarrow \nu \, \bar{\nu}$$

it might be possible to use that reaction to infer something new about the neutrino sector of physics. Invisible decay of the Higgs boson is defined by detecting evidence in proton-proton collisions that a Higgs *should* have been there, but when you add up the energy and momentum of the collision you fail to find particles that would have come from the decay of the Higgs. It's as if the Higgs boson had disappeared, leaving only indirect signs in the detector (like footprints on a sandy beach) but no direct trace in the detectors (the equivalent of watching the beach-walker make the footprints).

Since neutrinos are so ghostly, we don't expect them to interact in either the ATLAS or CMS detectors; there's simply not enough material in those instruments to entice neutrinos to leave traces of themselves. Therefore, if a Higgs decayed in that manner, it would seem to disappear entirely, leaving footprints, but no beachcomber. LHC physicists can infer the presence of the Higgs without actually observing its decay—a kind of "art" in collider physics—one of the more difficult, yet refined, uses of the detector.

Searches for the invisible decay of the Higgs boson have been conducted by both the ATLAS and the CMS Experiments. Neither experiment has yet reported a definitive observation of this phenomenon, but the sensitivity of such searches has been limited by the small amount of data produced and the early methods of conducting those searches. The techniques will continue to be refined and more and more data will be collected over the next ten to twenty years. We are in the infancy of these efforts, and there is much excitement and anticipation about what they will reveal as they go on.

Experiments in the U.S. and in Japan continue to map the landscape of the neutrino. Within the next number of years, physicists expect that they will have mapped all the known parameters of the neutrinos as defined by the neutrino mixing models. Of course, there will probably be surprises—measurements that don't comport with what is expected, or new properties not foreseen for those models. Experiments such as T2K (Tokai-to-Kamioka) in Japan and NOvA in the United States are advancing their particle accelerator and detector efforts with the aim of collecting large data sets to make more greatly detailed maps of the neutrino landscape. Future experiments, like the internationally sponsored DUNE (Deep Underground Neutrino Experiment) facility, will go further, deeply refining measurements in order to definitively map the neutrinos. This multi-decade program is an exciting frontier meant to lead the way to our understanding of the neutrino.

There are also complex efforts underway to directly measure the neutrino's mass using a variation of the beta decay process that led to the original prediction of the neutrino. If neutrinos have a Majorana spinor nature, then it is expected that these highly sensitive experiments will observe direct evidence for that mass and enable mapping their mass landscape of the three neutrinos. This work is independent of the particle accelerator-based experiments mentioned earlier, and complements the neutrino mixing work of those efforts. Experiments such as COBRA, CUORE, EXO, GERDA, KamLAND-Zen, MAJORANA, and XMASS use various materials and techniques in pursuit of measuring neutrino behavior.

What is the connection, if any, between the neutrino and the Higgs boson? Only patience, time, and the revelations of data will tell. Perhaps there is no connection. Perhaps the Higgs readily, and for unexpected reasons, decays into something neutrino-like. Guided by decades of experience in searching for predicted and new phenomena, the experimental physics community is open to surprises like this. They would welcome evidence linking the Higgs boson to this mysterious sector of the standard model. If the Higgs boson is a path to a deeper understanding of the neutrino, it is perhaps, also, a path out of the familiar territory of the standard model and into more exciting "terra incognita" elsewhere.

Higgs and Supersymmetry

In the earlier chapter on supersymmetry and later chapters on superstring theory, there has been almost no mention of the Higgs boson, yet the Higgs plays an important role in theories involving supersymmetry. If the standard model is to be extended to include supersymmetry, one is also forced to extend the Higgs sector beyond one physical Higgs boson to one that yields many more.

The "twinning" of the entirety of the matter and force particle of the standard model comes at a much larger cost in the Higgs portion of the theory. A single Higgs boson is insufficient to also explain the masses of the superpartners of the known fermions and bosons. One is forced to extend the Higgs quantum field in some way, such as postulating that there are additional Higgs quantum fields tacked on to those already in the standard model.

Adding fields comes at a cost. When the dust clears the smallest extension of the standard model, the MSSM, has at least five Higgs bosons. There is a lightest neutral Higgs boson, one that may be identified with the one we have already discovered, denoted in the standard model as H^0. In the MSSM, the lightest Higgs boson is denoted h^0, the lowercase h suggesting a lower mass compared to others in the theory. Then there is a heavy cousin of this lightest state, denoted in the MSSM by a capital "H," H^0. So far, the added Higgs bosons are familiar to physicists. They behave much like the lone standard model Higgs boson, with the exception that there are two of them, and one is heavier than the other.

Now we enter the stranger corner of the MSSM Higgs family. First, there is an odd-ball in the family, the A^0. While "odd-ball" is meant to be humorous, it is also somewhat literal, because the A^0 wave function behaves differently when certain mathematical symmetries are considered. Its pattern of decay is expected to be quite different from h^0 and H^0. So this cousin is, both literally and figuratively, odd.

And things get stranger still when we meet the remaining two new cousins of the standard model higgs boson. These two have something that no other Higgs boson has. It is a chimera, the mythic beast having the head and body of a lion, the head of a goat rising from its back, and the tail of a snake. Like the mythical chimera, the last two cousins are a new hybrid of the electroweak interaction—they possess both non-zero electric charge and weak hypercharge. These bosons are not predicted by the standard model. Discovering an electrically charged spin-0 particle would be unambiguous evidence of having found something beyond the standard model and would be a strong indicator in support of the supersymmetry hypothesis.

The discovery of any other Higgs-like boson would be sufficient evidence to indicate an extended sector of interactions beyond what is described by the standard model. The patterns of any such Higgs cousins would tell us something about the structure of that new physics, but physicists would need to discover more than just one new particle if they expect to learn what would be needed to know about the structure of a larger theory of nature. The MSSM predicts that, were you to find a second Higgs boson, then there should be three more to find.

As you might expect, the possibility of detecting more Higgs bosons has been an important discussion for the LHC experiments (and experiments prior to the

LHC—co-author SS earned his Ph.D. searching for a process that could be enhanced by the existence of an electrically charged Higgs boson). Both the ATLAS and CMS experiments list programs to search for cousins of the Higgs boson, using the properties of the known Higgs boson to direct the searches for additional Higgs bosons. To date, no compelling evidence for an additional Higgs boson has been discovered, but it is still quite early in the LHC's multi-decade program, and the design energy of the collider is still in the future. There are huge opportunities for new Higgs bosons, possibly to be found in the presence of supersymmetry, to be discovered through the use of this frontier instrumentation.

What about superpartners? Can the Higgs boson we have already discovered aid us in the search for possible superpartners, like stop squarks (a top quark's superpartner)? Indeed, SUSY is such a predictive framework that it suggests that there are opportunities to use the Higgs boson to pin down the existence of superpartners themselves. For instance, in the standard model there is a Higgs production process that involves the co-production of a Higgs boson and a pair of top quarks (expressed as $t\,\bar{t}$, a top and anti-top). This can be represented as

$$pp \rightarrow t\,\bar{t}\,H^0.$$

In the MSSM, there is also the possibility for its corollary process to occur:

$$pp \rightarrow \tilde{t}\,\tilde{\bar{t}}\,h^0$$

where the stop squark $(\tilde{t})$ is produced alongside a stop anti-squark $(\tilde{\bar{t}})$ and the lightest Higgs boson of the MSSM. In fact, it is possible for this latter process to happen more often than the standard model process. So, the hunt is on for signatures like this, and many others.

SUSY, however, constrained by its non-observation for thirty years, continues as a theory having a vast and unknown space of possible parameter values. This means that there is not strong guidance on where experimentalists should look to rule it out or to discover it. It pays, therefore, to be as open-minded as possible in these searches. A balancing act of openness and focus is constantly evolving in experimentation like the LHC. Experimentalists must trade off the certainty provided by specific fingerprints of SUSY with the knowledge that being too specific might cause one to miss a major discovery if one becomes over-focused on SUSY or a specific SUSY model.

The Higgs, only discovered in 2012, has already become a tool in the hunt for SUSY. What, if anything, will be found during the long program of LHC physics is not clear. SUSY particles have certainly not "rained from the collider" as was originally hoped. However, perhaps the Higgs and its strange nature—a spin-0 particle, a manifestation of a quantum field that, itself, defines the symmetry-broken vacuum state of the cosmos—will provide a path forward in the hunt for, or the elimination of SUSY.

The Meta-Stable Cosmos

The Higgs boson has served not only as a tool for discovering connections to a possible new physics (such as the origin of neutrino mass and supersymmetry), the Higgs boson also has served as a predicter of things to come in the cosmos. Cosmologists consider the balance of energy, matter, radiation, and space-time curvature because these are said to define the fate of the universe. Will it expand forever? Will expansion slow? Might it reverse? The Higgs boson presents physicists with a new possibility in this discussion. With its mass of 125 GeV/c^2, the Higgs may be signaling that the universe sits in a not-wholly stable location within its own vacuum state. This could mean that the final fate of the universe may not be determined by energy content, matter content, radiation content, or space-time curvature, but by the state of our universe in empty space and whether it is precariously perched to fall down to its true, lowest-energy point.

We have skirted this topic before, but it's time to consider it with a bit greater fullness. The focus is on that of "potentials." Potential here refers to an amount of energy stored by a force field, representing the field's ability to affect different kinds of particles. Quantum fields are easy to imagine if one considers the lines of force emanating from a bar magnet (i.e., its magnetic field). This provides us with an image of what a "field of force" looks like. Pursuing this further, underlying every field of force is a more fundamental structure—*potential*—that is, the source of the field and, thus, the associated force. Maxwell's equations for electric and magnetic fields describe how potentials shape reality.

An example of force fields and potentials is shown in Figure 15.2. The left illustration visualizes the electric force field around a positive electric charge (such as a proton) when considering how that proton pushes on another proton using its electric field. The right side shows us the electric potential associated with that electric field—the positive charge is located at the center of the spike and represents the strongest location of the potential.

All forces described by the standard model have a potential that generates them. This includes the electroweak force. Let's look at the potential associated with the electroweak force, as shown in Figure 15.2. It is this potential that gives rise to the origin of mass. This potential is defined, not in a physical space (as in Fig. 15.1, where the strength of the electric field and potential for a proton decreases with increasing spatial distance from the charge), but in an abstract space that has no physical analogy. The shape of the electroweak potential is in fact controlled in the standard model by the properties of the Higgs quantum field. Mathematics allows us to visualize and consider this shape. We see that it has a strange form compared to the electric potential, one that has been described as a "sombrero" or that it looks like the bottom of a wine bottle. (See Figure 15.3.)

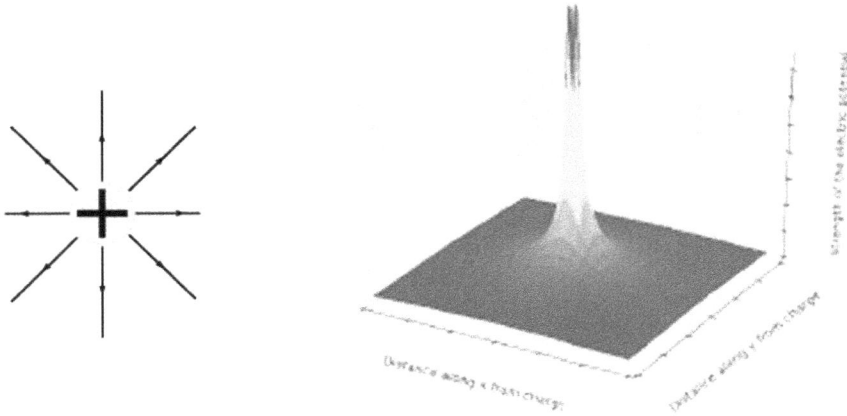

Figure 15.2   On the left is an illustration of a "field of force" around a proton, using arrows to show how the force that this charge exerts on other protons will act (protons push away other protons). The *origin* of this force field is shown on the right, showing the "electric potential" around the charge. How the shape of the potential changes with distance determines the strength of the associated force field. Potentials are more fundamental than fields.

When the electroweak symmetry of the universe is broken, the universe's foundational energy state is supposed to fall to the lowest point in electroweak potential that is pictured in Fig. 15.2. In all previous potentials described by mathematical theory, the lowest energy location always corresponds to the place where the field related to that potential also has zero strength. However, as we see from the shape of the electroweak potential, the lowest-energy position is not located where the Higgs field strength is zero (which occurs at the "top of the hill" at the center of the potential). Instead, the lowest energy point is offset from the place where the Higgs field has its zero strength. This is what makes the electroweak theory strange, yet crucial to the universe as we know it. This strangeness enables the weak bosons, the W and the Z, to acquire mass from the Higgs field. This strange disconnect between the places of lowest energy and the strength of the Higgs field is the key feature of the standard model, and remains to be fully explained.

This description of the foundational energy state of our cosmos is essential to understanding the stability of the universe. Imagine walking down a flight of stairs. Where would you consider the safest, most stable place on the stairway to be? At the bottom of the stairs, of course! On reaching the bottom of the stairs, there are no more stairs on which to trip on and fall. At the bottom, you can be sure that you have reached the lowest energy location that is possible. Should you stop on a higher step to accidentally fall down the rest of the stairs,

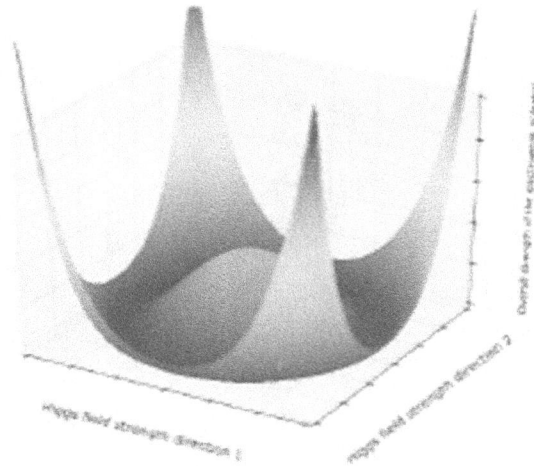

Figure 15.3   The illustration shows the structure of the electroweak potential as described in the standard model. It is this "wine bottle" or "sombrero" shape that defines the fundamental energy structure of empty space, and represents the properties of the Higgs force field in our universe.

you are more likely to injure yourself than after you have carefully reached the bottom.

So, if potentials are like staircases, and the bottom is the best place to be to assure a more permanent safety and stability, what if the shape of the Higgs potential, as described in the standard model, is wrong? What if the place we think is the bottom of the staircase is really a landing, and there are more steps down which the cosmos could fall? Studying the Higgs boson and other properties of the universe will test this idea. If we find that the universe is perched on a landing, with more cosmic steps to go down to the foundational energy level of the universe, it could be that the entire cosmos is perched on a landing from which it might someday fall. Falls have consequences!

Let's explore these consequences in the quantum realm. Let us consider the atom for a moment. Picture an electron orbiting around the central nucleus. Electrons can only exist in specific orbits; in-between orbits are forbidden. To jump from one orbit to another orbit takes energy. To go up in orbit requires energy to be input, and to fall down an orbit emits energy. You would expect that especially in complex atoms having lots of electrons, the electron cloud would always be at its lowest-possible total number of orbits unless something were pumping energy into the atom to make it otherwise.

It turns out that there is a special class of atoms that are "meta-stable"—its electrons are way, way up in orbit around the atom, far from the nucleus and so shielded from that nucleus by their sibling electrons, that they can remain in higher orbits for long periods of time even though lower orbits are available to them. These atomic electron configurations are defined as "kind of stable," or "meta-stable." When such an outlying electron does makes the leap down to a lower orbit, it emits energy during the process in the form of a photon, emitting one photon per leap.

So in the quantum realm, examples do exist that portray the staircase analogy. Some atoms hang out for a long time on a landing, rather than go to the bottom of the staircase. The landing is spacious, maybe with some attractive feature like a comfortable chair to dissuade continuing down the stairs. However, in the quantum realm, if something is possible it *will* happen, even when the chance is small. The electron is *eventually* going to complete the journey down and energy will be released. Can we apply this thinking back to the electroweak potential?

If the electroweak potential is the bowl in which the entire foundational energy state of the universe is described, what would happen if the universe were not in the true, lowest energy state, and could someday spontaneously slip down to it? Could it be that in all of the 13.8 billion years of the universe, that this has not already happened? The laws of physics state that energy must be conserved, so this drop would come at a cost: a flood of new input energy, likely in the form of high-energy photons, would occur. The release of such energy would be so tremendous that it could wipe out all matter, returning the universe to a soup of disconnected, ionized particles.

Can we determine that the final step has not yet been reached? Since the discovery of the Higgs boson, many theoretical physicists have been investigating this. Their efforts seek to understand whether the universe has lived at its most stable, lowest point of the electroweak potential since near to the beginning of time (when all the forces of nature should have been unified), to remain now, and hopefully forever, stable into the future. It turns out that the stability of the cosmos is much dependent on three parameters of the standard model: the mass of the Higgs boson, the mass of the top quark, and the interaction strength of the strong nuclear force.

As an example, physicists Sergey Alekhin, Abdel Djouadi, and Sven-Olaf Moch, have concluded from existing data that among the landscapes of possible stabilities for our universe—stable, meta-stable, and unstable—the current data slightly prefers a meta-stable universe, one that is not quite at the bottom of the energy staircase. However, the data flirts too with the possibility that the universe is actually already stable. The picture is not wholly clear, even though there appears to be a slight preference for meta-stability.

How can we interpret a universe in a meta-stable state, and what are its implications in determining the fate of the cosmos? We can understand the concept of meta-stability as it applies to the electroweak potential by considering Figure 15.4. It might be okay to be in a meta-stable state like that one. After all, to reach the very bottom, doesn't something have to "flick" the universe up

Our
Current
Universe

True
Minimum
Energy

Electroweak
Vacuum
Potential

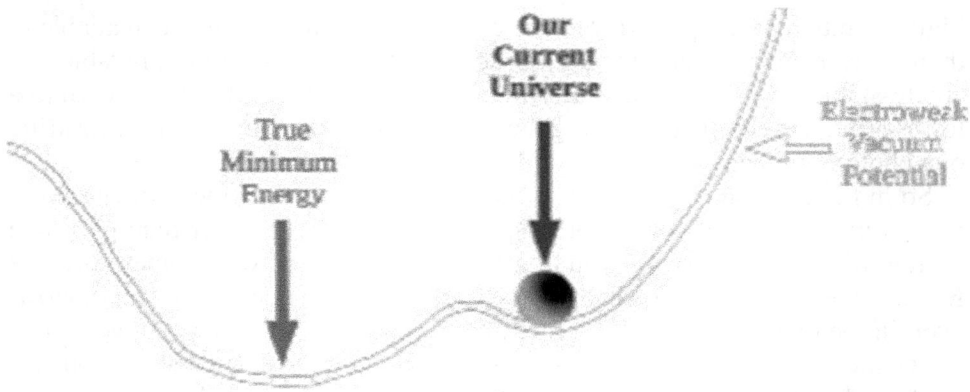

Figure 15.4 This illustration shows what a "meta-stable" electroweak potential might look like. The current state of the universe might be as though it is perched on a step (or ledge) just above the true lowest-energy position in the potential, meaning that one day the whole universe might spontaneously make a final leap down.

and over the little bump between there and the true minimum energy location? That's not going to happen . . . right?

The problem is that the universe is governed by quantum mechanics, and there is always that non-zero chance that at some time the universe may spontaneously surmount the barrier and land on the other side. Atoms and subatomic particles go beyond seemingly impenetrable barriers all the time. It's a well-known phenomenon called "quantum tunneling." It is used not only to probe the details of quantum physics, but to provide us with practical and useful technology, thus demonstrating its reality. Most solid-state devices, including transistors (and the microchips made from them) employ quantum tunneling either as a feature or, in some cases, a design choice. Quantum tunneling is so well understood that it even explains one of the most ubiquitous types of nuclear radiation, alpha emission, wherein two protons and two neutrons spontaneously break free from the "prison" of their parent atomic nucleus, taking with it energy from that nucleus. The quantum physics of tunneling is so well-described and precise that the exact rate at which alpha decay happens can be readily calculated for many atomic nuclei!

We know that tunneling is real; that the possibility of our entire universe spontaneously descending from its current level of electroweak potential is possible and; universal minimal energy level may not be zero. This is the definition of a meta-stable cosmos.

There is one saving grace in this: the data are not all in on the Higgs boson and the top quark, suggesting that our understanding of these particles is not complete enough. While the data flirts with supporting a stable cosmos, our knowledge continues blurry. More important than that, we know from all we have discussed that the standard model is not the final answer to the question, "How did the universe arise and what describes all of its features?" The standard model is thought to be only a part of a more complete description of nature, making the electroweak description of empty space not, perhaps, the final computational tool. Maybe supersymmetry or superstring theory or extra dimensions will complete our knowledge of the vacuum energy of the universe and solve this riddle. Perhaps the universe truly is in a stable state, but we haven't the knowledge yet to state that.

Maybe the Higgs boson serves as a lighthouse perched on the shore of the standard model, shining its beacon into the great unexplored sea of the cosmos. Perhaps, by watching in the many directions in which its light sweeps out, we spot a distant shore to sail to where we will gain a deeper understanding of the universe. Or, maybe in a negative construct, the Higgs is an indicator that our cosmos has not settled into its most comfortable low-energy state.

The Higgs and Dark Matter

Let's look in a different direction past the standard model to consider the role that the Higgs boson might play in understanding dark matter. Dark matter plays a huge role in the structure of the cosmos, by guiding the formation of galaxies and galaxy clusters through its gravitational reach, but its makeup is unknown. The Higgs boson offers the possibility of explaining how to detect dark matter, if it happens that dark matter interacts with the Higgs boson. After all, the Higgs boson is connected to mass by being able to create it, and, if dark matter can be explained by at least one kind of stable particle having mass, then perhaps it can speak (even if feebly) to normal matter through interactions that involve a Higgs boson.

Let's begin by looking at how experiments have been working to experimentally detect dark matter. This has been done by providing astrophysical dark matter (the dark matter around us in our galaxy) a target with which it can interact to reveal its presence. These experiments follow a regular set of approaches, although the methodology varies by experiment. The basic idea is illustrated below. (Figure 15.5.) A target material is selected based on its ability to: (1) be dense enough to coax dark matter to interact, even a little; (2) be well-understood enough that one or more clear signals will be observed when something interacts with this material; and (3) be affordable enough to, over time, scale up the size of the detector without having its costs run away. Experiments, responding to these factors, have proposed a variety of materials for searching for dark matter. Let's look at a few of the experiments presently operating, or preparing to operate.

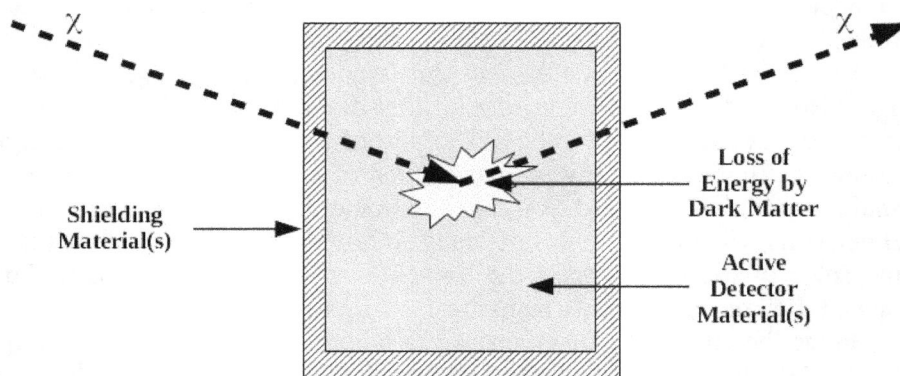

Figure 15.5 An illustration depicting the idea of a dark matter detector. Dark matter must interact rarely else it would have already been seen. The active parts of the detector are therefore heavily shielded from the surrounding world so that there is a greater chance of seeing dark matter deposit energy in the experiment.

There are experiments that use selected liquid versions of atomic noble gases as their detection medium. The most sensitive projects presently using Xenon are the XENON100 and XENON1T international experiments, located at the Gran Sasso Underground Laboratory in Italy, and the LUX and LZ international experiments, located at the Sanford Underground Laboratory at the Homestake Mine in the United States. Xenon is preferred because it can be scaled up to large volumes and is able to provide clear, well-characterized signals in the form of light pulses that indicate when a particle has interacted with a xenon atom.

There are experiments that have chosen semiconductor materials as their detection medium. The most sensitive instruments operating now are the international SuperCDMS SNOLAB, EDELWEISS-III, and CRESST experiments, at the Soudan Underground Laboratory in the U.S., the Modane Laboratory in France, and the Gran Sasso Laboratory in Italy, respectively. While solids are more difficult to scale up than liquids, they offer a more dense interaction medium; plus, semiconductors have a proven record of industrial manufacturing methods that are synergistic between the research and industrial communities.

There are many other detection technologies also in use, such as liquid argon, bubble chambers, light-emitting crystals, and gas-filled detectors. They each offer advantages and disadvantages in the search for cosmic dark matter. These technologies all focus on the same idea, that dark matter should scatter off the nucleus of atoms and, in doing so, leave a trace of its presence behind. That trace should be sufficient to determine the mass of the dark matter constituents.

But, where to look? How does one optimize the design of a dark matter detector when the mass of dark matter particles is not known? Or, how they interact (besides gravity)? A bit of guidance from theoretical physics is known as the "WIMP Miracle." "WIMP" is an acronym for "Weakly Interacting Massive Particle," a generic class of new particle that could have very feeble interactions with normal matter and would thus be a compelling candidate to explain dark matter. Let's look at what the "WIMP Miracle" tells us and how it gives guidance on the possible mass of dark matter.

If dark matter is present, as current evidence suggests, and if it was present during the very early universe, as evidence from the Cosmic Microwave Background experiment suggests, then we can propose that it is carried by a particle that came into existence shortly after the big bang. Let's denote this WIMP particle by the Greek letter, χ (pronounced "ky"). If there are interactions between these WIMPs and normal matter (standard model) particles in the early universe, then it's possible that WIMPs had been exchanging energy with standard model particles. This has implications for the structure of the universe, and for how much dark matter would be left after the universe cooled down. If you want to get the "right" amount of dark matter versus normal matter observed in the universe today, you have to tune the interactions of dark matter particles to be "just right." This tuning leads to the "WIMP Miracle." If it is the weak nuclear force that is the connection between WIMPs and standard model particles, either through the Z boson or the Higgs boson, then the mass of the WIMPs is constrained to lie in the neighborhood of about 100 GeV/c^2.

This idea has been used for more than a decade to guide the design of many dark matter detectors. However, the possibility that the Z boson is the connection between dark matter and normal matter has largely been ruled out. as these interactions would probably have been observed by now. So the hypothesis that the Z boson mediates the interaction between dark matter and normal matter is largely disfavored by observational evidence.

What about the Higgs boson? This predicted space of interactions is not quite as ruled out as the Z boson case, but a tremendous amount of progress has been made both by direct measurement of the Higgs boson and its properties, and by the direct search for WIMPs, that suggests that experiments are very close to either discovering or ruling out this means of interacting. Within the next decade or so, direct search experiments should achieve the sensitivities required to greatly cover the space of possible places where WIMP interactions could lie in the data.

Of course, perhaps the "WIMP Miracle" is just a coincidence that, in light of a more complex dark matter sector (for example, what if dark matter is not made from a single kind of particle, but a variety of particles, just like the standard model?) would be moot. Maybe dark matter really is a single particle, but lighter or heavier than the "WIMP Miracle" would imply. Experiments have been planned to explore this possibility. In addition, if there are more fundamental forces in nature than the electroweak, strong, and gravitational forces, then perhaps we've missed an important method for understanding dark matter.

Theoretical physicists have been considering these possibilities for a long time, and what has emerged from that are ideas about a "dark sector" of the universe. Perhaps there is a kind of "dark standard model," a shadow of the known standard model, with ways of bridging between the standard model particles and the dark sector particles that could be probed by particle colliders such as the LHC.

Indeed, LHC experiments have been searching for evidence that the Higgs decays to invisible final states, as mentioned earlier during the discussion of neutrinos and mass. If the Higgs is observed to readily decay to undetectable final states, perhaps it is explained, not by neutrinos, but by something else. It might be connected to dark matter!

Perhaps the dark matter mass is such that the Higgs can decay to a pair of these particles. This is something that fits nicely into these dark sector ideas. Searches have been conducted at the ATLAS, CMS, and other experiments for evidence of such dark sector aspects of the Higgs interaction. No compelling evidence has been found in favor of these ideas, but it is still quite early in the LHC program and experimental and theoretical physicists will continue to improve the techniques for searching for dark sector physics. After all, if dark matter was easy to detect, it would have been detected long ago. Indeed, all experiments—direct searches and the LHC alike—expect that dark matter will first be detected at the edge of sensitivity before being revealed in greater detail, with subsequent measurements.

Thus, perhaps, the Higgs may be a pathway to understanding dark matter. The efforts we have described here, and many others, are under way to assess this idea. Because the Higgs has such an intimate connection to mass, there is a belief that the Higgs and dark matter might play a fundamental role together. However, this idea may not be true and the light of experiment continues to probe the dark corners of the cosmos to see what is true about dark matter and what is not. The Higgs might serve as a sign post on the road into the frontier, but its utility is still not entirely clear.

Wrap Up

Okay then! What the heck's the Higgs?

It is a spin-0 particle whose quantum field, and thus quantum potential, defines the foundation of the electroweak theory. Its pattern of interaction with other quantum fields and particles yields the masses we observe in nature. But, what sets the pattern in the first place? The standard model, while describing the properties of this particle quite clearly, is mum on one of the most fundamental questions about the structure of the universe.

The Higgs may be more than just a means to understand the mass of quarks, gauge bosons, and charged leptons in the standard model. Can it shed light on the new mysteries associated with neutrino mass? What is it that sets neutrino masses to be *so much smaller* than even the smallest mass known before, the mass of the electron? Why would nature want things this way? Again, the standard model is mute. But, perhaps the Higgs boson can provide a means to see forward in this question, paralleling the tremendous efforts of dedicated neutrino experiments to shed more and more light on the properties and behavior of the neutrinos.

What is the fate of our universe? Will matter and the forces as we know them continue to gently evolve the universe for billions of years into the future? Or, might the Higgs mass serve as a warning that the universe has not quite settled on its foundations yet? If so, there is a small chance that, at some point in the future, the universe might quantum tunnel to its true lowest-energy state, dissolving the matter of the universe, resetting it to a hot and unstructured state. Or, maybe we're making dire predictions with too little information. What if there is new physics between the settled lands of the standard model and the too-distant island of the Planck scale? Maybe supersymmetry or superstring theory can gallop in here to settle this issue, helping us to understand whether or not the universe is settled on its foundation. Indeed, we can see how complex the Higgs sector of physics becomes when one adds supersymmetry to the standard model. Can the Higgs light the way to some new, intermediate land that soothes this question?

Dark matter is a central mystery of the cosmos today. What are its constituents? Are they heavy, weakly interacting particles? If so, are those interactions truly "weak" in the standard model-sense, governed by Z and H bosons? A great deal of that idea has been ruled out by the steady upward march of sensitivity in the design of direct detection dark matter experiments, but there is room left for this idea to triumph. Ah, but perhaps dark matter is as complex as the standard model, with a myriad of its own matter and force particles. Could the Higgs serve as a "portal" between the standard model and the dark sector?

And what of the Higgs and superstring theory? The discovery of the Higgs boson with a mass of approximately 125 (in appropriate units of measure) is tantalizing from the point of view, not of Superstring/M-Theory per se, but from the point of view of the existence of supersymmetry as an accurate description of our universe. Supersymmetry is a kind of balance in our cosmos between the types of particles that are bosons/fermions on one hand versus those that are carriers subject to the fundamental forces on the other.

Mathematical consistency allows for the mass of the Higgs boson to be much larger, say between 500 and 600 on our appropriate scale. Therefore, the one discovered at 125 GeV/c2 is considered "light." But we've seen earlier how quantum mechanics, if not hand-tuned carefully by the theoretical physicist, could keep driving the expected Higgs mass higher and higher. Yet such ferocious corrections are not seen in the laboratory. Perhaps it is supersymmetry that answers the question of "why not?"

The Higgs boson can be regarded as being analogous to a tightrope walker dealing with blowing winds. If the wind blows too hard from left to right, the performer risks being blown off on the right side of the tightrope. If the wind blows too hard from right to left, the performer risks being blown off on the left side of the tightrope. However, if a wind coming from both directions blows equally hard, then the walker stands a good chance to successfully complete the rope walk.

The mathematical condition of supersymmetry is equivalent to the winds from both directions blowing equally hard. So long as it is in force the Higgs boson could maintain its light mass.

So, is the light mass of the Higgs an effect of Superstring/M-Theory or not?

The answer is confusing. It could be an indication inasmuch as almost all consistent string theories involve supersymmetry. However, it is possible to mathematically construct equations that possess the property of supersymmetry in extensions of the standard model, yet are totally independent of any string theory. So the presence of supersymmetry, while encouraging for string theory, does not provide iron-clad evidence for string theory.

Finally, whether or not the Higgs boson is a "portal" to dark matter is another tantalizing possibility on the horizon. In order to have a mathematically consistent set of equations that include the standard model and supersymmetry, five Higgs bosons are required as well as an equal number of superpartners. So it is conceivable that within this zone there exists a particle type that could explain dark matter.

In truth, the question of "What the heck's the Higgs?" drives a complex and diverse program of physics set to operate for decades to come. The LHC experiments will continue into the 2030s, and there are already plans underway to construct a new facility, perhaps in Japan or in China (or both), to collide electrons and positrons to study the Higgs employing alternative means than presently. These experiments would allow the properties of the Higgs boson to be known to better than 1% precision, the level at which we presently understand many of the other parameters of the standard model.

There is the hope that, with lessons learned from the construction and operation of the LHC, a new proton collider will be built in this century that achieves a new level of proton collision energy at the level of 100 TeV, as compared to the present 14 TeV design of the LHC. We know from past experience that each new increase in energy is accompanied by new discoveries. However, what we learn at the LHC about the Higgs and its interactions may not be sufficient to light the way forward, and we may have to work on a new *discovery machine* that can achieve that step.

The discovery of the Higgs boson was a first step in a much grander journey that will be executed by thousands upon thousands of physicists, theoretical and experimental alike, along with engineers, technicians, and data scientists working on collider and non-collider physics experiments. We now have a Higgs boson, yes, but the final answer to a most basic question, one that has driven the subtitle of this book, lies ahead of us. Perhaps one of our young readers will be inspired to take up this question as their life's work—to be the one who finally cracks it wide open to expose to full light all those myriad forms that continue to cast their questioning shadows upon the cave's wall.

## Chapter 16

Seeking Reality

Among the most often-asked questions humans pose are: "Who am I?" "Why am I here?" and "What is happening around me?"

The scientist's challenge in seeking to answer these questions is to create an as-accurate-as-possible description of our universe without appealing to supernatural causation. Using technology, humans observe the heavens, the stars, planets, comets, and galaxies to find that this array of objects fits into a scheme of things. Over the millennia this method has worked well, enabling a continuing progress in understanding. As more is learned about the universe and its workings, we often discover that the mysteries are deeper than we anticipated—shadows that cast themselves upon the walls of our catalog of knowledge.

Technology reveals new information about the universe and enables our understanding to become broader, deeper, and more nuanced while making it more comprehensive. Continued observation produces new information about the universe and its behavior. In our present information and computation age, a single physics experiment can produce more data than is available in all of the libraries that have ever existed on earth. In turn, this new knowledge introduces further questions about how these observations relate to what is already known. A cycle of refinement, often punctuated by unexpected discovery, defines the evolution of knowledge. A side effect of this process explains why society supports and continues to support this way of improving knowledge. The obtaining and evolutionary refinement of knowledge drives parallel technological evolution, thereby providing us a more bountiful quality of life.

We now explore what it means to refine a mathematical theory to better explain the universe that we know. We will use a particular example to show this, reminding the reader that progress is often made, not by a single person, but by a community of people in competition or in cooperation.

Our universe is complex. Explanations of it, while sometimes based on simple ideas, often also present new complexity. Adding complexity to explain complexity may offer new discoveries waiting to be made, or they may push the promise of discovery out of reach. This produces the need for careful balancing. The goal of a theoretical physicist is to model nature, but in modeling nature closely, one must be mindful that one may push the testable features out of the reach of observational methods presently available. One cannot propose a model of the universe that provides ad hoc excuses for failure. Such models are not falsifiable, and thus fall outside the rigors of the scientific method. In the pursuit of a deeper understanding of nature, we must be mindful that nature itself is the final arbiter of a proposed physical theory.

The Pursuit

Superstring/M-Theory, in introducing a new slant on quantum theory, has created a mystery about how it fits into cosmology. Theorists (of course, not *all* of them) generally agree that the mathematical structure of string theory mimics the standard model in its broadest context and is the only known method that appears to do this using mathematics that is consistent with the standard model. For string theory to become physics, its mathematics must consistently describe a quantum theory of gravitation that is supported by the cosmology we observe with our telescopes and satellites, while also coinciding with the behavior of the fundamental matter and forces that we see in nature (e.g. as can be demonstrated within particle colliders).

Unable to experiment directly on string theory, scientists continued to work in the manner as they always had and turned to modeling the universe in the hope that this would produce predictions from the mathematics that corroborates string theory's ability to attain a unified cosmology of the universe. What has been achieved so far has helped to provide greater definition in models that describe what we observe in nature. Superstring/M-Theory models are further attempts to produce high-quality predictions that can be benchmarked against known behaviors of the universe.

Branes, as an alternative to, or as a subset of, Superstring/M-Theory, can also be used to depict the behavior of our universe because branes are able to support particles and forces that we experience in our everyday lives. However, a brane, like all mathematical constructions, may be a device that we have rationalized. These may be yet more shadow figures as we have no proof that branes, like strings, actually exist.

Let us return to that most basic of all questions: "What is the universe?" We might turn, for a first impression, to what the sum of human perception tells us as is summarized in a dictionary. The dictionary defines it as, "All things that exist—the earth, the heavens, the galaxies, and all else therein, considered together as a whole; the cosmos. . . ."

The characteristics of branes can contain all of the forces, particles, and objects (even we humans) that are the observed cosmos. In that ability, branes have been proposed as a fundamental groundwork for the model-building process. As we saw while discussing the building and testing of models, branes are useful in model-building because they are a powerful alternative tool to help understand the behaviors and interactions of forces and particles (strings, for example) in describing the scenarios implied by the standard model. As in all human endeavors, it is a good practice to have at least two ideas in order to spur competition and for each to hone the other. The arena of theory works in many ways like all other areas of human competition. Branes, because they contain all of the descriptors of the universe, can be used as a framework to describe and analyze specific behavioral activities of the universe. Branes have become the foundation for modeling an entire universe, visible and invisible—a *braneworld.*

To see how an idea can be used to point the way forward, even absent evidence from a current generation of experiments, we will explore how braneworld models are important in suggesting future particle collider-based tests. Branes provide tools for describing what can be expected to occur during particle-force interactions. Consequently, branes possess the potential to become significant contributors in establishing physically realizable experiments.

It is important to recognize that braneworld models are not the same as Superstring/M-Theory. While it is true that certain types of branes exist in this context, this is generally not the case. For example, the Randall-Sundrum braneworld employs the concept of hidden dimensions, but is agnostic on the suggestion that matter and force particles arise from an underlying filament-like structure. The brane is also silent on the incompatibility between quantum theory and gravitation, even as it leaves open the possibility for gravity to be weakened by the separation created by bulk space. Finally, the brane is not subject to quantum mechanical laws.

At the technical level, there remain unresolved issues as to whether the Randall-Sundrum brane might actually be a mathematically consistent solution to the equations of Superstring/M-Theory. It may be that this theory relates to the less expansive supergravity theory, a theory that goes only so far as to combine supersymmetry with general relativity, independent of being a superstring theory.

Branes were the focus of intense study in the 1990s, which research activity continues to this day. The importance of brane theory has been broadly accepted by many physicists (but, again, by no means all) who ask whether branes exist in the real world. Are branes real, or are they another shadow on the wall? If they do exist, what does their existence imply? Accepting the possibility that branes are a part of the makeup of the universe, they have become an integral part of braneworlds, real-world models that combine branes with strings and bulk dimensions in a way that pursues a detailed probing of the behaviors of our universe.

When the role of higher dimensions is compared in string and brane approaches, the braneworld allows setting up mathematical models that are closer to the standard model when comparing real-world assessments of proposed scenarios. Always in the context of the standard model, computer simulation becomes a powerful method for identifying collider experiments that can provide real-world comparisons, as we saw earlier about mathematical and computational modeling.

Let's retrace: Mathematical modeling and experimental simulation are used to examine brane behavior so that real-world assessment of known and postulated laws can be evaluated. After 1995, branes became a new tool for model-building, allowing researchers to experiment mathematically with different configurations such as hidden dimensions. The universe will remain a shadowy mystery until this idea or some as-yet-to-be-defined idea is confirmed by an experimental test.

Brane scenarios introduce interesting new possibilities about the global nature of space-time. If standard model particles are found on a brane, then humans also must reside there as humans are composed of these particles and we and everything we know experience these forces. Not all particles and forces are required to be on the same brane. Branes can be connected together in many configurations using open strings. Such combinations of branes and strings appear to be able to produce effects that accurately mimic the standard model. For instance, explaining the hierarchy of interaction strengths from the strong nuclear force down to gravity can be described by branes. This increases the possibility for arriving at a braneworld model that mimics our real world.

Let us look at some of the specific models that have shown promise in explaining the universe. These models build on ideas already presented in this book—that the particles and forces of the standard model can be confined to a low-dimensional space (3-dimensional) while that space is embedded in a higher-dimensional space; that the extra dimensions beyond those on our brane are known as bulk space; and that communication between branes is possible and mechanism-dependent. A brane can have more than the three spatial dimensions we experience, but the dimensional extent of the bulk must always be higher than or equal to that of the highest dimension of any brane that resides in that bulk space.

The first model we will explore employs the interactions of a specific string type, the heterotic string. The second model will look at a universe that stresses larger dimensions (yet still quite small by our standards) than those we have previously considered as well as extra dimensions. (By "larger dimension," we refer here to the possibility that the extra dimensions in brane models do not have to curl up into the *very* small sizes [like Calabi-Yau spaces] that would be required in order to be consistent with a string-type theory.)

An important factor to keep in mind, a remnant of superstring theory ideas, is what the conception of the quantum of gravity will look like. This quantum, the graviton, is considered a closed string. Because it attaches only to itself (unlike an open string, which may attach its ends each to a different brane), the graviton is free to travel in bulk space.

As we move forward in this chapter, we again recall the analogy of the two simultaneous basketball semi-final games, where one game represents one brane and the other game a second brane. In order to picture a model of our universe in this manner, let us first simplify our own observable universe by imagining that our three spatioal dimensions can be represented by two spatial dimensions. Picture our universe as a sheet (2-D) instead of a volume (3-D). To imagine additional spatial dimensions, imagine placing our sheet into a box in which the extra dimsnsion of the box (moving out from—above—or down from—beneath—the sheet) represents the bulk (perhaps even more than one of them). This is a bit like imagining our universe as a fish tank—the observable portion being a glass sheet placed vertically into the tank. Our sheet, a 2-D brane, is embedded in the larger space, the fish tank. The fish tank is all that there is, while the sheet is all that we know about for certain. We might even imagine that somewere else in the tank there is another universe like ours—another glass sheet—separated from us by the bulk space.

By now it should be apparent that there are likely many types of braneworld models. Each model can be shown to include the forces and particles of the standard model on one or more branes that reside in a bulk space, allowing communication between them via closed strings and other particles such as the graviton. Physicist Lisa Randall's book, *Warped Passages,* presents several kinds of braneworlds, highlighting particular characteristics found in each of them. For example:

1. That dimensions can be much larger than physicists had previously thought possible (although still very small, as we will see).
2. That space-time can be warped or curved so that we can expect objects in the bulk to be affected (e.g. mass).
3. That the curvature of space-time might itself hide the number of extra dimensions (which could be infinite) and that space-time might even appear to have different numbers of dimensions in different places.

These behavioral attributes were carefully chosen by Randall and Sundrum as suitable examples to display different universe scenarios that can become the bases for collider experiments. They are distinct possibilities that remain to be proven and demonstrate that they are not shadows of the real world!

We are about to learn that larger-sized dimensions are possible and extra dimensions can give rise to Kaluza-Klein particles (which are fingerprints of higher dimensions) that could be made detectable in the LHC. However, in spite of now having a few years of data from the LHC, these extra dimensions have not revealed themselves. Consequently, there is also the need to get a better handle on the cosmological implications of braneworld models to determine whether we can see their effects in the cosmos even if they are not yet revealed in the collider. For instance, perhaps the presence of extra dimensions might reveal themselves in gravitational wave observations by instruments like LIGO or Virgo.

Because there are a great number of braneworld models possible as mathematical constructions, each with different interactions and behaviors, many such models will need to be filtered out in order to find the ones that best describe the behavior of the real universe. Each model can be a useful tool to define experiments that will demonstrate whether these effects exist and which will be most important to determine whether extra dimensions and other behavioral characteristics can be made visible and how sizeable they can be. We must separate the realities of our universe from the mathematical shadows and mysterious behaviors that cannot be supported.

The Hořava-Witten (HW) Model

We will first describe the Hořava-Witten (HW) model, a braneworld model in which particles and forces are divided between exactly two branes in a bulk space. It is possible in this configuration to encounter particles that experience different forces and interactions than we experience in our world. The particles and forces we regularly observe may be only a small part of a much larger universe, one not shown by the observations we have made thus far. Therefore, it is possible that what we see is but a limited portion of the entirety of our universe. Branes separated by large distances can experience interactions only through other branes or strings traveling in the bulk, thus conducting communication from brane to brane. This separation may preclude our becoming aware of other configurations of branes because of the inability for branes to communicate across the bulk space even as branes are just "next door" to each other.

We have already met Ed Witten, one of the two physicists whose name is lent to this model. Let us meet the other creator of this model.

Petr Hořava is a string theorist, originally from the Czech Republic, who is presently a physics professor at the University of California-Berkeley. He wrote articles with Ed Witten, starting in 1995. They discovered that if one begins with an 11-dimensional theory (ala M-theory) and assume that one of the dimensions can have edges (called domain walls), this produces one of the known superstring theories on that boundary. That superstring theory is the heterotic string based on a mathematical group of symmetries called $E_8 \times E_8$. A heterotic string is a closed string that has both superstring and bosonic string properties, making it a hybrid (the word "heterotic" is a medical term referring to the marked growth or vigor exhibited by cross-bred animals or plants). This discovery provided crucial support for the possibility that all string theories could arise as varying limits of a single higher-dimensional theory (M-Theory).

The Hořava-Witten braneworld model was the first braneworld model ever conceived. The strength of this model is that it is a natural consequence of a unified theory of gravity and other forces, symbolized by the key actor in the theory: the heterotic string. Thus, the heterotic string has the potential to unify the forces of nature. This model, in many ways, provides the ability to describe

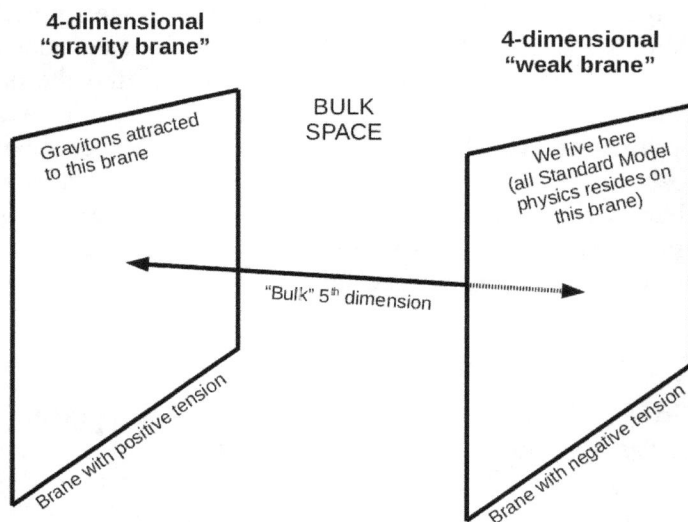

Figure 16.1 A depiction of a universe with a four-dimensional brane containing the forces and particles we already know about, separated from a brane where gravity is strong. The gulf between them is bridged by at least one extra dimension.

the low-energy universe we observe with our best tools while also being a string theory.

A generic braneworld model described by Hořava and Witten looks like the illustration in Figure 16.1. A generic model is a model that has a set of particles and forces that are commonly found in a world like ours, having the ability to be modified in order to examine the circumstances known to exist in the observable universe. The generic model configuration in this figure has two branes, each containing a dimensional description bounding a bulk space of higher dimensionality.

The heterotic string is a closed superstring that has left-moving and right-moving vibrations along its circumferential length. The two opposing directions interact differently from each other. This string was chosen because it looked mathematically like it could unify the force of gravity with the other three forces of nature. The heterotic string can produce both closed- and open-string vibrations. Because of this, Hořava and Witten saw an opportunity to transmigrate its capabilities into that of a multi-dimensional brane. Consequently, this model can treat both open- and closed-string interactions.

There are known drawbacks to this model. In particular, its higher dimensional behavior does not realistically describe all behaviors and interactions of the standard model, which is four-dimensional in nature. Thus, the Hořava-Witten model is a shadow of the real world. While it does produce behaviors typical of those of the standard model, it does not mimic the real world as we see it around us. This model, however, can be useful to help to define experiments to look for real-world behaviors characteristic of these models. For example, experiments could be devised to search out Kaluza-Klein (KK) particles that might be accessible in the bulk space of this model.

In the Hořava-Witten model, the heterotic string is extended to provide two branes, each having nine spatial dimensions located at both ends of a bulk space. The two branes bound the higher-dimension bulk space of ten dimensions plus a time dimension. The combination of two branes bounding an 11-dimensional bulk makes it possible to incorporate a supergravity model. Each of the branes contains force and matter particles that are consistent with those we see in our universe. The brane on the left contains half of the forces having the left-moving characteristics of the heterotic string; the right brane contains the other half of the forces, providing the right-moving string characteristics.

There are enough forces and particles on each brane that either of them could provide the necessary ingredients of the standard model that are needed for humans to reside there and survive. Hořava and Witten assumed that one of the branes confines the standard model particles and forces of our universe, while the bulk and the second brane provide gravity and the other forces and particles needed to complete the behavioral characteristics of the universe. Some of these behaviors have not yet been seen in our universe, but if any unexpected new particles are discovered at the LHC or another particle collider, it might reflect the complexity described in this model

Even in this specific braneworld model, there is the opportunity for both large dimensions and small, curled up dimensions. For instance, the six extra dimensions required to exist on each brane in the model could be hidden away in Calabi-Yau spaces so as to not disturb the physical phenomena already observed in our apparent 4-dimensional universe. This is a rich picture of a cosmos, and predicts an equally rich spectrum of features for the universe. This is what has made it an attractive model to this day.

Perhaps the most intriguing thing about a model like the Horava-Witten approach has to do with gravity. The bulk between the two branes shown in Figure 16.1 provides an 11-dimensional space-time that allows the two branes to interact via the gravity force (the graviton), the communication being provided by closed strings in the bulk. Unlike the gauge fields of the standard model (and, therefore, the photon, the W, the Z, the gluon, and the Higgs boson) it may be that the graviton is the only force-carrier capable of moving freely through the bulk. Gravity might thus leak between all the dimensions of the universe.

Gravity could, in fact, be a very strong force (referred to as Planck strength) in the 11-dimensional universe, but as it only encounters our brane by leaking through the bulk, its strength appears diminished in the process. This would mimic what we observe in our universe while preserving the Planck strength in the bulk. It should be noted that this feature is not unique to this model. It is found in many other models, including those of Randall and Sundrum that would arise in the late 1990s and early 2000s. We will come to them shortly. The leaking of gravity into our brane, belying its true strength in the bulk, is a quite popular way of seeking to resolve this particular hierarchy problem.

In fact, this idea might work so well in describing the various hierarchy (and other) problems seen in the standard model that one of the original saviors of the standard model, supersymmetry, might not be needed at all. Let's take a look at that possibility.

The ADD Model

So far, the HW model has been useful in examining such things as the occurrence of extra dimensions. We have described them from the standpoint of examining what expectable effect higher dimensions would have on experimental results. Extra dimensions make it possible to describe string behaviors more readily. The size of the extra dimensions is ever a point of discussion in these efforts, and we've see that both kinds, large and small, could be possessed by a single model.

Let us consider what it would mean for an extra dimension to be big, that is, far larger than the curled up size of dimensions that would enable them to hide safely from all conceivable future experiments. If dimensions were larger than 10^{-33}cm (Planck length), the curled-up dimensions we have spoken about, how large would a dimension need to be to be detected?] We've discussed the impracticality of building a collider to probe the Planck scale (be it length or energy, both measure too far beyond the detection level of any conceivable technology). If known particles can be given enough energy that their wavefunctions can travel in the extra dimensions, and if those dimensions are Planck-scale in size, the corresponding KK particles would be so heavy that they, too, would be well out of range of experiments conceivable in colliders for decades to come, and possibly for centuries to come.

But, what if extra dimensions are bigger than we have guessed and KK particles are lighter? We introduced the idea earlier that measuring the pattern of KK particle masses for, say, the KK excitations of the electron, would tell us about the structure of extra dimensions if you remember the discussion of the de Broglie wavelength, that energy is inverse to size. The higher the energy of a particle, the smaller the wavelength it possesses.

This is how accelerators probe short distances. The lightest KK particle and its mass will tell us something about the size of the extra dimension(s) in which it travels, because that mass is energy, and energy is inversely proportional to size.

Consider the simplest assumption, that there is one extra dimension and that it is rolled up into a circle. In that case, the mass of the lightest KK particle would be larger than its corresponding standard model particle's mass by a factor inversely proportional to the extra dimension's size. Since the dimensions of larger and larger sizes would give rise to lighter and lighter KK particles, this allows for collider physicists to bound the size of extra dimensions, should they exist at all. Experiments have searched for KK particles, making assumptions about the number and size of extra dimensions, and have observed no evidence for such particles up to a mass of 1000 GeV/c² (about one thousand times the mass of the proton).

Non-observation of a phenomenon is not a bad thing. As a model makes a prediction about what you *should* see, if you don't see it you can turn that non-observation around and constrain the model. Since KK particles present themselves as signatures of higher dimensions, not seeing them tells experimental physicists that extra dimensions cannot be very large or are nonexistent. Current experimental constraints indicate that extra dimensions cannot be larger than 10^{-18}cm (one-millionth of a trillionth centimeter) In other words, current experiments suggest that the largest possible size of a hidden dimension is one thousand times smaller than the size of a helium nucleus! This is a very small size indeed, and one that, with currently available technology, challenges the best instruments and physicists, were they to look directly for smaller-sized hidden dimensions. This limit to the size of an extra dimension is about a hundred times smaller than the length scale at which the weak force becomes important.

Nonetheless, while 10^{-18}cm is small, it is huge compared to 10^{-33}cm (the Planck length scale), which is the distance at which physicists expect you absolutely have to have a quantum theory of gravity. There are still fifteen powers of ten between where physicists can now probe (albeit indirectly) the size of extra dimensions and the scale at which they would absolutely *have to be* in order for superstring theory to work.

This means that extra dimensions could be much bigger than the Planck scale length and still evade detection. A contemporary physicist, Ignatius Antoniadis, was one of the first to anticipate that extra dimensions could be far larger than the Planck length, in the range of the length scale associated with the weak nuclear force. If extra dimensional theories come to rescue the standard model by providing a larger theoretical framework that includes what we know, but extends it by adding extra dimensions, this may have implications for particle colliders. For example, if those extra dimensions have sizes that correspond to energies of several hundred or a few thousand GeV—the weak energy scale—then physicists would expect the LHC to see *something*.

What if not seeing KK particles has no bearing on the size of extra dimensions? What if their dimensions are large but KK particles continue to remain hidden from collider experiments? At first, this sounds like an *ad hoc* excuse for a failure to observe anything. As the history of science tells us, theoretical physicists should take cues from the observations and non-observations found by experiment. The lack of the observation of a predicted phenomenon is a roadsign from nature telling the theory to go another way.

The possibility that large extra dimensions exist but remain hidden has been an issue since the late 1990s. Dimensional characteristics are interwoven with gravity and the hierarchy problem, as we hope we have conveyed. In this context, a very interesting idea came forth in 1998 at the annual Supersymmetry Conference (SUSY 1998). The topics discussed at this conference do not always focus on the predictions of SUSY theory. In fact, that year's conference was quite different due to a Stanford physicist, Savas Dimopoulos, who presented a fascinating concept.

He reported on a collaboration with his colleagues, Nima Arkani-Hamed and Gia Dvali (collectively referred to by the initials, "ADD"). They presented the concept of extra-large dimensions. Extra-large dimensions would mean that dimensions could be much larger than 10^{-33} cm. Dimopoulos proposed to the audience that extra dimensions, rather than supersymmetry, could be the physical theory underlying the standard model, quite a turn for a SUSY-themed conference! He went on to say that, if this were true, experimenters could expect to find evidence of extra dimensions (rather than supersymmetry) when they explore energies above 100 GeV, easily achieved with the Tevatron and LHC.

The ADD model summarizes the principal results that Dimopoulos presented at the conference in 1998, which introduced a novel view about how extra-large dimensions might explain the weakness of gravity. This idea was discussed earlier in the context of the Horava-Witten Model—that gravity, spread out in extra dimensions, is weakened in any subset of dimensions by its ability to travel in all of them.

However, while resolving the problem of the hierarchy of strengths in fundamental forces these models create a new kind of hierarchy problem: one must explain why the extra dimensions are so large compared to the Planck length. There is a companion question that ADD asks: without contradicting experimental results, how big can rolled-up extra dimensions be if standard model particles are confined to a brane and are not free to travel in the bulk? The answer from using the ADD approach was most extraordinary. At the time the paper was written, it appeared that extra dimensions could be as big as one millimeter. One millimeter is enormous relative to a particle! Particle behavior in collider experiments is generally measured at far smaller dimensions.

We can compare and contrast the HW model we discussed earlier and the concepts embodied in the ADD model. This enables us to see how adjustments to the assumptions of the model allow for a universe like our own while suggesting new consequences for the observable cosmos. The ADD approach confines all standard model particles and forces to a single brane, placing the extra dimensions in curled-up form (Figure 16.2). In the illustration, the cylinder represents the curled-up extra dimensions. The standard model's contents reside on the single brane confined to that space. The brane is represented by the single dotted line. It has all the characteristics of a brane, but this brane does not bound space as was the case for the HW model. A brane that is said to "bound space" means that there are no directions of space that are outside of the brane itself. For example, a tabletop bounds a two-dimensional space

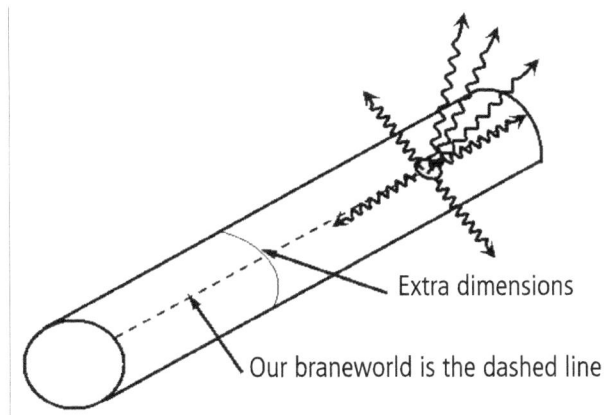

Extra dimensions

Our braneworld is the dashed line

Figure 16.2   An ADD braneworld, where our universe is in the presence of a large, curled up extra dimension of space. Gravity is capable of moving in the large extra dimension, but other particles and forces are not.

as there are two independent directions for motion that can be performed on objects on the table, but it does not bound a three dimensional space since one can move an object off the table. Space in ADD models, depending on their make-up, contains two, three, or more additional curled-up dimensions.

These dimensions are rolled-up and large. As we noted, they could be as big as a millimeter. So, how can extra dimensions be large and still remain hidden? And what, if any, are the consequences of gravity's ability to, here too, traverse the bulk dimensions?

Clearly, if this were the case, then one would expect some observable activity to constrain this claim. Indeed, one thing to do is assess the strength of gravity on the millimeter scale. The ADD model predicts that gravity's strength will change when probed at or below the scale of the extra dimensions since its strength is tied to its ability to move through extra dimensions. One can think of this effect in the following way: that gravity's strength diminishes by a factor of four each time one doubles the distance between two massive objects (the inverse-square law), but this diminution in strength is really due to the fact that every time the distance is doubled, more space occurs between the masses and provides more places where gravity can leak into the bulk space. Close the distance and fewer leaks occur. One should, at the distance scale of the extra dimension or smaller, detect that gravity's strength disobeys the inverse-square law as its greater strength is revealed.

But gravity is far, far, far weaker than other forces as the distance scales shrink, making it very difficult to use an experiment to investigate gravitational attraction in isolation. To conduct a table-top gravitational attraction experiment between two masses at a distance scale of a millimeter is greatly complicated by the fact that the masses are made of atoms, which carry electric charges that need to be accounted for and will strongly influence experimental results. Errant charges such as static electricity and cosmic ionization easily appear. Their effects must be considered and taken into account to avoid inaccuracies.

To get a sense of how weak the gravitational force is, two half-pound masses separated by approximately three feet experience a gravitational attraction that is about equivalent to the weight of a few blood cells. In comparison, stray electric charges present on the two masses (say, an imbalance of charge of just 0.1%) would experience a force ten times greater than that due to the multiplicity of electromagnetic effects that cloud the measurement of gravity.

To really drive home the point that gravity is a super-weak force in comparison to the other known forces, the gravitational force between two electrons is 10^{43} times weaker than the electromagnetic force. The gravitational force of the earth dominates the force of attraction on all earthbound bodies because the net electric charge on these bodies is nearly zero. However, note that you don't fall through the earth even though gravity is pulling you down. This is because the repulsion of electrons between atoms in the floor and atoms in your feet prevents you from being yanked through the earth. Electromagnetism wins again! As atoms press closely together, the minor imbalances in electric neutrality (the electron clouds in your feet are a little closer to the electron clouds in the floor than are the corresponding neutralizing nuclei, exposing more of the electromagnetic field and leading to mutual repulsion) are sufficient to prevent you from falling thorugh the floor.

Because of the sensitivity to other very small forces, bodies that attract (via electric and magnetic forces) while in the presence of gravitational forces must be totally isolated and made insensitive to those small forces, while at the same time being sensitive to the force of gravitational attraction, which is *very very* small. Recall the LIGO experiment, which, while it is capable of detecting gravitational waves rippling through space-time having sizes just a millionth of the size of a proton, could do so only after decades of work to refine the design and reduce its sensitivity to local "noise" and other sources of error.

Although gravity is a most familiar force (we all experiment with it, whether we want to or not, when learning to walk), precise measurement of its value is exceedingly difficult because it is so weak. Newton's gravitational constant, a measure of the strength of the gravitational interaction, is not predicted by any theory so far known and must be evaluated by experimental observation. While independent measurements of this constant have been made, they only agree to about 1 part in 1000 because of the challenge in measuring this number precisely. By comparison, electromagnetic interaction strength is known to 1 part in a billion!

Gravity, while it has been verified in its behavior on grand and cosmic scales, had not really been tested at the millimeter scale even in the late 1990s. ADD provided an opportunity to test extra-dimensional ideas in other than the tunnels of a collider. Let us look at how one experiment achieved this kind of test and constrained the ADD ideas.

Measuring Gravity at Small Scales

Devising a test to measure gravity's effects poses a problem that is not solely related to there being no suitable test instrumentation able to measure the effects between two or more bodies on the earth. As gravity is so weak, all manner of tiny effects that are reported by the instrumentation resultant of non-gravitational forces (friction, heat, etc.) must be fully taken into account. While this is a problem on the earth, small gravitational effects have been measured elsewhere by using a physically large apparatus. Satellites in space, separated from the earth by several hundred miles, use the earth's gravitational attraction to sense a gravity gradient (a difference in the level of gravity's effect) and use that gradient to stabilize the satellite's orientation relative to the earth. By creating a distance separation between two relatively large objects on a satellite, thereby sensing the gravity differential between them while it is in orbit, the electric and magnetic forces on the satellite have a net value of zero so that the small gravity force can be detected. Such a technique, called gravity gradient measurement, is used today in some satellites and is illustrated schematically in Figure 16.3. This technique requires little added equipment.

The magnitude of the attracting gravitational force depends on the distance between the satellite and the earth's center, counterbalanced by its centrifugal force. This technique depends on the law of gravity as first discovered by Isaac Newton. A satellite is pointed continuously toward the earth along the line

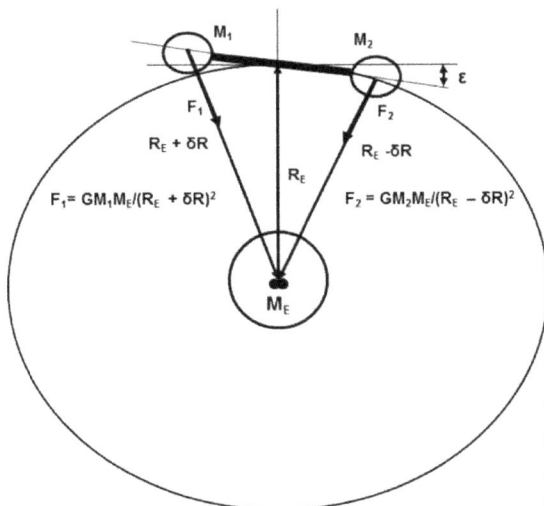

Figure 16.3 A depiction using two masses to probe the changes in strength in a gravitational field. In this case, changes in a planet's gravitational field is felt by two carefully positioned masses in orbit.

passing through the earth's instantaneous local vertical as it circles the earth. Its azimuth and pitch orientation is also stabilized in orbit.

Now, suppose that a disturbance causes the satellite to rotate away from the local vertical orientation, such as being hit by a bit of debris in space. When this occurs, the satellite tilts from its local vertical attitude, in general, in the pitch, yaw, and roll axes. The inertial gyros correct its yaw and roll orientation, while the two balls of the satellite "dumbbell" sense the gravity gradient, causing a differential restoring force that rights the satellite axes to its nominal orientation attitude.

The two balls produce a differential restoring force. With reference to Figure 16.3, that force is produced by a tilt of the satellite away from its local vertical attitude, producing a force gradient (F_1-F_2) caused by the unbalanced forces that result when the two balls are tilted from the local vertical by the angle, ε. This gravity force gradient induces a correction force that restores the spacecraft to its original orientation at the instantaneous local vertical position. The restoring force is negative because the equations assume that F_1 is smaller than F_2. F_1 is smaller than F_2 because the left ball is farther away from the earth's center than the right ball, making its attraction force weaker than that of the nearer ball. The satellite is designed in the manner of a dumbbell, each ball of the dumbbell having sufficient mass to sense the gravity gradient resulting from the satellite's tilt away from its instantaneous local vertical direction.

This same technique can be used to sense the gravitational attraction between two bodies on the earth's surface if there are no other forces large enough to compete. Proper design states that the differential force of attraction is twice the attractive force from one of the objects. The two forces on the satellite are still very small, but they are measurable, given reasonable orbit dimensions and reasonable masses of the two balls. On the earth's surface the use of this technique would be much less sensitive to gravity because of other, stronger disturbances such as extraneous vibration, frictional forces, extraneous charges, and magnetic forces that would swamp the gravity gradient force.

To accomplish a small-scale measurement of gravity's strength, custom instrumentation, called Eöt-Wash, was built at the University of Washington in Seattle by a group of physicists led by Eric Adelberger. This experiment was designed to make it possible to measure the net force of attraction between two objects, their distance separation being no more than a fraction of a millimeter.

The Eöt-Wash Experiment is a modern precision-engineered variation on an older experiment conducted by Count Loránd Eötvös (1848–1919) using a device called a torsion balance. Such devices have been used for centuries to measure very weak forces (Charles-Augustin de Coulomb, a French physicist, used one prior to 1784 to determine the law for the electrostatic force, a law that later became enshrined in Maxwell's equations. Eöt-Wash measures the gravitational effect on a pair of masses spaced very closely together.

Every possible consideration was incorporated into the measuring instrumentation and its setup to make certain that the two objects being measured were fully isolated from all known sources of force. The team that developed and ran the Eöt-Wash Experiment was able to test the inverse-square law to better than a half-millimeter. For an ADD model with two extra dimensions, this implies that extra dimensions cannot have a size larger than about 0.15mm! The scale of large extra dimensions had to be less than that if they are to affect Newton's gravitational constant and the force law that accompanies it, but still defy detection via such experiments.

Alone in Bulk Space

One of the key distinguishing features of the ADD model, as compared to something like Horava-Witten, is that the only KK particles we would be likely to see are those caused by gravitons in the bulk space. In the ADD vision of the universe, the standard model really is trapped on a brane. There is no ability to move in the extra dimensions for electrons, photons, or Higgs bosons. Everything would look four-dimensional. Study standard model particles all you like, you will not see direct evidence of extra dimensions. This would make living on this brane, alone in bulk space, quite a lonely existence. While a graviton can have KK partners, the graviton interacts so weakly in space (again, consider the relative weakness of two electrons' attraction by gravity as opposed to they being repelled by electromagnetism) that these partners, and the graviton itself, would be almost impossible to detect directly.

There is a danger in this model, one that we hinted at in the opening of the chapter, that it might not be falsifiable. Science seeks to make testable, reproducible explanations of the natural world. A key feature of scientific explanation is that there be a test that can falsify it. Absent a means to test gravity at shorter and shorter scales, the ADD vision of the universe is approaching a point where it cannot be falsified, making it permanently safe from testing (akin to Glashow's challenge, a scientific criticism of M-Theory). Theoretical physicists worry about this problem when choosing ideas with which to model the universe; avoiding assessment is akin to creating supernatural explanations for natural things.

Let us look at an alternative to the models of the universe we have reviewed so far. In doing so, we will come upon a model that is not only tuned in to the universe, but capable of being adjusted to accommodate future ideas (such as a superstring or an M-Theory) as well as being capable of predicting the outcomes of experiments we can conduct today.

A Light at the End of the Tunnel

It is always useful to review what we know about the universe. As we have oft repeated, it is wise for theory to be guided by experimental observation. We've explored the quantum realm and its adherence (so far) to the description of the standard model. There are twelve matter particles and three distinct forces that rule the atomic and subatomic realm. Separately, there is gravity, described accurately at scales above a millimeter by general relativity. The cosmos at its grandest scale can be thought of as a volume filled with various components whose densities in that volume control the overall structure and fate of the cosmos. There is the shape of space-time, observed to be quite flat in our four dimensions. There is matter, 15% of which is of the kind described by the standard model and 85% of which is unidentified and interacts, so far, only by gravitational attraction. The latter is dark matter. Together, these two forms of matter make up only about 30% of the energy density of the universe. The remainder appears to be an unidentified dark energy whose role in the present cosmos seems to be to accelerate the expansion of the universe.

As we have emphasized, any model of the cosmos must minimally include these features. The forward movement of the history of physics is one of incremental progress punctuated with unexpected discoveries that drive a rapid explosion of new ideas and, often, a winning idea that stands the long test of experimental observation. Newton's laws of mechanics were not cast aside after 1916, when Einstein came to recognize that Newton's original ideas were but a part of a larger picture of space, time, energy, and matter. Rather, Newton's laws remain as a part of the more complete laws of both general relativity and quantum physics (the standard model). Maxwell's equations tell us about the electrical and magnetic forces, and are a part of the standard model that continues to work today for new applications of electricity. The old laws of physics have not been thrown out. Rather, they are recognized as part of a larger picture, and as time goes by, physicists see more of this picture and come to understand that the old ideas are often movements toward more general concepts. These laws, although partial explanations, they work fine within a range of application.

We seem presently to be in a period when there are many unexpected observations: dark matter, the accelerated expansion of the cosmos, the non-zero tiny mass of neutrinos, to name a few, prompting a huge number of ideas to bubble up in the effort to explain them (and more).

Before we close this look to explain how extra dimensions might provide a model for the universe, let's look at a set of ideas that might allow for the universe we know, explain some of the mysteries we face, and predict outcomes that we can test.

The Warped Cosmos

No matter how often we think about the various problems in understanding the cosmos, one question recurs as a large target upon which all theories of the universe set their sights: the hierarchy problem. Why is the Higgs boson mass so different from the Planck energy scale? Why are the masses of the known fermions so different from each other, with neutrinos having masses that are about a trillion times smaller than that of the top quark? What, if anything, fills the apparent gap between the energy domain ruled by the standard model—energies below about 1000 GeV—and the much higher energy scale where gravity would seem to rule and perhaps might be governed by its own quantum theory?

It is the hierarchy problem, and its many aspects, that attracts the many ideas that form in the minds of theoretical physicists. Solving this problem while explaining other features of the universe has much appeal.

We've mentioned Randall-Sundrum-style models with branes embedded in bulk space. Let's take a look at the specific models that have been adopted to attempt to solve this problem.

The first is the "RS-1" model, consisting of a pair of branes with three spatial dimensions and flat space-time geometry. Branes, like superstrings, can have tension, and in this model the tension of one brane is set to achieve a cosmological constant that explains the accelerated expansion of the observable universe. The choices reproduce the cosmos-level space-time features of the universe we know. The branes are embedded in a higher-dimensional bulk space with one extra dimension. In this model the universe is 4+1 dimensional. The presence of two branes allows for a standard model-like space on one of them (the one with negative tension) and other physics on the second brane, creating a hierarchy. Thus, the hierarchy problem is explained by the geometric properties of the model. This is depicted in Figure 16.4, which duplicates Figure 16.1

You can see the basic principles at work. First, minimize the number of constraints imposed on the model's configuration that cause it to look and behave like our real world; second, eliminate the hierarchy problem and; third, guide the identification of experiments that the LHC is then able to perform.

But what if the universe is not 5-dimensional? Superstring theories require on the order of eleven dimensions. The RS-1 model has room. Should one need to add more dimensions to accommodate M-Theory, one can simply tack them on as tiny, curled-up dimensions.

Look at the geometry of this universe, the brane on which the standard model resides is known as the weakbrane (the right-hand brane in Figure 16.4). The other brane (the left-hand brane in Figure 16.4), is where the unknown-Planck-scale physics resides and is called the gravitybrane. On the gravitybrane, gravity is a strong force equal to that described in the standard model.

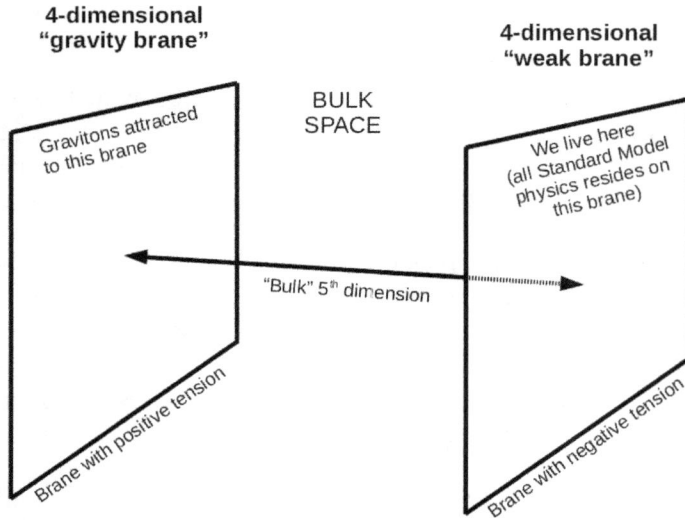

4-dimensional "gravity brane"

4-dimensional "weak brane"

BULK SPACE

Gravitons attracted to this brane

We live here (all Standard Model physics resides on this brane)

"Bulk" 5th dimension

Brane with positive tension

Brane with negative tension

Figure 16.4

We experience its gravitational content on our brane (the weakbrane) via the leakage through the bulk between branes. Unlike the weakbrane, the gravitybrane has a negative tension.

A key aspect here is that the extra spatial dimension is also structured. It is warped such that the 4+1 dimensional system has a cosmological constant, part of which resides on our brane and provides the accelerated expansion of space-time. The warping of the extra dimension allows escape from the constraints on such dimensions that existed when the model was developed, around 1999. Such warped extra dimensions could be quite large. This type of space has not been sensed or observed in our universe, but its effects have been observed. It is called inflationary acceleration.

Energy, and where it can be stored, is important to the RS-1 model, as it is in any model of the universe. Einstein's theory of general relativity is based on the existence of energy and geometry, both of which induce the gravitational field. Branes that carry energy curve space-time even though that space is observed to be spatially flat. The positive tension of the gravitybrane attracts gravitons, causing them to congregate near it, while the negative tension of the weakbrane tends to repel them. Recall that Einstein's equation describes a gravity force field induced by the matter-energy density in the universe. This quantity, in turn, induces an energy density in the universe. That energy density curves space-time. Dark energy would be quite neatly explained by this feature, without the need to introduce new contributions to quantum vacuum energy (which the standard model greatly over-predicts, as we noted earlier).

Were it not for gravity in the bulk, each brane would look like a four-dimensional universe. Gravity is not restricted to a brane. It leaks into the branes from the bulk, where gravitons carry the gravitational force. Gravity is felt everywhere in the bulk, but it is not felt everywhere with equal intensity. Space-time warping of the bulk dimension(s), along with tension on the branes, alter gravity in the bulk space.

The large extra-dimensional component of the ADD model discussed earlier takes advantage of the fact that branes hold particles and forces, but neglect the energy that the branes carry because of their tension. Einstein's theory assumes that energy curves space-time, but it is not clear in the RS construct how a universe with an extra dimension would behave as regards gravity. The additional sources of energy could radically alter gravity's behavior.

This leads to the question, how does gravity change as we travel through the bulk from the gravitybrane to the weakbrane? Does it change uniformly? Does it remain constant in strength? Does it fluctuate in some unexpected way? The answer is not obvious.

The Shape of Gravity

While one may be familiar with curved surfaces, one might have little familiarity with the mathematical definition and significance of a curved surface. When physicists and mathematicians discuss curvature it is not at all in the same terms as it is described by everyday language. A cone is a perfect example. Most people would say that the shape in Figure 16.5 is curved, but, if a mathematician or physicist were asked about the two-dimensional surface shown, the answer is unexpected, as a perfect cone is actually mathematically flat except at its tip. So, one must exercise care in the interpretation of curved surfaces and objects.

When a mathematician or physicist refers to the flatness of a surface, they are asking about its intrinsic curvature. This is a precise way to answer the following question. "If a tiny being was confined to the surface under examination, how could that being tell if it was curved?" Using this definition and its associated mathematics, one can prove that a perfect cone is flat except at its tip. Using those same mathematics, it is shown that a sphere is not flat.

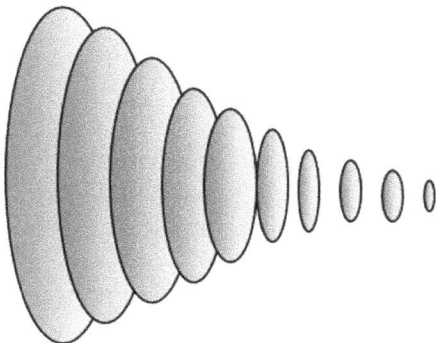

Figure 16.5   A depiction of gravity's changing strength closer to the gravity brane (left side of the illustration) or weakbrane (right). The disks are like slices of gravity's strength. That strength, represented by a disk's area, diminishes as one looks at gravity's pull when closer to the weakbrane.

A spherical surface, when represented in two dimensions, is clearly a curved surface—both to non-scientists as well as to mathematicians and physicists. Latitude and longitude determine one's location on the surface of a sphere; and its curvature is very noticeable. Even the ancient Greek philosopher Eratosthenes (c. 276 B.C.–194 B.C.), deduced this curve and calculated the earth's circumference to be about 25,000 miles (impressively close to the true value).

Curvature affects the gravitational field in the universe. If the bulk space of the RS-1 model is curved, what does this mean? It means that the gravity strength in the bulk will vary as a function of the distance from, say, the gravitybrane.

The graviton is the only force-carrying particle in the bulk space. Were gravitons to clump up in various places in the bulk, gravity's intensity would vary as one travels from the gravitybrane to the weakbrane; but, were the gravitons to congregate nearer the gravitybrane, the intensity there would be strongest, and become progressively less intense as one moves toward the weakbrane. To visualize this, see Figure 16.5.

Assume that gravity is strongest on the left, where the disks are large near the gravitybrane; the larger slices imply greater gravitational strength. Gravity is weakest at the narrow end of this representation. This analogy, however, is not accurate because each slice is portrayed as flat, the only curvature shown is that of its outer funnel-shaped surface. The curvature of the overall funnel shape implies the changing strength of gravity with distance in the bulk, while in each slice of the bulk one can think of that as a space where the gravitational interaction has a specific strength, as it does on either of the two branes. In a warped space, the slices of the space are physically flat, but the measuring rod that measures dimensions like length, speed, mass, and energy changes at each point within the slice, causing variation across the slice. Although each slice is physically flat, its gravitational interaction strength is warped at each point throughout the slice.

This is an analogy. The characteristics of warped space cannot adequately be drawn on the flat surface of a book's page, nor, even, can it be described well in a 3-dimensional space. It is a special kind of effect that we described earlier in the discussion of physics inside a black hole, a special curved space known as anti-de Sitter (AdS) space. Recall from that discussion that its negative curvature causes its shape to appear as a manikin-like object.

While the bulk in this model has a warped geometry that behaves like an ADS space, there is more to the story. There are, in this model, other effects resultant of the gravity behavior of this unique geometry. There are several ways to describe this. The description that the RS model uses is based on how gravitons are distributed in the bulk because that analogy is easily understood. This distribution is echoed by the way energy is distributed on the branes and in the bulk.

In a flat space-time, even one involving more than three spatial dimensions, gravitons would be equally likely to be found anywhere in the bulk. We would say that the probability function is constant throughout the bulk. The slices we described earlier would all have the same gravity strength. However, in a warped geometry this would not be true. The curvature of the bulk space is equivalent to a probability function that drives the most likely place to find a graviton, and in this picture the weakbrane is the place of lowest probability while the gravitybrane is the place of highest probability. This would, in turn, make gravity stronger on the gravitybrane than on the weak brane, even though there really is only one kind of gravitational interaction in the 5-dimensional universe.

Thus, the RS-1 model tells us that in a warped space-time we would expect to find a hierarchy between standard model-like interactions bound as they are to one brane (the weakbrane) and the gravitational interaction, which, while it is observable on the weakbrane, is actually exposed at its full strength on the gravitybrane. Having to travel through the warped space between branes is a game of chance, diminishing the effect of gravity on the weakbrane. Geometry defines the observed strength of gravity. A hierarchy problem is a consequence of this situation.

This leaves a significant question to be answered: how large do these extra dimensions have to be to dilute gravity, and how many dimensions does it take? In addition, how far apart in bulk space must the branes be for this principle to begin to work? The distance between the two RS-1 branes needs to be as large as the Planck length (10^{-33}cm). This is a very small distance, and not a useful lower limit because such distances cannot be tested for directly. If one considers the warping of the extra dimension as the engine that drives the hierarchy problem, then it is estimated that one needs a change in the curvature corresponding to 16 powers of 10, or 10^{16}. This is consistent with the ratio in energy between Planck energy and where the electroweak unification occurs. We have traded a problem in energy scales for a geometric warping factor, which is much easier to consider.

Unification and Warped Extra Dimensions

It is tempting that the hierarchy problem could be so easily resolved in this picture, trading the problem of energy scales at which point physics models take over for the simple warping of an extra dimension. One might then wonder what this picture can do for the unification of forces? That has been the aesthetic quest of particle physics ever since the success of the standard model and its ability to unify the electromagnetic and weak interactions. There is as yet no satisfying model that cleanly unifies the electroweak and QCD interactions (although supersymmetry alone can do it, this opens up other problems, the biggest of which is that there is no evidence for supersymmetry!). Can the RS-1 picture manage to solve this problem? And even if it does, how do we test it?

There are scenarios that would allow for unification to be naturally explained in models like RS-1. These scenarios have to maintain consistency with what is known: that fermions and gauge bosons in the standard model, with perhaps the exception of neutrinos, acquire mass from the Higgs field (interactions with the Higgs boson). We know from experimentation that, as one cranks up the energy at which one probes matter, the interaction strengths of the weak, electromagnetic, and strong interactions change in such a way as to approach each other. Perhaps what is happening is that the gauge fields of the standard model, represented by the particle excitations (the photon, the W and Z, and the gluons) are also able to travel in the bulk space. They would then enjoy that change in strength that gravity experiences thanks to the warping of the extra dimension. Could this explain not only the changing strength of fundamental forces at higher energies, but their possible and eventual unification?

This has been found to be plausible, and physicists, such as Alex Pomerol at the University of Barcelona, have been exploring this kind of augmentation to the original assumptions of the RS-1 model. Such an idea is capable of reproducing the known features of our universe so long as the breaking of electroweak vacuum symmetry, the origin of masses for gauge fields like the W and Z and for quarks and charged leptons, occurs on the weakbrane and not elsewhere. This would require that the Higgs field be pinned to the weakbrane. But, if one conceives of scenarios like this, the unification of forces again becomes a feature of the ability to move in a warped bulk space.

In string theories, one has a tremendous amount of freedom to place particles and fields anywhere in the possible dimensions. Observational evidence is a means to constrain what otherwise seems like an infinite number of possiblities. The choices we have discussed here, suggested by physicists actively working to better understand extra-dimensional ideas, are some that can achieve the goal of being guided by observation. The question, of course, is: how does one test this idea, and is it falsifiable? Is there a test to reveal features of this model (e.g. the existence of bulk space), or is there at least a test that rules it out and tells us that this is not a useful idea?

Experimentation, Proof Positive

String theory, extra dimensions, warped geometries, unification, the solution of the hierarchy problem, the behavior of braneworld models, these are each resultant of research by many over a long period. But we must remember that they are, all of them, conjecture. Until experimentation can corroborate the mathematics, this work is wholly unproven. It is not physics. Its reality remains in the shadows.

Mathematics has painted physically realizable models that behave in ways consistent with the observations of our universe while mimicking the standard model, but experiments must now take these ideas to task in the laboratory, or in the daily workings of the cosmos outside the laboratory. Only through such experimentation can it finally be determined that these conjectures have merit. The possibility for experimentation now presents itself at the forefront of the field. The RS-1 model, and related models, open the door to this experimentation.

We have discussed how the wavefunction of a particle, like an electron, can travel in extra dimensions. This momentum in a space hidden from us will appear to us as if the electron has gained mass, and for every oscillation of its wavefunction in the extra dimension, a quantum unit of mass is added to the state. These are KK particles.

While KK particles can exist in models like RS-1, they no longer tell us anything about the size of the dimensions in which the wavefunction is traveling. For instance, gravitons moving in the bulk will occasionally have zero momentum in our 3-dimensional space when they are on our weakbrane. This manifests as a massive particle in our brane, thus there are detectable consequences. KK particles may not be as light as they were once thought to be (owing to, perhaps, a large size of the extra dimension corresponding to a lower energy required to travel through that extra dimension), but they should exist. Such KK particles would be expected to decay to photon pairs and, in general relativity light is bent by curved space-time. If we zoom into that bending on the quantum level it might look like a photon has repeatedly scattered off of gravitons. Predictions at the LHC of KK particles being produced on the weakbrane place the masses of such particles at the level of 1 TeV/c^2. While this is not as low as one would like, it is within the grasp of the LHC.

What have LHC experiments like ATLAS and CMS observed based on the predictions of RS-1 and related models? To find the key signature of a pair of photons produced by a heavy particle, one need only look at the results of searches for di-photon resonances at ATLAS and CMS. Both experiments look for these with each new release of data from the LHC. In 2010, the first search came up empty-handed and, as a result the experiment, was able to rule out a KK graviton with masses below about 0.5-1.0 TeV in certain parameter spaces of the RS model (assuming that the graviton's interaction strength was above a certain range).

Those searches continue, and there was even a bit of excitement in the search for particles decaying to photon pairs in early 2016. ATLAS and CMS reported that there was a hint in the data of a "mass bump." Such bumps can indicate that a new particle is being produced by the collider. Both experiments saw suspicious activity at a mass of about 750 GeV/c^2. However, the hints were really not very significant. Despite the lack of promise, the theoretical community went on a paper-writing storm, with everyone positioning to have a correct explanation of this particle, should it turn out to be real. By the time the summer of 2016 arrived, both ATLAS and CMS had greatly increased their data sets and the hints disappeared. What had seemed to be weak bumps in the data had been coincidental statistical fluctuations. The community quickly lost interest in the subject.

Using the data they had brought forth in 2016, ATLAS and CMS were able to determine that there was no evidence for a KK graviton up to masses of about 3.5 TeV/c^2 within a limited range of interaction strength between photons and gravitons. So far, there is no strong evidence for the existence of KK gravitons consistent with the RS-1 and associated models of physics.

The Dilemma of Extra-Dimensions

The RS-1 is not the only way to model an observable universe like ours. For instance, are we certain that two branes are needed? There is an alternative model, RS-2, that moves the gravitybrane and weakbranes apart by an infinite distance in the extra dimension, making it appear as if there is only one brane in the universe. This preserves many of the desirable features of RS-1 without the need for a second brane.

While these are clever ideas, many questions need to be addressed. For instance, nowhere do we see how dark matter might emerge, nor the origin of neutrino mass. Supersymmetry, if it is incorporated into RS-1 or RS-2 scenarios, might solve the dark matter question (we earlier discussed the "neutralino" of SUSY as a compelling candidate). One could also assume that there may be some other as-yet-undiscovered kind of matter out there, perhaps described by a new "dark standard model." This leaves neutrino mass as an open question. Perhaps neutrino mass is separately generated by a mechanism in supersymmetry in that new dark standard model.

Another important consideration is that string theorists have been saying for forty years that strings require a minimum of ten dimensions plus the gravity force. Investigations of models that include an RS-2-like system, but with many more extra dimensions, have revealed that there can be other branes that live on dimensions higher than five. These branes could interact with gravity and geometry to create a localized gravity behavior that could have five, six, seven, and even eleven additional dimensions. Such structures could produce a four-dimensional behavior of our brane similar to the behavior of RS-2. Or, perhaps, the dissident view that superstring theory can reside (with mathematical challenges) in fewer than 10 or 11 dimensions, could resolve the point.

Perhaps RS-2 and similar braneworlds are islands in a higher bulk space, as was suggested by theoretical physicist Ed Witten in connection with string-types, each one having its own unique behavior, depending on its force and particle interactions. This can be seen as a future direction of the community's effort.

Alternatively, maybe what we are researching is but a microcosm of a system of parallel universes. We must anticipate that, if there are extra dimensions and branes, why can't there also be multiple universes? If there are multiverses, how will we find them?

While physicists once believed that the work of identifying and describing our universe would be completed by the end of the twentieth century, a reliable, singular, and testable description still evades us. While much has been unveiled about nature's laws through mathematical constructs and experimental evidence, it remains to be seen whether these more encompassing models reflect real-world behavior.

Thus, while there has been a great deal that has been unveiled, the universe remains *a giant shadow*. We realize that there are many questions remaining to be answered about where we came from and where we are going.

Wrap Up

We have shown the struggle of the theoretical physicist (indeed, a community of theoretical physicists) to refine mathematical ideas in the quest to better describe reality. Using various braneworld schemes (Horava-Witten, ADD, and RS-1) we have explored what it means to refine a model to make it more realistic and the consequences that come from making adjustments. We have also shown how experimental tests have been conducted to assess key claims of these models. So far, there is no evidence in support the verity of these models.

Absence of evidence is not scientific evidence that they are absent in nature, but science cannot generally provide a universal negative. In this case there is no such thing as extra dimensions. As we have seen, it is possible to push the detectable effects of extra dimensions to be just out of reach of the current generation of experiments. But, if this is done for too long, it begins to feel like a model-adjustable ad hoc excuse for failure is being created. Such a feature is a hallmark of poor scientific theory. It is important that theoretical physicists working on models consider how far away they can push evidence for their ideas and continue researching a meaningful theory of nature. At what point does theoretical physics become pure mathematics or philosophical musing?

Extra dimensions are a compelling idea. They offer a grand view of the universe, akin to the wonder of being trapped on our little blue planet yet able to discern laws of nature that affect worlds countless light-years away. Trapped on our little weakbrane, we may be just beginning to probe the true dimensions of the universe. We can only hope for evidence of the success of an idea as the search continues each day for the true structure of reality.

A Glimpse
into the Near Future

The physics community will continue its focus on resolving the many un-answered questions about our universe. While this quest has not yet yielded a complete answer as to how the universe operates, as indeed it may never so do, with each new step into the light this sheds, the community encounters even more fundamental and even deeper questions than were earlier revealed.

Even with our detailed present understanding of ordinary matter, there are so many unsettled subjects. Unanswered questions are opportunities for people to make breakthroughs and discoveries in the future. As we have seen, even the mature ideas of string theory and the braneworld concept have unfinished business, and cannot yet yield an accurate picture of reality. We wish to leave you with a sense of wonder at the power of humanity to understand the universe, but we also wish to leave you with the reality that there are many, fundamental, unsolved problems. Each of these is an opportunity, not a crisis.

This era we live in right now is very similar to the one that occurred at the end of the 1800s. That era, too, had reached a moment of wondrous un-derstanding of the universe, having resolved many successful ideas: Newton's Laws, the Laws of Thermodynamics, and Electromagnetism (using Maxwell's Equations). Still, there were puzzle pieces that didn't fit into the then-accepted explanations of nature. Why did atomic spectra exhibit a structure of bright bands and dark bands? Why do bodies that absorb all frequencies of electro-magnetic radiation emit a spectrum of radiation so different from that predicted by thermodynamics? Why are some atoms unstable, emitting energy spontane-ously? And, of course, there was the mystery of the speed of light: why did it not change when the motion of the source (or the observer) changed? And just what was light? Was it a wave or a particle?

Those mysteries unsettled many physicists. It turns out that these puzzle pieces that didn't quite fit into accepted mathematical explanations were the gateways to new discoveries. These days, we (physicists and non-physicists alike) look romantically back on the early 1900s as a renaissance in physical thought, but a renaissance was possible only because there were unsettled issues.

Let us deal with the question of fundamental mass for the matter found in the universe. While many exploratory ideas about matter bubbled up across the decades, there is but one (so far) that wins the day—the existence of the Higgs boson makes two features of nature possible. It explains why the weak force is so short in range while the electromagnetic force is infinite in range, as well as providing an origin for fundamental mass in matter. The Large Hadron Collider and the thousands of physicists who perform experimental physics using this frontier machine provided the crucial evidence that demonstrated the existence of the Higgs boson. Having concluded nearly fifty years developing this idea, we now begin a decades-long program to further define this new particle.

There is clearly much more needing understanding about the cosmos. Let us again consider ideas like superstring theory and braneworlds. While they are presently far removed from the kind of now-proven reasoning that led to the prediction of the Higgs boson, these ideas have begun to yield experimentally testable consequences that can be probed in current and future experiments. They are ideas within the grasp of near-term research efforts that will be explored during the next decade. This work will seek to reveal a fuller picture of the cosmos—a hidden cosmos that shapes our observable one. That picture will be a grand one—if it is true!

As the writing of this book concludes, the program to understand the Higgs boson is in its early stages. We have pictured that the Higgs boson is a kind of road sign that points on to potential pathways to take us into farther frontiers of human understanding. The Higgs boson provides a foundation upon which physicists will build to increase the understanding of realities that continue for the moment to remain in the shadows. There could be other versions of the Higgs boson, subatomic cousins that await discovery at the LHC that will paint a fascinating "family portrait" hinting additional roles in the cosmos, its possible connection to quantum gravity and the graviton, and even to mathematical symmetries that lie outside the standard model. The discovery of a Higgs boson that is curiously low in mass encourages physicists to think that these kinds of discoveries may be just around the corner.

For decades, one of the most widely explored ways of achieving the prospect for a unified theory of all of nature has been string theory. While mathematically consistent (as all explanations of nature must be to be acceptable), it needs provable evidence showing its accuracy in describing nature—at the smallest scales seen in collider data, at the largest scales seen in the light left over from the big bang, and by astrophysical and cosmological observation. It is of note that *only* string theory, among the unified theories, currently attempts to include the force of gravity in its *quantum* aspects. We must recall Sheldon Glashow's criticism that there is a danger that such models are permanently safe from falsification. We reinforce this criticism by noting that there is presently the inability even to count all of the possible ways to compactify the small extra dimensions in M-theory.

Braneworld models are less complex than are string theory models, thereby supporting an effective means for study through near-term experimentation. While a few braneworld models have been discussed in this book, other iterations aimed at a better description of nature are possible. Many models can be described that synthesize reality, but they will continue to be speculative so long as their predictions remain unverifiable by experiment. Warped extra dimensions are a key player required in this landscape of higher dimensional thinking. The LHC will provide the ability to probe for such features of nature during the next decade.

Particle collider experiments at the LHC will search, within controlled conditions and at higher energies, as technology permits it to do so, aspects of the subatomic realm that are predicted by models built upon the ideas of extra dimensions, supersymmetry, or superstrings. However, colliders are not the only way to do this.

The kinds of research performed with colliders may seem very different from that performed with telescopes or other instruments designed to view the largest structures in the cosmos. However, if physics has taught us anything it is this: there is a harmony among all the scales at which one can view the universe. The structure of the nucleus is the result of the details of the strong force, quarks, and gluons, viewed from a scale far larger than any of those ingredients. The atom is a representation of the details of the nucleus, electrons, and the electromagnetic force working together and viewed from a scale larger than each of those parts. Molecules, and even the stuff of life, are but the details of how atoms behave when viewed from distances far larger than the size of atoms. Matter that is on the scale of human beings (things sized similar to ourselves) has its properties determined by all of the aforementioned microscopic details added together into vast structures we call "the everyday world." The solar system is but the structuring of matter by gravity, as viewed on scales much larger than the human scale, going on to view from ever-larger scales, until we can imagine the entire cosmos—the details of the whole are dictated by the details of the tiniest parts. Viewed in this way, there is no difference between the research conducted at a collider and the research conducted at a telescope. To understand the universe we need to combine all of these pursuits to make progress.

Astrophysical observation of objects such as black holes, neutron stars, supernovas, and gamma-ray bursts provide an excellent means to search for evidence of new phenomena and synergies in the natural world. A collection of large telescopes, telescope arrays, and satellites will observe the sky for decades to come, capturing huge amounts of data at finer and finer scales, searching for hints of anything beyond what we already know about the cosmos. Experiments that hunt for dark matter's constituents will probe the universe in other ways, searching for new particles beyond those encoded in the standard model. The dark matter that shaped the evolution of galaxies, clusters, and superclusters should be all around us; we need only cleverness and patience to figure out from what it is made. If it is made from something other than matter, something of which we have yet to conceive, then we need mathematical cleverness to ascertain its possible natures and devise new tools to study it.

Although they have very tiny masses, neutrinos have performed a role in shaping the cosmos as it is today. If they had been even a little heavier than they are now, the sum total of their effect on the early cosmos will have left a measurable imprint on the cosmic microwave background and, indeed, on the distributions of galaxies in the sky. It is crucial that we use ever more defining experiments to map the properties of neutrinos to further enable a complete understanding of the universe.

Some neutrino experiments will seek to determine the exact mass of each kind of neutrino; some will attempt to determine the nature of its wave behavior; yet others will hunt for signs that matter and anti-matter neutrinos behave differently from each other.

Gravitational wave observatories, having embarked on a new land-based form of astronomy, will read messages in space-time—some of these messages that come to us are viewed as the subtle squashing and stretching of our planet as space-time wobbles emanating from the event horizons of colliding black holes. Gravitational waves will allow us to make "sonograms" of the universe.

Data from any of these methods, along with observational methods not yet developed, will provide unexpected support for one or more of the various mathematical constructions presently extant—braneworlds, supersymmetry, superstring/M-theory—and others yet to be conceived.

Some braneworlds possess properties that may explain the unification of forces. The features of these models (large extra dimensions, warped extra dimensions, or the presence of heavy graviton Kaluza-Klein particles, that could manifest in our 4-dimensional space-time) are intriguing signatures to be searched for with existing or anticipated near-future experiments. Braneworlds have apparent deficiencies—there is no clear limit as to how long these models can continue to evade detection before they are deemed truly falsified.

A Glimpse into the Farther Future

Future research, extending across many decades, will focus in a variety of areas. The list is long: increased understanding of the Higgs boson; continued searching for unpredicted structures and patterns in ultramicroscopic realms; exploring the frequency spectrums of radio—infrared, visible, ultraviolet, x-ray, gamma-ray; knowing more about gravitational wave astronomy; dark matter; dark energy; supersymmetry (and other possible symmetries); the possibility of extra dimensions and its associated mathematics; additional Superstring/M-theory development; and, perhaps, some areas not yet identified by mathematics.

Following are some snapshots of experiments that will soon come on line or were actively being conducted at the time of this book's publication (and are expected to continue to run for years onward). Collider physics will continue to be dominated by the LHC well into the 2030s. While a future collider project might ramp up during that period, construction of such facilities requires decades of planning, building, and shake-down before the first well-understood data becomes available. For example, the LHC was conceived of in the early 1980s and provided its first collisions in 2010, a span of about thirty years.

The present LHC program will focus on the study of the Higgs boson and its connection to new models of physics; on supersymmetry and other scenarios; and the study of nuclear collisions, which can teach us about a "quark-gluon fog" that should have existed during the period immediately after the big bang occurred. These and many other areas of study will be accessible to specialized and multi-purpose particle detector experiments at the LHC. The low mass of the Higgs boson has provided a boost for the expectations of the supersymmetry community by pointing indirectly to the possibility for extending the standard model using SUSY. The community continues to look for direct evidence of the existence of this long-sought and intriguing but elusive symmetry of nature.

Astronomy is engaged in a diverse portfolio of experiments. The James Webb Space Telescope, successor to the Hubble Space Telescope, is expected to launch in Spring, 2019. It will bring us new and incredible images from deep space that will be well beyond the technical abilities that were the Hubble's. The Large Synoptic Survey Telescope (LSST), a land-based telescope in northern Chile, is scheduled to be built and begin operations in early 2022. Capable of scanning large sections of the sky very quickly (attributable to its extremely high resolution of 3.2 gigapixels), LSST will build a huge catalog of astronomical objects meant to better understand the evolution of the universe and the physical nature of astronomical phenomena.

A series of telescopes, both ground-based and balloon-borne, will probe the light from the big bang and the cosmic microwave background looking for evidence of gravitational waves that would have been imprinted on that light dating from the birth of the cosmos. Gravitational wave interferometer experiments, such as LIGO and Virgo, are the vanguard of a new generation of space-time telescopes. So far, they have heard the chirps in space-time that emanate from colliding black holes. As more such amazing phenomena are observed, one can only speculate on the new information that will be learned about the universe during the coming decades.

In the area of high-energy particle astronomy, the AMS-02 experiment located on the international space station is a major instrument for studying particles arriving from outer space. Gamma ray satellites such as the Fermi Gamma-Ray Space Telescope will continue to look at the most cataclysmic phenomena in the cosmos, teaching us about the deaths of stars and even larger objects, including the feeding behavior of supermassive black holes. Neutrino telescopes like IceCube at the south pole or ANTARES in the Mediterranean Sea are capable of detecting the highest-energy neutrinos in the universe, and are thus expected to teach us about the kinds of extreme phenomena capable of producing these neutrinos.

The dark matter search community will be very busy for at least one or two decades. The next generation of dark matter search experiments, such as LUX-ZEPLIN (LZ), SuperCDMS SNOLAB, XENON1T, and ADMX Gen2, will push the boundaries of sensitivity while searching for ultra-low-mass dark matter (millionths of the proton mass), low-mass dark matter (less than ten times the proton mass), and heavy dark matter (greater than ten times the proton mass). Eventually, these experiments will become so sensitive to particle interactions that even the very-low-energy neutrinos emitted by our sun may be easily spotted by these instruments, but that is probably still a decade away. Orbital and ground-based astronomy experiments will also be on the hunt for dark matter in the cosmos, looking for interactions where dark matter might clump in our universe—at the centers of galaxies and, perhaps, even in the hearts of stars. The LHC will also participate, looking for the production of dark matter particles in the collider.

Neutrino experiments are expected to be of great interest over the coming decades. In addition to presently operating experiments like T2K in Japan and NOvA in the United States, there is the expectation of a next-generation program called DUNE/LBNF (Deep Underground Neutrino Experiment/Long-Baseline Neutrino Facility). These international programs are studying neutrinos very carefully, mapping out their properties as they observe mixing among their various kinds. Over the coming decades, the neutrino will be coaxed to yield its secrets. It has already provided many surprises. What, for example, might it tell us about the origin of mass?

Let us recall a subject we have looked at in detail from multiple perspectives: the implications for experiments as to whether extra spatial dimensions actually exist. What if these dimensions are large? The most popular reason for invoking extra dimensions is to explain the relative weakness of gravity as compared to the other forces of nature. If gravity does travel in these extra dimensions, thereby weakening its influence in the space and time dimensions by spending little time in our own brane, this can be detected by studying how gravity changes strength on small scales; or, by searching for evidence that particles of gravity—gravitons—travel in the extra dimensions and appear to us, in our four dimensions, as though they are a zoo of new heavy particles (Kaluza-Klein particles). Neither of these effects have been observed, but they continue to be hunted. This has constrained the original breadth of these ideas, limiting possible detectable effects. Kaluza-Klein particles might just be heavier than our present ability to produce them at the LHC. Or, maybe large extra dimensions are not as large as we hoped they might be, or that they are more numerous than we would prefer. If there are more than just one or two extra dimensions, there are more places for Kaluza-Klein particles to have extra momentum, raising the smallest masses they would appear to possess in our 4-dimensional space-time and putting them further from the reach of the LHC. Or, perhaps they don't exist at all! Experiment has ruled out the simpler possibilities, but has not ruled out the idea entirely.

What would it mean if experiments like those at the LHC or in the dark matter search community were to fail to observe anything beyond what is presently and comfortably explained in the standard model? The absence of SUSY in the realm of elementary particles would likely make most superstring concepts less attractive for study (although we note that the failure to detect SUSY since its prediction in the 1970s has not thus far wholly dissuaded the superstring theory community from its pursuit). Alternately, perhaps SUSY particles are trapped on a brane we cannot yet (or ever) access with experiments, making them untestable and thus unsuited to the physical description of nature. Finding new SUSY constructs would require a fresh look at how 4-dimensional and extra-dimensional versions of string theory would avoid their own demise. Because SUSY is a framework that can accommodate many models, its data cannot be used to provide an iron-clad rigorous argument against SUSY. There is an omnipresent danger that the only tenable models of SUSY or extra dimensions will be those that allow for our universe to be as it exists in nature while all the models' novelties (the traits that distinguish them) are not directly provable. In such a case, they would become more a kind of mathematical philosophy than a physical theory.

We remind the reader again that the true test of any good description of the natural world is through experiment and observation; failure to detect direct evidence for an idea means that that concept may be wrong, or, at the least, not relevant. Recall that the expectation for and the implications for the existence of a Higgs boson were conceived of in the early 1960s. The standard model incorporated these ideas and *indirect* evidence for the Higgs boson was collected from the early 1980s until the early 2000s, yet the Nobel Prize (we use the Nobel here as a proxy for a community having truly accepted an idea as "correct") was not awarded for this mechanism until it was definitively detected and confirmed during 2012–2013, the prize having then been awarded in 2013. Direct evidence is the cornerstone of any successful theory of nature.

We also remind the reader that models—specific mathematical constructs designed to allow for near-term experimental testing—are more easily refuted than are the frameworks upon which they are built. Braneworlds are built upon the notion of extra dimensions; failure to detect evidence of a specific braneworld model is not the same as falsifying the existence of extra dimensions. While it might be possible to rule out the RS-1 or RS-2 braneworld models by the end of the LHC program, that is not evidence against extra dimensions. The theoretical physics community will have to work very hard, as they have in building other frameworks (such as the standard model), to find the clear boundaries of these mathematical ideas in which absolute falsifiability is then achieved. This is probably the most difficult task the community of mathematical explorers will have to undertake in the coming decades.

Competition among ideas is crucial to this process. Let us consider superstring theory as one arena in which competition may prove extremely useful at separating valid from useless ideas. If there are only five dimensions, not ten or eleven, the mathematics of strings cannot (in its present form) describe our universe. If there are ten or eleven dimensions, we need to know where they exist—if they are curled up and hidden, or perhaps they disappeared after the big bang. Maybe, as we warned in earlier chapters, the problem of dimensionality is a red herring, a misleading folly, since the inception of bosonic string theory in the 1970s. Multiple dimensions were, in many ways, the primary problem of the theory and the concept on which tremendous intellectual effort was spent trying to explain the non-observation of so many necessary extra dimensions.

Perhaps this was a false flag from nature, a curiosity in the mathematics that distracted us from a better course, causing us to seek to make superstring theory work in the four space-time dimensions we know for certain do exist. As we have illustrated, the casting of such dissident concepts (this one has been championed, as we have noted, by one of the authors of this book, SJG) into the intellectual ecosystem are crucial, if only to spur on a community of thinkers to get out of its rut. Who can say which idea will be the right idea? Only experiment can sort that out, but if not all of the potential ideas have been trotted out, then one won't know to test for the one that may be the correct one.

If we were to observe generalities for a moment, we would suggest that the discovery of new particles in nature (wherever such a discovery might arise, whether in dark matter detectors, the LHC, astrophysical observations, or any other) appears to be crucial to further progress in answering the big questions that face the physics community. If we want to explain why the standard model over-predicts the energy of empty space (vacuum energy, a possible source for dark energy), there may need to be more particle interactions to cancel out those in the standard model. This is also true if we want to understand the relative lightness of the Higgs boson mass or if we want to understand dark matter. Kaluza-Klein excitations of the graviton are a key prediction of warped extra dimensions—these are another class of new particle looking to be found, if they exist. It seems that in most of the places we look, where there exist mature theoretical notions for a more generic mathematical description of nature, new particles are required as a path to completion of those ideas.

Another intriguing possibility that we will leave for the reader to consider is the possibility for the disappearance of some particles at high energy levels. If high energies are required to observe extra dimensions, we might first discover that the disappeared particles are traveling in extra dimensions by noting their absence during collider experiments. We looked briefly at this concept in the later chapter on the Higgs boson, while discussing the idea of invisible Higgs boson decay. Search strategies could reveal that particles at the highest energies are regularly missing from the detectors, even though we expect them to be found there. While hunting for extra dimensions, the disappearance of something old might be just as important as the appearance of something new.

Looking back deeper in time, by studying the light from the big bang (or maybe one day, when the technology permits, neutrinos that are left over from the big bang), is intended to give us more information about the first instants of time when the universe came into being. Mathematically based braneworld models offer the possibility to explain how the big bang might have occurred in the first place—by the collision of two nearby branes. Can the fingerprints of such occurrences be detected as we become able to peer closer and closer to the first instant of time?

A model of this kind of brane-induced big bang was proposed by Justin Khoury, Burt Ovrut, Paul Steinhardt, and Neil Turok in 2001. Their "ekpyrotic model" moved braneworlds in a new direction. They found that, after such a collision, there can be derived a set of mathematical equations that describes a universe much like the one we live in! As with any new idea in physics, serious technical challenges have been raised about this model, but still, one can see how branes have taken physicists on a path to consider new perspectives about what it means to be "in a universe." If branes can collide once, why not more than once? And what if *multiple* branes are capable of colliding? What happens when they *do* collide en masse? Many new questions emerge to join the shadows! It would not be improbable to continue carrying further questioning into even deeper, darker corners of the shadows.

Until positive proof arising from direct observation and experiment occurs, superstring/M-Theory, braneworld scenarios, and whatever other ideas will emerge from the fertile imaginations of theorists, these ideas must continue each to be borne into a shadow of reality. Each is an enigma to be resolved in the expectation that we are bringing them a light-step forward, rather than casting them into new darkness in the cosmos. Observation is the true test of hypothesis. When observation yields direct evidence, we will have met the challenge of bringing reality out of the shadows.

About the Authors

S. JAMES GATES, Jr's (Jim Gates) interest in science began at age four when his mother, Charlie, brought her children to a science fiction movie when the family was living at Ft. Pepperell near St. John's, Newfoundland. Four years later at Ft. Bliss, near El Paso, Texas, Sylvester James Gates, Sr., his father—a member of the U.S. Army and a veteran of WWII—gave him books about the coming of the space age that solidified an early interest in science (and kindled his wish to become an astronaut). Science fiction, comic book superheroes, a fantastic high school physics teacher, Mr. Freeman Coney, and the Orlando Public Library served as the launch pad for a life in physics.

Gates became the Ford Foundation Professor, Physics and Affiliate Professor of Mathematics at Brown University in Providence, Rhode Island after retiring in 2017 as a University System Regents Professor; Center for String and Particle Theory Director; Distinguished University Professor; John S. Toll Professor of Physics; and Affiliate Professor of Mathematics at the University of Maryland—College Park.

He received the 2011 National Medal of Science, the 2006 Public Understanding of Science & Technology Award from the American Association for the Advancement of Science, the 2003 Klopsteg Award for excellent physics teaching from the American Association of Physics Teachers, and the 1994 Bouchet Award of the American Physical Society. He is a member of the National Academy of Science, the American Association for the Advancement of Science, the American Academy of Arts and Sciences, the American Physical Society, and the American Philosophical Society.

Gates has contributed to the mathematical foundation of supersymmetry since authoring M.I.T.'s first Ph.D. written on the subject in 1977. Among his discoveries has been four dimensional string theory using the mathematics of the standard model and connections to graph theory, information theory, and indications of the possibility that following the big bang there might have been an "inchoate epoch" during which processes similar to evolution acted on the mathematical laws that describe our universe. His research continues to expand the understanding of supersymmetry in unique and innovative ways.

In 2017, he completed forty-five consecutive years as a college instructor in physics and/or mathematics. He has appeared in many TV science documentaries, on-line videos, and in 2006 completed *Superstring Theory: The DNA of Reality,* a video series of twenty-four half-hour non-technical presentations.

FRANK BLITZER received his B.S.E.E from Purdue University with a second major in mathematics. After graduating, he continued his studies toward the Master's Degree in Electrical Engineering. Frank spent fifty years working for several major companies in the aerospace industry on such things as the design of the B-52 bomber and various missiles and space systems. He facilitated operations with inertial guidance, communications, and surveillance systems. He contributed to the design of systems for the Lacrosse Missile, Patriot Missile, the APOLLO Manned-Space Program, and the Strategic Defense Initiative Program. He developed and patented several missile guidance and control systems, and space systems, as well as pattern recognition and surveillance systems, in which fields he is published. He received the Honeywell Top Performer Award in 1992.

STEPHEN JACOB SEKULA's parents, Annetta and Stephen, fed Steve's early science habits with dangerous chemistry sets, providing his writing interest with endless pads of paper, typewriters, and, eventually, word processing software, as well as intellectual criticism of that writing. His sister, Kate, was essential in dragging him back to reality when he spent too much time in the same shadows this book explores.

SJS is Associate Professor of Experimental Particle Physics at Southern Methodist University in Dallas, Texas. He earned his Ph.D. from the University of Wisconsin–Madison and his B.S. from Yale University.

He has been involved in particle collider experiments since his undergraduate days; first at Fermilab, then at SLAC National Accelerator Laboratory, and presently at CERN. He led a team within the BaBar collaboration in 2008 that discovered a state of matter, the lowest energy configuration of bottom quark and its antimatter counterpart, that was first predicted to exist in 1977. He participated in the discovery and measurement of the Higgs Boson in 2012 and 2013, culminating in the announcement of its discovery on the Fourth of July in 2012. He continues still to be facinated by the potential of this particle to explain even more about the presently unknown origins of our universe.

He is a recipient of the 2017 SMU Altshuler Distinguished Teaching Professor Award and he received the Texas Section American Physical Society's Robert S. Hyer Award for excellence in the supervision of undergraduate research.

www.ingramcontent.com/pod-product-compliance
Lightning Source LLC
Chambersburg PA
CBHW082134210326
41599CB00031B/5981